# Advances in Dimension Theory, Fractal Functions and Measures

# CONTEMPORARY MATHEMATICS

825

# Advances in Dimension Theory, Fractal Functions and Measures

AMS Special Session on Fractal Geometry
and Dynamical Systems
Spring Eastern Virtual Sectional Meeting
April 1–2, 2023

Virtual Conference on Functional Analysis and Fractals
Organizer: Indian Institute of Information Technology
Allahabad (IIIT-A), India
February 16–18, 2024

Saurabh Verma
María A. Navascués
Amit Priyadarshi
Editors

AMERICAN MATHEMATICAL SOCIETY

2020 *Mathematics Subject Classification.* Primary 26A18, 28A20, 28A75, 28A78, 28A80, 37C30, 37D05, 42C05, 11K55, 60B05.

---

**Library of Congress Cataloging-in-Publication Data**

Names: Verma, Saurabh, 1995– editor | Navascués, María A., 1957– editor | Priyadarshi, Amit, 1981– editor | American Mathematical Society. Spring Eastern Sectional Meeting, Special Session on Fractal Geometry and Dynamical Systems (2023 : Online)

Title: Advances in dimension theory, fractal functions and measures / AMS Spring Eastern Sectional Meeting, Special Session on Fractal Geometry and Dynamical Systems April 1–2, 2023, virtual; Saurabh Verma, María A. Navascués, Amit Priyadarshi, editors.

Description: Providence, Rhode Island : American Mathematical Society, [2025] | Series: Contemporary mathematics, 0271-4132 ; Volume 827

Identifiers: LCCN 2025013554 | ISBN 9781470477844 paperback | ISBN 9781470481018 ebook

Subjects: LCSH: Measure theory | Dimension theory (Algebra) | AMS: Real functions – Functions of one variable – Iteration | Measure and integration – Classical measure theory – Measurable and nonmeasurable functions, sequences of measurable functions, modes of convergence | Measure and integration – Classical measure theory – Length, area, volume, other geometric measure theory | Measure and integration – Classical measure theory – Hausdorff and packing measures | Measure and integration – Classical measure theory – Fractals | Dynamical systems and ergodic theory – Smooth dynamical systems: general theory – Zeta functions, (Ruelle-Frobenius) transfer operators, and other functional analytic techniques in dynamical systems | Dynamical systems and ergodic theory – Dynamical systems with hyperbolic behavior – Hyperbolic orbits and sets | Harmonic analysis on Euclidean spaces – Nontrigonometric harmonic analysis – Orthogonal functions and polynomials, general theory | Number theory – Probabilistic theory: distribution modulo 1; metric theory of algorithms – Metric theory of other algorithms and expansions; measure and Hausdorff dimension | Probability theory and stochastic processes – Probability theory on algebraic and topological structures – Probability measures on topological spaces

Classification: LCC QC20.7.M43 A383 2025

LC record available at https://lccn.loc.gov/2025013554

Contemporary Mathematics ISSN: 0271-4132 (print); ISSN: 1098-3627 (online)

DOI: https://doi.org/10.1090/conm/825

---



10 9 8 7 6 5 4 3 2 1    30 29 28 27 26 25

# Contents

# Preface

This volume is based on the papers that were presented at the AMS Special Session on Fractal Geometry and Dynamical Systems at the AMS Spring Eastern Virtual Sectional Meeting (https://meetings.ams.org/math/spring2023e/meetingapp.cgi/Symposium/141), USA, April 1–2, 2023, and a virtual Conference on Functional Analysis and Fractals (https://cfaf.iiita.ac.in/), India, February 16–18, 2024, organized by us.

Fifty years ago, Mandelbrot created a new type of geometry called fractal. One of the novelties of this new mathematics is a systematic qualitative and quantitative approach to the concepts of irregular shapes and roughness. Galileo said that the universe is written in mathematical language, and its characters are triangles, circles, and other geometric figures. Mandelbrot masterly defined "other" geometric objects whose main property is self-similarity and coined the term "fractal" for them. This kind of model fits better complex patterns such as the circulatory system, the coastline of a littoral country or a stock market chart. One way of quantifying the complexity of such structures is the computation of their fractal dimension.

This book presents modern advances in the concept of dimension and its related notion of fractal measure. The volume is oriented to give an insight into the current research in the area and contains novel contributions of important scientists in the field. The volume deals with very diverse topics such as the Hausdorff dimension of a set of continued fractions, the dimension theory of inhomogeneous attractors, ergodic conjecture of falling balls systems, or Hausdorff measures to represent uncertainty in neural networks.

The first paper investigates some bounds of the fractal dimension of the fractional integral of a fractalized continuous function. Thereafter, approximation aspects of a continuous function preserving fractal dimension are elaborated.

The second paper determines the optimal sets of $n$-means and the $n$th quantization errors for all $1 \leq n \leq 6$ for two nonuniform discrete distributions with support the set $\{1, 2, 3, 4, 5, 6\}$ along with other results on optimal sets and quantization errors. It also discusses how to find a discrete distribution if the optimal sets are given.

The third paper focuses on an application of the model of coherent upper conditional prevision based on Hausdorff outer measures to represent uncertainty in neural networks when the distribution over data is unknown. For the group of all bi-Lipschitz self functions, the standard Borel group is introduced and coherent conditional prevision is defined with respect to a quasi-invariant measure if the conditioning event has positive and finite measure. Otherwise, it is defined with respect to $0-1$ valued finitely additive, but not countably additive, probability.

The fourth paper proposes Rakotch-type contraction mappings to obtain novel fractal interpolation surfaces with broad, essential, and profound meaning.

The fifth paper is expository in nature that discusses the dimension theory of inhomogeneous attractors, giving several examples and some open questions. It also focuses on the upper box dimension with emphasis on how to derive good estimates, and when these estimates fail to be sharp.

The sixth paper outlines a construction motivated by the class of IFS fractal measures and proves new duality results for Gaussian processes. Further, a necessary and sufficient condition is derived for the equivalence of pairs of probability measures which determine the corresponding two Gaussian processes.

The seventh paper generalizes two types of contractions proposed by Ciric to the framework of $b$-metric spaces. It provides some Collage Theorems for these mappings. Sufficient conditions on the ratios and the space for the existence of fixed points and common fixed points of two maps are obtained. Two iterative algorithms to approximate the fixed points are analyzed and their convergence and stability are studied.

The eighth paper gives some elementary "comparison theorems" for Hausdorff dimensions of sets associated with continued fractions, which sometimes allow one to prove that $\dim(E(F_1)) > \dim(E(F_2))$ without actually computing $\dim(E(F_1))$ or $\dim(E(F_2))$ for $F_1, F_2 \subset \mathbb{N}$.

The ninth paper contains a survey of recent works of Olsen. It provides a detailed study of the growth of the eigenvalues of multifractal transfer operator, significantly improving the current state of the art estimates by Bandtlow & Jenkinson from 2008. Also, a general theory of fractal and multifractal prime counting formulas for Graph-Directed Self-Conformal Iterated Function Systems is developed. As an application, a multifractal prime geodesic counting formula for classical Fuchsian Schottky groups is discussed.

The tenth paper explores different conditions on the matrix $M$ so that the measure $\mu_{M,D}$ is a spectrum for a specific four-element digit set $D$. Further, the measure $\mu_{M,D}$ is shown to be a spectrum for specific $M$ and general four-element digit set $D$.

The eleventh paper introduces a new general multifractal analysis, addressing the constraints associated with conventional multifractal models by incorporating specific functions. It focuses on the exploration and analysis of a comprehensive Hewitt-Stromberg measure, introducing a generalized multifractal structure based on the mentioned functions. It also includes the estimation of upper and lower bounds for the general Hewitt-Stromberg dimension, providing insights into the divergence of multifractal functions.

The twelfth paper presents an unconditional proof of Wojtkowski's Ergodicity Conjecture for almost every system of 1D perfectly elastic balls falling down in a half line under constant gravitational acceleration. By introducing a new algebraic approach, it is shown that almost every such system is completely hyperbolic and ergodic.

The thirteenth paper focuses on the construction of some classes of compact sets that are dense in the class of all non-empty compact subsets of the real line with respect to the Hausdorff metric. It also contains a weaker version of Hochman's theorem for the Hausdorff dimension of self-similar sets and also collects some remarks and examples for attractors generated by iterated function systems consisting of all bi-Lipschitz or affine mappings under various separation conditions.

Last but not least, we would like to extend our heartfelt gratitude to all contributors who submitted papers for this volume. Special appreciation goes to the reviewers for their invaluable suggestions and comments. We want to thank all participants and speakers at the AMS special session and the conference for making the events successful. We also express our sincere thanks to Christine Thivierge, Bridget Manzella, and Mike Saitas from the AMS and the entire Contemporary Mathematics Editorial Committee for their support, technical assistance, and the opportunity to publish this volume.

Saurabh Verma
María A. Navascués
Amit Priyadarshi

Contemporary Mathematics
Volume **825**, 2025
https://doi.org/10.1090/conm/825/16505

# Dimension preserving approximation and estimation: Fractal surfaces and Riemann-Liouville fractional integrals

Ekta Agrawal and Saurabh Verma

Abstract. We primarily aim to study the composition of the fractal operator associated with bivariate fractal interpolation (known as fractal surface) and the mixed Riemann-Liouville fractional integral operator, and vice versa. Subsequently, some bounds of the fractal dimension of the fractional integral of a fractalized continuous function are obtained. Thereafter, we elaborate the approximation aspects of a continuous function with respect to fractal dimension and its fractional derivative. In the end, we define the fractional convex functions on the rectangular domain and study their approximation.

## 1. Introduction

Barnsley [**3**] in his seminal work 'Fractal functions and interpolation' introduced the notion of fractal interpolation functions (FIFs) on closed and bounded intervals of $\mathbb{R}$, via the iterated function system (IFS) [**13**]. The resulting FIF, which is the invariant point of the Read-Bajraktarević operator, is continuous, albeit not necessarily smooth. This leads to a significant milestone in the realm of interpolation theory since conventional interpolation typically relies on smooth functions. Thereafter, Navascués [**25, 26**] parametrized fractal interpolation and presented the concept of $\alpha$-FIFs. Recently, Jahn and Massopust [**14**] proposed the integral RB operator and studied some of its properties. Motivated by the exploration in the univariate case, Massopust [**21**] initially formed a fractal interpolation surface (FIS) on the triangular domain of $\mathbb{R}^2$, providing the data set on the boundary is co-planar. Subsequently, several other formulations are executed by imposing some restrictions on the data set; for details, see [**4, 9, 20, 24, 41**]. Further, Ruan and Xu [**29**] proposed a general framework of FISs on a rectangular domain of $\mathbb{R}^2$ with no restriction on the data set nor the scaling factor. Subsequently, Verma and Viswanathan [**36, 38**] proposed $\alpha$-FIS and elaborated several properties of the fractal operator. Recently, Massopust [**22**] proposed fractal interpolation over domain with a non-linear partition. To fulfill a preparatory role, we briefly mention Ruan and Xu's formulation in Section 2.

---

2020 *Mathematics Subject Classification.* Primary 28A80; Secondary 26A33, 47B38.
*Key words and phrases.* Fractal surfaces, Fractal operator, Box dimension, Hausdorff dimension, Riemann-Liouville fractional integral and derivative.
The first author was supported by IIIT Allahabad, India.
The second author was supported by SEED grant project of IIIT Allahabad, India.

The geometrical interpretation of the Riemann-Liouville (RL) fractional calculus of any continuous function (see [**28**]) attracted a lot of researchers. We refer readers to [**19, 31**] for the fundamentals of fractional calculus. Liang et al. [**15–17**] studied the dimension of fractional calculus for continuous functions. Wang et al. [**39**] have shown the existence of a linear relationship between the order of fractional calculus and the Hausdorff dimension of the Weierstrass function. Later, Chandra and Abbas [**5**] initiated the evaluation of the fractal dimension of the RL fractional integral of a continuous function defined in the rectangular region. In this article, we investigate several properties of the composition of the fractal operator and the RL fractional integral operator. Subsequently, the bounds of the fractal dimension of the function obtained from the composition are studied in a bivariate setting. There are several fractal dimensions that are introduced in the literature. Here, we deal with Hausdorff and box dimensions, which are discussed in [**8**].

The approximation of a real-valued continuous function on a compact interval by a sequence of polynomials, or in particular Bernstein polynomials, has gained a lot of attention in the literature. In 2005, Navascués [**25**, Theorem 3.9] initiated the approximation of a continuous function on a compact interval via fractal functions and explored that the set of $\alpha$-fractal polynomials with $\alpha \neq 0$ is dense in the space of continuous functions over a compact interval and generalized the result in [**26**]. Verma and Massopust [**35**] studied the approximation of real-valued continuous functions on $[0, 1]$ that preserve fractal dimensions and its derivative. Subsequently, they also approximated continuous functions via fractal function preserving dimension in a univariate setting. Agrawal et al. [**1**] generalized the work done in [**35**], and deduced approximation results for continuous functions defined on a rectangular domain. Since fractional calculus is advanced compared to classical one, Yu and Liang [**40**] investigated the approximation of a continuous function that preserves fractal dimension and its RL fractional derivative in the univariate case. The current article targets to investigate several approximation results for the continuous functions on the rectangular domain concerning fractal dimension and its fractional derivative. Following that, we proposed a fractional convex function and studied its approximation.

It is notable that Section 2 deals with all the required preliminaries, followed up by our new results in Section 3.

## 2. Preliminaries

DEFINITION 2.1 ([**31**]). Let $h : [a, b] \times [c, d] \to \mathbb{R}$ be a continuous function and $\nu = (\nu_1, \nu_2)$ with $\nu_1, \nu_2 > 0$. The mixed Riemann-Liouville (RL) fractional integral of $h$ of order $\nu$ is defined as

$$\mathcal{I}^{\nu} h(x, y) = \frac{1}{\Gamma(\nu_1)\Gamma(\nu_2)} \int_a^x \int_c^y (x - s)^{\nu_1 - 1}(y - t)^{\nu_2 - 1} h(s, t)\, ds dt.$$

Let $\mu = (\mu_1, \mu_2)$ with $0 < \mu_1, \mu_2 < 1$. The mixed RL fractional derivative of $h$ of order $\mu$ is defined as

$$\begin{aligned}\mathcal{D}^{\mu} h(x, y) &= \frac{\partial^2}{\partial x \partial y} \mathcal{I}^{(1-\mu_1, 1-\mu_2)} h(x, y) \\ &= \frac{1}{\Gamma(1 - \mu_1)\Gamma(1 - \mu_2)} \frac{\partial^2}{\partial x \partial y} \int_a^x \int_c^y (x - s)^{-\mu_1}(y - t)^{-\mu_2} h(s, t)\, ds dt.\end{aligned}$$

NOTE 2.2 ([**31**, Pg. 36]). *Let* $-1 < \tau_1, \tau_2 < 0$. *Consider* $h(x,y) = (x-a)^{\tau_1}(y-b)^{\tau_2}$, *for all* $(x,y) \in [a,b] \times [c,d]$. *Then*

$$\mathcal{D}^{\mu} h(x,y) = \frac{\Gamma(1+\tau_1)\Gamma(1+\tau_2)}{\Gamma(1+\tau_1-\mu_1)\Gamma(1+\tau_2-\mu_2)}(x-a)^{\tau_1-\mu_1}(y-b)^{\tau_2-\mu_2}.$$

*It follows that*

$$\mathcal{D}^{\mu} h(x,y) = 0, \text{ if } h(x,y) = (x-a)^{\mu_1-1}(y-b)^{\mu_2-1}.$$

NOTE 2.3 ([**27**]). *Some properties of RL fractional derivative and integral are outlined as follows. Consider* $h_1, h_2 \in \mathcal{C}([a,b] \times [c,d])$ *and* $\kappa_1, \kappa_2 \in \mathbb{R}$.

(1) $\mathcal{D}^0 h_1 = h_1$.
(2) $\mathcal{D}^{\mu}(\kappa_1 h_1 + \kappa_2 h_2) = \kappa_1 \mathcal{D}^{\mu}(h_1) + \kappa_2 \mathcal{D}^{\mu}(h_2)$, *provided* $\mathcal{D}^{\mu}(h_1)$ *and* $\mathcal{D}^{\mu}(h_2)$ *exist.*

THEOREM 2.4 (Fundamental theorem of fractional calculus, [**31**, Theorem 2.4]). *Let* $f : [a,b] \times [c,d] \to \mathbb{R}$ *be a continuous function, and* $\mu = (\mu_1, \mu_2)$ *with* $0 < \mu_1, \mu_2 < 1$. *Then for every* $(x,y) \in (a,b) \times (c,d)$, *we have*

(1) $\mathcal{D}^{\mu}\mathcal{I}^{\mu} f(x,y) = f(x,y)$.
(2) $\mathcal{I}^{\mu}\mathcal{D}^{\mu} f(x,y) = f(x,y) - \kappa_f (x-a)^{\mu_1-1}(y-b)^{\mu_2-1}$, *where* $\kappa_f \in \mathbb{R}$ *is some fixed constant depending on the function* $f$.

THEOREM 2.5 (New fractional mean value theorem in one variable, [**10**, Corollary 3.3]). *Suppose that* $\eta \in (0,1)$, $f \in \mathcal{C}([a,b])$ *with* $\mathcal{D}^{\eta}[(x-a)^{\eta-1} f] \in \mathcal{C}([a,b])$. *Then there exists* $\xi \in [a,x]$ *such that*

$$f(x) - f(a) = \frac{\mathcal{D}^{\eta}[(x-a)^{\eta-1} f(x)]\Big|_{x=\xi}}{\Gamma(\eta+1)}(x-a).$$

DEFINITION 2.6 ([**7**]). The spectral radius of any operator $T$ is defined as

$$\rho(T) = \sup\{|\lambda| : \lambda \in \mathbb{C} \text{ and } T - \lambda I \text{ is not invertible}\}.$$

We have a well-known Gelfand's formula to compute the spectral radius as $\rho(T) = \lim_{n\to\infty} \|T^n\|^{\frac{1}{n}}$.

THEOREM 2.7 ([**7**]). *Let* $T$ *be a bounded linear operator. Then* $\rho(T) \le \|T\|$.

DEFINITION 2.8 ([**8**]). Let $K$ be a non-empty bounded subset of $\mathbb{R}^n$ and $N_\delta(K)$ be the smallest number of sets of diameter at most $\delta$ that cover $K$. Then the box dimension of $K$ is defined as

$$\dim_B(K) = \lim_{\delta\to 0^+} \frac{\log N_\delta(K)}{-\log\delta},$$

provided the limit exists. If the limit does not exist, then the upper and lower box dimensions, respectively, are defined as

$$\overline{\dim}_B(K) = \limsup_{\delta\to 0^+} \frac{\log N_\delta(K)}{-\log\delta}, \qquad \text{and} \qquad \underline{\dim}_B(K) = \liminf_{\delta\to 0^+} \frac{\log N_\delta(K)}{-\log\delta}.$$

DEFINITION 2.9 ([**8**]). Let $K$ be a subset of $\mathbb{R}^n$ and $r$ be any non-negative number. Then the $r$-dimensional Hausdorff measure of $K$ is defined as

$$\mathcal{H}^r(K) = \lim_{\delta\to 0^+} \mathcal{H}^r_\delta(K),$$

where $\mathcal{H}_\delta^r(K) = \inf\{\sum_{i=1}^{\infty} |U_i|^r : \{U_i\}$ is a $\delta$-cover of $K\}$ and $|U_i| = \sup_{x,y\in U_i} \|x - y\|_2$. The Hausdorff dimension of $K$ is defined as

$$\dim_{\mathcal{H}}(K) = \inf\{r \geq 0 : \mathcal{H}^r(K) = 0\}.$$

Further, it is shown in [**8**] that for any non-empty bounded subset $K$ of $\mathbb{R}^n$, following inequality holds:

$$\dim_{\mathcal{H}}(K) \leq \underline{\dim}_B(K) \leq \overline{\dim}_B(K).$$

THEOREM 2.10 ([**37**, Corollary 3.13]). *Let $g : [a,b] \times [c,d] \to \mathbb{R}$ be a Hölder continuous function with exponent $\sigma$ with $0 \leq \sigma \leq 1$, and $\mathcal{G}(g) = \{(x,y,g(x,y)) : (x,y) \in [a,b] \times [c,d]\}$ is the graph of $g$. Then*

$$2 \leq \dim_{\mathcal{H}}(\mathcal{G}(g)) \leq \underline{\dim}_B(\mathcal{G}(g)) \leq \overline{\dim}_B(\mathcal{G}(g)) \leq 3 - \sigma.$$

Mauldin and Williams [**23**] estimated the Hausdorff dimension of nowhere differentiable functions. Later, the dimensional evaluation of the Weierstrass function is explored in [**12, 32**]. Songping and Guolong [**33**] investigate the box dimension of Besicovitch functions. Several recent works on the dimensional analysis of constructed fractal functions and $\alpha$-FIFs can be studied in [**34**]. The dimension estimation of fractal functions using function spaces are studied in [**2, 6**]. Now onwards for any $s \in \mathbb{N}$, we consider $\Sigma_s = \{1, 2, \ldots, s\}$, $\Sigma_{s,0} = \{0, 1, \ldots, s\}$, $\partial\Sigma_{s,0} = \{0, s\}$ and $\operatorname{int} \Sigma_{s,0} = \{1, 2, \ldots, s-1\}$.

**2.1. Revisit to $\alpha$-fractal surfaces.** Let the interpolation data set be given as $\{(x_i, y_j, z_{ij}) \in \mathbb{R}^3 : (i,j) \in \Sigma_{N,0} \times \Sigma_{M,0}\}$ such that $x_0 < x_1 < \cdots < x_N$ and $y_0 < y_1 < \cdots < y_M$. Suppose $I = [x_0, x_N]$, $J = [y_0, y_M]$, $I_i = [x_{i-1}, x_i]$ and $J_j = [y_{j-1}, y_j]$, for all $i \in \Sigma_N$ and $j \in \Sigma_M$, respectively. Further, for $i \in \Sigma_N$, define $u_i : I \to I_i$ be a contractive homeomorphism such that

$$\begin{cases} u_i(x_0) = x_{i-1},\ u_i(x_N) = x_i, & \text{if } i \text{ is odd}, \\ u_i(x_0) = x_i,\ u_i(x_N) = x_{i-1}, & \text{if } i \text{ is even}. \end{cases} \tag{2.1}$$

On the same line, for any $j \in \Sigma_M$, $v_j : J \to J_j$ is defined. Again, for convenience in notation, define $\omega : \mathbb{Z} \times \{0, N, M\} \to \mathbb{Z}$ as follows

$$\omega(i,N) = \omega(i,M) = \begin{cases} i, \omega(i,0) = i-1, & \text{if } i \text{ is odd}, \\ i-1, \omega(i,0) = i, & \text{if } i \text{ is even}. \end{cases}$$

It is notable that

$$u_i(x_k) = x_{\omega(i,k)},\ \forall\, (i,k) \in \Sigma_N \times \partial\Sigma_{N,0},$$
$$v_j(y_k) = y_{\omega(j,k)},\ \forall\, (j,k) \in \Sigma_M \times \partial\Sigma_{M,0}.$$

Let $\mathcal{K} = I \times J \times \mathbb{R}$. Subsequently, for any $(i,j) \in \Sigma_N \times \Sigma_M$, define $F_{ij} : \mathcal{K} \to \mathbb{R}$ to be a continuous function satisfying end-point conditions given as

$$F_{ij}(x_k, y_l, z_{kl}) = z_{\omega(i,k),\omega(j,l)},\ \forall\, (k,l) \in \partial\Sigma_{N,0} \times \partial\Sigma_{M,0},\ \text{and}$$
$$|F_{ij}(x,y,z) - F_{ij}(x,y,z_*)| \leq \eta_{ij}|z - z_*|,\ \forall\, (x,y) \in I \times J,\ z, z_* \in \mathbb{R},$$

where $0 < \eta_{ij} < 1$ is a contraction factor. Recently, Gurubachan et al. [**11**] showed the existence of a new class of fractal functions without the end-point conditions. Therefore, $\{\mathcal{K}; \mathcal{W}_{ij} : (i,j) \in \Sigma_N \times \Sigma_M\}$ is an iterated function system (IFS), where $\mathcal{W}_{ij} : \mathcal{K} \to \mathcal{K}$ is specified as

$$\mathcal{W}_{ij}(x,y,z) = \big(u_i(x), v_j(y), F_{ij}(x,y,z)\big),\ \forall\, (x,y,z) \in \mathcal{K}.$$

Let $\mathcal{C}_0(I \times J) = \{g \in \mathcal{C}(I \times J) : g(x_i, y_j) = z_{ij}, \forall (i,j) \in \Sigma_{N,0} \times \Sigma_{M,0}\}$, where $\mathcal{C}(I \times J)$ is the set of all real-valued continuous functions defined on $I \times J$. Then $\mathcal{C}_0(I \times J)$ being a closed subset of $\mathcal{C}(I \times J)$ with respect to sup-norm, is complete. Define an operator $\mathcal{T}$, commonly referred to as Read-Bajraktarević (RB) operator, on $\mathcal{C}_0(I \times J)$ as follows:

$$\mathcal{T}(g)(x,y) = F_{ij}(u_i^{-1}(x), v_j^{-1}(y), g(u_i^{-1}(x), v_j^{-1}(y))), \; \forall\, g \in \mathcal{C}_0(I \times J),$$

where $(x,y) \in I_i \times J_j$, $(i,j) \in \Sigma_N \times \Sigma_M$. It is shown in [**29**, Theorem 3.1] that

- there exists a unique function $h \in \mathcal{C}_0(I \times J)$ such that $\mathcal{T}h = h$, that is,

  $$h(x,y) = F_{ij}(u_i^{-1}(x), v_j^{-1}(y), h(u_i^{-1}(x), v_j^{-1}(y))),$$

  for all $(x,y) \in I_i \times J_j$, $(i,j) \in \Sigma_N \times \Sigma_M$. The equation mentioned above is commonly termed a self-referential equation.
- $\mathcal{G}(h)$ is the invariant set or attractor of the IFS, that is,

  $$\mathcal{G}(h) = \bigcup_{ij} \mathcal{W}_{ij}(\mathcal{G}(h)),$$

  where $\mathcal{G}(h) = \{(x, y, h(x,y)) : (x,y) \in I \times J\}$ is the graph of $h$.

The function $h$ is referred to as a fractal interpolation function (FIF) and $\mathcal{G}(h)$ is the required fractal interpolation surface (FIS) for the IFS. In a recent study, Liang and Ruan [**18**] proposed recurrent FIS and evaluated its box dimension. Now onwards, we mention the formulation of $\alpha$-fractal surface.

Let $f$ be a continuous function on $I \times J$, referred to as the germ function. Assume the interpolation data generated by $f$ is $z_{ij} = f(x_i, y_j)$, for all $(i,j) \in \Sigma_{N,0} \times \Sigma_{M,0}$, see [**36**]. Further, for $(i,j) \in \Sigma_N \times \Sigma_M$, the functions $u_i$, $v_j$ and $F_{ij}$ specified above can be chosen as

$$\begin{aligned} u_i(x) &= a_i x + b_i, \\ v_j(y) &= c_j y + d_j, \\ F_{ij}(x,y,z) &= \alpha(u_i(x), v_j(y))z + f(u_i(x), v_j(y)) - \alpha(u_i(x), v_j(y))Lf(x,y), \end{aligned}$$

where constants $a_i, b_i \in \mathbb{R}$ are determined using equation (2.1) as follows

$$\text{(2.2)} \qquad \begin{cases} a_i = \frac{x_i - x_{i-1}}{x_N - x_0}, b_i = \frac{x_N x_{i-1} - x_0 x_i}{x_N - x_0}, & \text{if } i \text{ is odd}, \\ a_i = \frac{x_i - x_{i-1}}{x_0 - x_N}, b_i = \frac{x_0 x_{i-1} - x_N x_i}{x_0 - x_N}, & \text{if } i \text{ is even}. \end{cases}$$

Similarly, $c_j, d_j \in \mathbb{R}$ are determined. The function $\alpha \in \mathcal{C}(I \times J)$ is called scaling function with $\|\alpha\|_\infty < 1$, and $L : \mathcal{C}(I \times J) \to \mathcal{C}(I \times J)$ is a bounded linear operator such that for all $h \in \mathcal{C}(I \times J)$, we have $Lh(x,y) = h(x,y)$, where $(x,y) \in \partial\Sigma_{N,0} \times \partial\Sigma_{M,0}$. The corresponding $\alpha$-FIF, $f_L^\alpha$ is the invariant point of the RB operator stated as

$$\mathcal{T}^\alpha(h)(x,y) = f(x,y) + \alpha(x,y)\big[h(u_i^{-1}(x), v_j^{-1}(y)) - Lf(u_i^{-1}(x), v_j^{-1}(y))\big],$$

for $(x,y) \in I_i \times J_j$ and $(i,j) \in \Sigma_N \times \Sigma_M$, and obeys the self-referential equation as

$$\text{(2.3)} \qquad f_L^\alpha(x,y) = f(x,y) + \alpha(x,y)\big[f_L^\alpha(u_i^{-1}(x), v_j^{-1}(y)) - Lf(u_i^{-1}(x), v_j^{-1}(y))\big],$$

for all $(x,y) \in I_i \times J_j$ and $(i,j) \in \Sigma_N \times \Sigma_M$. Further, $\mathcal{G}(f_L^\alpha)$ is the required $\alpha$-FIS. Consequently, there exists a linear operator $\mathcal{F}_L^\alpha$, referred to as the $\alpha$-fractal operator on $\mathcal{C}(I \times J)$ defined as

$$\mathcal{F}_L^\alpha(f) = f_L^\alpha, \; \forall\, f \in \mathcal{C}(I \times J).$$

Some properties of $\mathcal{F}_L^\alpha$ are outlined below which will serve as a prelude to our coming results in the next section.

THEOREM 2.11 ([**36**, Theorem 3.2]). *Let $\Im$ be the identity operator on $\mathcal{C}(I \times J)$ and $\|.\|$ denote the operator norm. Then we have the following:*

(1) *The fractal operator $\mathcal{F}_L^\alpha$ is a bounded linear operator with respect to sup norm on $\mathcal{C}(I \times J)$ and*

$$\|\mathcal{F}_L^\alpha\| \leq 1 + \frac{\|\alpha\|_\infty \|\Im - L\|}{1 - \|\alpha\|_\infty}.$$

(2) *If $\|\alpha\|_\infty \neq 0$, then the fixed points of $L$ are the fixed points of the fractal operator $\mathcal{F}_L^\alpha$ as well.*

## 3. Main results

PROPOSITION 3.1. *The set $A := \{f \in \mathcal{C}(I \times J) : f(x_0, y_0) = 0\} \subseteq \mathcal{C}(I \times J)$ is a closed invariant subspace of operators $\mathcal{F}_L^\alpha$ and $\mathcal{I}^\nu$.*

PROOF. It is straightforward to validate that $A$ is a closed subspace of $\mathcal{C}(I \times J)$. Let $g \in A$, then $\mathcal{F}_L^\alpha g \in \mathcal{C}(I \times J)$. Also, with the reference to self-referential equation (2.3), we have $\mathcal{F}_L^\alpha g(x_0, y_0) = 0$. Therefore, $\mathcal{F}_L^\alpha(A) \subseteq A$. Again, by definition of RL fractional integral, we obtain

$$\mathcal{I}^\nu g(x_0, y_0) = \frac{1}{\Gamma(\nu_1)\Gamma(\nu_2)} \int_{x_0}^{x_0} \int_{y_0}^{y_0} (x-s)^{\nu_1 - 1}(y-t)^{\nu_2 - 1} g(s,t)\,ds\,dt = 0.$$

It follows that $\mathcal{I}^\nu(A) \subseteq A$, establishing the claim. □

COROLLARY 3.2. *Both $\mathcal{F}_L^\alpha \circ \mathcal{I}^\nu$ and $\mathcal{I}^\nu \circ \mathcal{F}_L^\alpha$ have a closed invariant subspace.*

PROOF. The proof is immediate from Proposition 3.1, hence it is omitted. □

COROLLARY 3.3. *$\mathcal{F}_L^\alpha + \mathcal{I}^\nu$ has a closed invariant subspace.*

In general, $\mathcal{F}_L^\alpha \circ \mathcal{I}^\nu \neq \mathcal{I}^\nu \circ \mathcal{F}_L^\alpha$. An example to support the statement is mentioned below.

EXAMPLE 3.4. Consider the interval $[0,1]$ and its partition $\{0, 0.25, 0.5, 0.75, 1\}$. Thus, we get a rectangular grid on $[0,1]^2$, and $N = M = 4$. Further, for $i \in \Sigma_4$, the affine maps $u_i$ are given as follows:

$$u_1(x) = 0.25x, \qquad u_2(x) = -0.25x + 0.5,$$
$$u_3(x) = 0.25x + 0.5, \qquad u_4(x) = -0.25x + 1.$$

Suppose $\alpha\,(\neq 0) \in (-1,1)$, and $L$ is an operator on $\mathcal{C}([0,1]^2)$ chosen as

$$Lh(x,y) = (x-2)h(x,y) - x^2 + 9,\ \forall\, h \in \mathcal{C}([0,1]^2),\ (x,y) \in [0,1]^2.$$

The fixed point of $L$ is obtained as $g(x,y) = x + 3$, for all $(x,y) \in [0,1]^2$. Subsequently, applying Theorem 2.11, $g$ is also the fixed point of fractal operator $\mathcal{F}_L^\alpha$, that is, $\mathcal{F}_L^\alpha(g) = g = g_L^\alpha$. Thus, using equation (2.3), $g_L^\alpha$ satisfies

$$\begin{aligned} g_L^\alpha(x,y) &= g(x,y) + \alpha\big[g_L^\alpha(u_i^{-1}(x), v_j^{-1}(y)) - Lg(u_i^{-1}(x), v_j^{-1}(y))\big] \\ &= x + 3, \end{aligned}$$

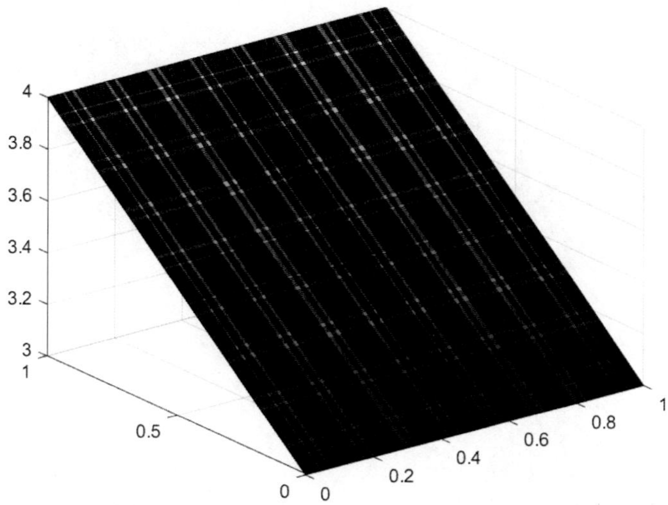


FIGURE 1. $\mathcal{G}(g_L^\alpha)$ when $\alpha = 0.5$.

for all $(x, y) \in I_i \times J_j$ and $i, j \in \Sigma_4$. Figure 1 represents the graph of $g_L^\alpha$ for $\alpha = 0.5$. Now, for $\nu = (\nu_1, \nu_2) = (\frac{1}{2}, \frac{1}{2})$, we obtain

$$\begin{aligned}\mathcal{I}^\nu \circ \mathcal{F}_L^\alpha(g(x,y)) = \mathcal{I}^\nu g(x,y) &= \frac{1}{\Gamma(\frac{1}{2})^2}\int_0^x\int_0^y (x-s)^{\nu_1-1}(y-t)^{\nu_2-1}g(s,t)dsdt \\ &= \frac{1}{\Gamma(\frac{1}{2})^2}\int_0^x\int_0^y (x-s)^{\frac{-1}{2}}(y-t)^{\frac{-1}{2}}(s+3)dsdt \\ &= \frac{4}{3\Gamma(\frac{1}{2})^2}x^{\frac{1}{2}}y^{\frac{1}{2}}(2x+9) = 0.42x^{\frac{1}{2}}y^{\frac{1}{2}}(2x+9).\end{aligned}$$

Again, the fractal perturbation of $\mathcal{I}^\nu g$ is given as $\mathcal{F}_L^\alpha(\mathcal{I}^\nu g)$, represented in Figure 2 for different value of $\alpha$. Consequently, exerting equation (2.3), $\mathcal{F}_L^\alpha(\mathcal{I}^\nu g)$ satisfies

$$\begin{aligned}\mathcal{F}_L^\alpha(\mathcal{I}^\nu g)(x,y) &= \mathcal{I}^\nu g(x,y) + \alpha\big[\mathcal{F}_L^\alpha(\mathcal{I}^\nu g)(u_i^{-1}(x), v_j^{-1}(y)) - L\mathcal{I}^\nu g(u_i^{-1}(x), v_j^{-1}(y))\big] \\ &= 0.42x^{\frac{1}{2}}y^{\frac{1}{2}}(2x+9) + \alpha\big[\mathcal{F}_L^\alpha(\mathcal{I}^\nu g)(u_i^{-1}(x), v_j^{-1}(y)) - \big(u_i^{-1}(x) - 2\big) \\ &\quad 0.42(u_i^{-1}(x))^{\frac{1}{2}}(v_j^{-1}(y))^{\frac{1}{2}}\big(2u_i^{-1}(x)+9\big) - u_i^{-2}(x) + 9\big],\end{aligned}$$

for all $(x, y) \in I_i \times J_j$ and $i, j \in \Sigma_4$. Observe that for $\alpha = 0$, $\mathcal{F}_L^\alpha(\mathcal{I}^\nu g) = \mathcal{I}^\nu g$, represented in (**a**) of Figure 2. Hence, $\mathcal{F}_L^\alpha \circ \mathcal{I}^\nu(g) \neq \mathcal{I}^\nu \circ \mathcal{F}_L^\alpha(g)$.

REMARK 3.5. It is observed that $\mathcal{I}^\nu$ is a bounded linear operator on $\mathcal{C}(I \times J)$. Furthermore, we get

$$\|\mathcal{I}^\nu\| \leq \frac{(x_N - x_0)^{\nu_1}(y_M - y_0)^{\nu_2}}{\Gamma(\nu_1+1)\Gamma(\nu_2+1)},$$

where $\nu = (\nu_1, \nu_2)$ with $\nu_1, \nu_2 > 0$.

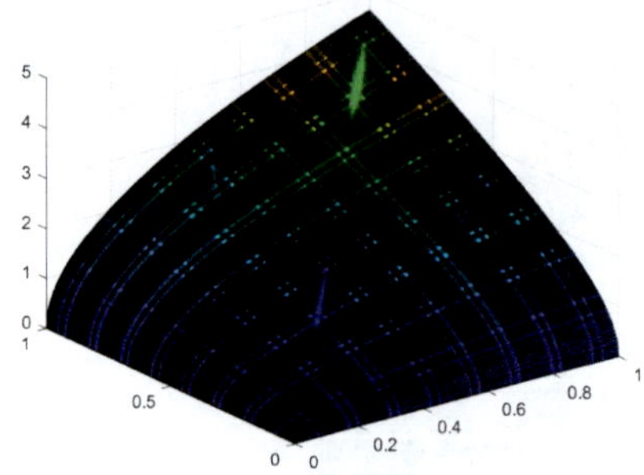


**(a)** $\mathcal{G}(\mathcal{F}_L^\alpha(\mathcal{I}^\nu g))$ when $\alpha = 0.00$

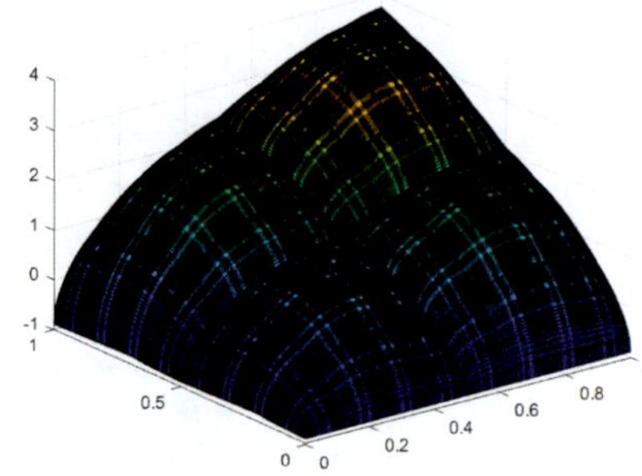


**(b)** $\mathcal{G}(\mathcal{F}_L^\alpha(\mathcal{I}^\nu g))$ when $\alpha = 0.09$

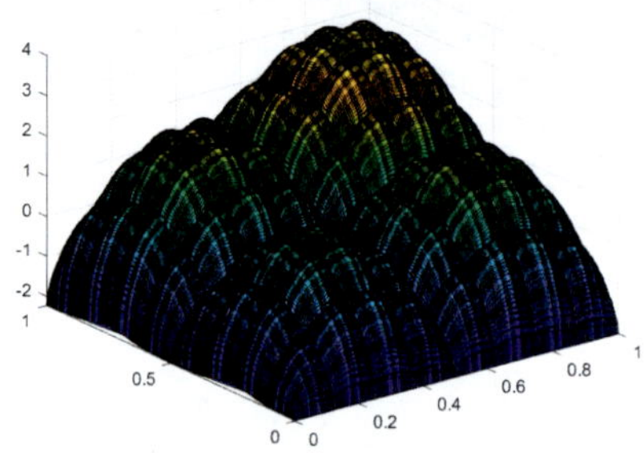


**(c)** $\mathcal{G}(\mathcal{F}_L^\alpha(\mathcal{I}^\nu g))$ when $\alpha = 0.2$

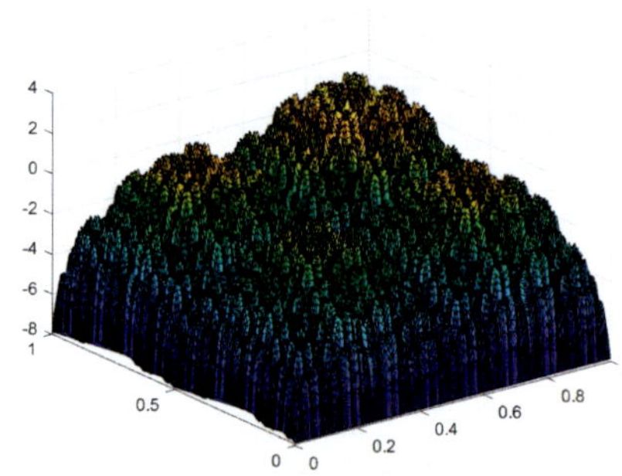


**(d)** $\mathcal{G}(\mathcal{F}_L^\alpha(\mathcal{I}^\nu g))$ when $\alpha = 0.5$

FIGURE 2

It is witnessed from Theorem 2.11 and Remark 3.5 that $\mathcal{F}_L^\alpha \circ \mathcal{I}^\nu$, $\mathcal{I}^\nu \circ \mathcal{F}_L^\alpha$ and $\mathcal{F}_L^\alpha + \mathcal{I}^\nu$ are bounded linear operators on $\mathcal{C}(I \times J)$. Moreover, we have

$$\|\mathcal{F}_L^\alpha \circ \mathcal{I}^\nu\| \le \|\mathcal{F}_L^\alpha\|\|\mathcal{I}^\nu\| \le \Big(1 + \frac{\|\alpha\|_\infty \|\mathfrak{I} - L\|}{1 - \|\alpha\|_\infty}\Big)\Big(\frac{(x_N - x_0)^{\nu_1}(y_M - y_0)^{\nu_2}}{\Gamma(\nu_1+1)\Gamma(\nu_2+1)}\Big),$$

$$\|\mathcal{I}^\nu \circ \mathcal{F}_L^\alpha\| \le \|\mathcal{F}_L^\alpha\|\|\mathcal{I}^\nu\| \le \Big(1 + \frac{\|\alpha\|_\infty \|\mathfrak{I} - L\|}{1 - \|\alpha\|_\infty}\Big)\Big(\frac{(x_N - x_0)^{\nu_1}(y_M - y_0)^{\nu_2}}{\Gamma(\nu_1+1)\Gamma(\nu_2+1)}\Big), \text{ and}$$

$$\|\mathcal{F}_L^\alpha + \mathcal{I}^\nu\| \le \Big(1 + \frac{\|\alpha\|_\infty \|\mathfrak{I} - L\|}{1 - \|\alpha\|_\infty}\Big) + \Big(\frac{(x_N - x_0)^{\nu_1}(y_M - y_0)^{\nu_2}}{\Gamma(\nu_1+1)\Gamma(\nu_2+1)}\Big).$$

PROPOSITION 3.6. *Let $\nu = (\nu_1, \nu_2)$ with $\nu_1, \nu_2 > 0$. The bound for the spectral radius of the operators $\mathcal{F}_L^\alpha \circ \mathcal{I}^\nu$, $\mathcal{I}^\nu \circ \mathcal{F}_L^\alpha$, and $\mathcal{F}_L^\alpha + \mathcal{I}^\nu$ are estimated as follows:*

$$\rho(\mathcal{F}_L^\alpha \circ \mathcal{I}^\nu) \le \Big(1 + \frac{\|\alpha\|_\infty \|\mathfrak{I} - L\|}{1 - \|\alpha\|_\infty}\Big)\Big(\frac{(x_N - x_0)^{\nu_1}(y_M - y_0)^{\nu_2}}{\Gamma(\nu_1+1)\Gamma(\nu_2+1)}\Big),$$

$$\rho(\mathcal{I}^\nu \circ \mathcal{F}_L^\alpha) \le \Big(1 + \frac{\|\alpha\|_\infty \|\mathfrak{I} - L\|}{1 - \|\alpha\|_\infty}\Big)\Big(\frac{(x_N - x_0)^{\nu_1}(y_M - y_0)^{\nu_2}}{\Gamma(\nu_1+1)\Gamma(\nu_2+1)}\Big), \textit{ and}$$

$$\rho(\mathcal{F}_L^\alpha + \mathcal{I}^\nu) \le \Big(1 + \frac{\|\alpha\|_\infty \|\mathfrak{I} - L\|}{1 - \|\alpha\|_\infty}\Big) + \Big(\frac{(x_N - x_0)^{\nu_1}(y_M - y_0)^{\nu_2}}{\Gamma(\nu_1+1)\Gamma(\nu_2+1)}\Big).$$

PROOF. The proof follows from Theorem 2.7, hence it is omitted. □

In the forthcoming theorems, we investigate the impact of the operator $\mathcal{I}^\nu \circ \mathcal{F}_L^\alpha$ on fractal dimension. The definitions of the Hausdorff and box dimensions are mentioned in Section 2; for details, see Ref. [**8**].

THEOREM 3.7. *Let $\mathcal{I}^\nu \circ \mathcal{F}_L^\alpha : \mathcal{C}(I \times J) \to \mathcal{C}(I \times J)$ be the operator defined as above and $\nu = (\nu_1, \nu_2)$ with $0 < \nu_1, \nu_2 < 1$. Then*

$$2 \le \dim_{\mathcal{H}} \mathcal{G}(\mathcal{I}^\nu \circ \mathcal{F}_L^\alpha(f)) \le \underline{\dim}_B \mathcal{G}(\mathcal{I}^\nu \circ \mathcal{F}_L^\alpha(f)) \le \overline{\dim}_B \mathcal{G}(\mathcal{I}^\nu \circ \mathcal{F}_L^\alpha(f)) \le 3 - \min\{\nu_1, \nu_2\},$$

*for all $f \in \mathcal{C}(I \times J)$.*

PROOF. Let $f \in \mathcal{C}(I \times J)$ and $0 < \nu_1, \nu_2 < 1$. For $0 < x_0 \le x < x + h_1 \le x_N$, $0 < y_0 \le y < y + h_2 \le y_M$, $h_1, h_2 > 0$. Consider

$$\begin{aligned}
&|\mathcal{I}^\nu \circ \mathcal{F}_L^\alpha(f)(x + h_1, y + h_2) - \mathcal{I}^\nu \circ \mathcal{F}_L^\alpha(f)(x, y)| \\
&\quad = |\mathcal{I}^\nu(f_L^\alpha)(x + h_1, y + h_2) - \mathcal{I}^\nu(f_L^\alpha)(x, y)| \\
&\quad = \frac{1}{\Gamma(\nu_1)\Gamma(\nu_2)} \bigg| \int_{x_0}^{x+h_1} \int_{y_0}^{y+h_2} (x + h_1 - s)^{\nu_1 - 1}(y + h_2 - t)^{\nu_2 - 1} f_L^\alpha(s, t)\,dsdt - \\
&\qquad \int_{x_0}^{x} \int_{y_0}^{y} (x - s)^{\nu_1 - 1}(y - t)^{\nu_2 - 1} f_L^\alpha(s, t)\,dsdt \bigg| \\
&\quad \le \frac{1}{\Gamma(\nu_1)\Gamma(\nu_2)} (K_1 + K_2 + K_3 + K_4),
\end{aligned}$$

where

$$\begin{aligned}
K_1 &= \bigg| \int_{x_0}^{x} \int_{y_0}^{y} \Big[(x+h_1-s)^{\nu_1-1}(y+h_2-t)^{\nu_2-1} - (x-s)^{\nu_1-1}(y-t)^{\nu_2-1}\Big] f_L^\alpha(s,t)\,dsdt \bigg|, \\
K_2 &= \bigg| \int_{x_0}^{x} \int_{y}^{y+h_2} (x + h_1 - s)^{\nu_1 - 1}(y + h_2 - t)^{\nu_2 - 1} f_L^\alpha(s, t)\,dsdt \bigg|, \\
K_3 &= \bigg| \int_{x}^{x+h_1} \int_{y_0}^{y} (x + h_1 - s)^{\nu_1 - 1}(y + h_2 - t)^{\nu_2 - 1} f_L^\alpha(s, t)\,dsdt \bigg|, \\
K_4 &= \bigg| \int_{x}^{x+h_1} \int_{y}^{y+h_2} (x + h_1 - s)^{\nu_1 - 1}(y + h_2 - t)^{\nu_2 - 1} f_L^\alpha(s, t)\,dsdt \bigg|.
\end{aligned}$$

Since $f \in \mathcal{C}(I \times J)$, the obtained fractal perturbation $f_L^\alpha$ is also continuous on $I \times J$. Thus, there exists $\xi > 0$ such that

$$|f_L^\alpha(x, y)| \le \xi,\ \forall\,(x, y) \in I \times J.$$

Now, we estimate the bounds of $K_1, K_2, K_3$ and $K_4$. Consider

$$\begin{aligned}
K_1 &\leq \xi \int_{x_0}^{x}\int_{y_0}^{y} \left|(x+h_1-s)^{\nu_1-1}(y+h_2-t)^{\nu_2-1}-(x-s)^{\nu_1-1}(y-t)^{\nu_2-1}\right|dsdt\\
&= \xi \int_{x_0}^{x}\int_{y_0}^{y} \big|(x+h_1-s)^{\nu_1-1}(y+h_2-t)^{\nu_2-1}-(x+h_1-s)^{\nu_1-1}(y-t)^{\nu_2-1}+\\
&\quad (x+h_1-s)^{\nu_1-1}(y-t)^{\nu_2-1}-(x-s)^{\nu_1-1}(y-t)^{\nu_2-1}\big|dsdt\\
&\leq \xi \int_{x_0}^{x}\int_{y_0}^{y} \left|(x+h_1-s)^{\nu_1-1}\right|\left|(y+h_2-t)^{\nu_2-1}-(y-t)^{\nu_2-1}\right|dsdt+\\
&\quad \xi \int_{x_0}^{x}\int_{y_0}^{y} \left|(y-t)^{\nu_2-1}\right|\left|(x+h_1-s)^{\nu_1-1}-(x-s)^{\nu_1-1}\right|dsdt\\
&= \xi \int_{x_0}^{x}\int_{y_0}^{y} (x+h_1-s)^{\nu_1-1}\left[(y-t)^{\nu_2-1}-(y+h_2-t)^{\nu_2-1}\right]dsdt+\\
&\quad \xi \int_{x_0}^{x}\int_{y_0}^{y} (y-t)^{\nu_2-1}\left[(x-s)^{\nu_1-1}-(x+h_1-s)^{\nu_1-1}\right]dsdt\\
&= \frac{\xi}{\nu_2}\left[h_2^{\nu_2}+(y-y_0)^{\nu_2}-(y+h_2-y_0)^{\nu_2}\right]\int_{x_0}^{x}(x+h_1-s)^{\nu_1-1}ds+\\
&\quad \frac{\xi}{\nu_1}\left[h_1^{\nu_1}+(x-x_0)^{\nu_1}-(x+h_1-x_0)^{\nu_1}\right]\int_{y_0}^{y}(y-t)^{\nu_2-1}dt.
\end{aligned}$$

Further, it is notable that

$$\begin{aligned}
&\int_{x_0}^{x}(x+h_1-s)^{\nu_1-1}ds \leq \frac{(x_N-x_0)^{\nu_1}}{\nu_1}, \text{ and}\\
&\int_{y_0}^{y}(y-t)^{\nu_2-1}dt \leq \frac{(y_M-y_0)^{\nu_2}}{\nu_2}.
\end{aligned}$$

Also, it is observed that $(x-x_0)^{\nu_1} \leq (x+h_1-x_0)^{\nu_1}$, as a result

$$h_1^{\nu_1}+(x-x_0)^{\nu_1}-(x+h_1-x_0)^{\nu_1} \leq h_1^{\nu_1}.$$

Consequently, the estimation of $K_1$ reduces to

$$K_1 \leq \frac{\xi}{\nu_1\nu_2}\Big(h_2^{\nu_2}(x_N-x_0)^{\nu_1}+h_1^{\nu_1}(y_M-y_0)^{\nu_2}\Big).$$

Again, consider

$$\begin{aligned}
K_2 &\leq \xi \int_{x_0}^{x}\int_{y}^{h_2}(x+h_1-s)^{\nu_1-1}(y+h_2-t)^{\nu_2-1}dsdt\\
&\leq \frac{\xi}{\nu_1\nu_2}(x_N-x_0)^{\nu_1}h_2^{\nu_2}.
\end{aligned}$$

Similarly, we obtain

$$\begin{aligned}
K_3 &\leq \frac{\xi}{\nu_1\nu_2}(y_M-y_0)^{\nu_2}h_1^{\nu_1}, \text{ and}\\
K_4 &\leq \frac{\xi}{\nu_1\nu_2}h_1^{\nu_1}h_2^{\nu_2}.
\end{aligned}$$

Thus,

$$
\begin{aligned}
&|\mathcal{I}^\nu \circ \mathcal{F}_L^\alpha(f)(x+h_1, y+h_2) - \mathcal{I}^\nu \circ \mathcal{F}_L^\alpha(f)(x,y)| \\
&\leq \frac{\xi}{\Gamma(\nu_1+1)\Gamma(\nu_2+1)}\Big(2h_2^{\nu_2}(x_N - x_0)^{\nu_1} + 2h_1^{\nu_1}(y_M - y_0)^{\nu_2} + h_1^{\nu_1}h_2^{\nu_2}\Big).
\end{aligned}
$$

For suitable constant $C$, we obtain

$$
|\mathcal{I}^\nu \circ \mathcal{F}_L^\alpha(f)(x+h_1, y+h_2) - \mathcal{I}^\nu \circ \mathcal{F}_L^\alpha(f)(x,y)| \leq C\big(h_1^{\nu_1} + h_2^{\nu_2}\big) \leq 2C\|(h_1,h_2)\|_2^\eta,
$$

where $\eta = \min\{\nu_1, \nu_2\}$. The last step of the above inequality is obtained by exerting $h_1, h_2 \leq \|(h_1,h_2)\|_2$. Therefore, for suitable $C > 0$, we get

$$
|\mathcal{I}^\nu \circ \mathcal{F}_L^\alpha(f)(x+h_1, y+h_2) - \mathcal{I}^\nu \circ \mathcal{F}_L^\alpha(f)(x,y)| \leq C\|(x+h_1, y+h_2) - (x,y)\|_2^\eta.
$$

Consequently, exerting Theorem 2.10 completes the proof. □

The Hölder space $\mathcal{C}^\sigma(I \times J)$ is defined as follows:

$$
\mathcal{C}^\sigma(I \times J) := \{f \in \mathcal{C}(I \times J) : f \text{ is Hölder continuous with exponent } \sigma\}.
$$

THEOREM 3.8. *Let $\mathcal{I}^\nu \circ \mathcal{F}_L^\alpha : \mathcal{C}^\sigma(I \times J) \to \mathcal{C}(I \times J)$ be the operator defined as above with $|I \times J| \leq 1$, and $\nu = (\nu_1, \nu_2)$ with $0 < \nu_1, \nu_2 < 1$. Further, $\alpha, Lf \in \mathcal{C}^\sigma(I \times J)$ such that*

$$
\max_{ij} \frac{NM}{|a_i|^\sigma |c_j|^\sigma}(2\|\alpha\|_\infty + [\alpha]_\sigma) < 1,
$$

*where $[\alpha]_\sigma = \sup_{\substack{(x,y),(z,w)\in I\times J \\ (x,y)\neq(z,w)}} \frac{|\alpha(x,y)-\alpha(z,w)|}{\|(x,y)-(z,w)\|_2^\sigma}$. Also, assume $f \in \mathcal{C}^\sigma(I \times J)$ with $f(x_0, y_0) = 0$. Then*

$$
2 \leq \dim_{\mathcal{H}} \mathcal{G}(\mathcal{I}^\nu \circ \mathcal{F}_L^\alpha(f)) \leq \underline{\dim}_B \mathcal{G}(\mathcal{I}^\nu \circ \mathcal{F}_L^\alpha(f)) \leq \overline{\dim}_B \mathcal{G}(\mathcal{I}^\nu \circ \mathcal{F}_L^\alpha(f)) \leq 3 - \max\{\sigma, \eta\},
$$

*where $\eta = \min\{\nu_1, \nu_2\}$.*

PROOF. Let $f \in \mathcal{C}(I \times J)$ and $0 < \nu_1, \nu_2 < 1$. For $0 < x_0 \leq x < x + h_1 \leq x_N$, $0 < y_0 \leq y < y + h_2 \leq y_M$, $h_1, h_2 > 0$. Consider

$$
\begin{aligned}
&\mathcal{I}^\nu \circ \mathcal{F}_L^\alpha(f)(x+h_1, y+h_2) - \mathcal{I}^\nu \circ \mathcal{F}_L^\alpha(f)(x,y) \\
&= \mathcal{I}^\nu(f_L^\alpha)(x+h_1, y+h_2) - \mathcal{I}^\nu(f_L^\alpha)(x,y) \\
&= \frac{1}{\Gamma(\nu_1)\Gamma(\nu_2)}\Big(\int_{x_0}^{x+h_1}\int_{y_0}^{y+h_2} (x+h_1-s)^{\nu_1-1}(y+h_2-t)^{\nu_2-1} f_L^\alpha(s,t)\,ds\,dt - \\
&\quad \int_{x_0}^{x}\int_{y_0}^{y} (x-s)^{\nu_1-1}(y-t)^{\nu_2-1} f_L^\alpha(s,t)\,ds\,dt\Big) \\
&= \frac{1}{\Gamma(\nu_1)\Gamma(\nu_2)}(K_1 + K_2 + K_3 + K_4 - K_5),
\end{aligned}
$$

where

$$\begin{aligned}
K_1 &= \int_{x_0}^{x_0+h_1}\int_{y_0}^{y_0+h_2}(x+h_1-s)^{\nu_1-1}(y+h_2-t)^{\nu_2-1}f_L^{\alpha}(s,t)dsdt,\\
K_2 &= \int_{x_0}^{x_0+h_1}\int_{y_0+h_2}^{y+h_2}(x+h_1-s)^{\nu_1-1}(y+h_2-t)^{\nu_2-1}f_L^{\alpha}(s,t)dsdt,\\
K_3 &= \int_{x_0+h_1}^{x+h_1}\int_{y_0}^{y_0+h_2}(x+h_1-s)^{\nu_1-1}(y+h_2-t)^{\nu_2-1}f_L^{\alpha}(s,t)dsdt,\\
K_4 &= \int_{x_0+h_1}^{x+h_1}\int_{y_0+h_2}^{y+h_2}(x+h_1-s)^{\nu_1-1}(y+h_2-t)^{\nu_2-1}f_L^{\alpha}(s,t)dsdt,\\
K_5 &= \int_{x_0}^{x}\int_{y_0}^{y}(x-s)^{\nu_1-1}(y-t)^{\nu_2-1}f_L^{\alpha}(s,t)dsdt.
\end{aligned}$$

Consider

$$\begin{aligned}
&|K_4-K_5|\\
&=\Big|\int_{x_0+h_1}^{x+h_1}\int_{y_0+h_2}^{y+h_2}(x+h_1-s)^{\nu_1-1}(y+h_2-t)^{\nu_2-1}f_L^{\alpha}(s,t)dsdt-\\
&\quad\int_{x_0}^{x}\int_{y_0}^{y}(x-s)^{\nu_1-1}(y-t)^{\nu_2-1}f_L^{\alpha}(s,t)dsdt\Big|\\
&=\Big|\int_{x_0+h_1}^{x+h_1}\int_{y_0+h_2}^{y+h_2}(x+h_1-s)^{\nu_1-1}(y+h_2-t)^{\nu_2-1}f_L^{\alpha}(s,t)dsdt-\\
&\quad\int_{x_0+h_1}^{x+h_1}\int_{y_0+h_2}^{y+h_2}(x+h_1-s)^{\nu_1-1}(y+h_2-t)^{\nu_2-1}f_L^{\alpha}(s-h_1,t-h_2)dsdt\Big|\\
&\le\int_{x_0+h_1}^{x+h_1}\int_{y_0+h_2}^{y+h_2}\Big|(x+h_1-s)^{\nu_1-1}\Big|\Big|(y+h_2-t)^{\nu_2-1}\Big|\Big|f_L^{\alpha}(s,t)-f_L^{\alpha}(s-h_1,t-h_2)\Big|dsdt.
\end{aligned}$$

With reference to [**2**, Theorem 3.1], the fractal perturbation of $f$ is also Hölder continuous with Hölder exponent $\sigma$, that is, $f_L^{\alpha}\in\mathcal{C}^{\sigma}(I\times J)$. Consequently, the above estimation reduces to

$$\begin{aligned}
|K_4-K_5| &\le C\|(h_1,h_2)\|_2^{\sigma}\int_{x_0+h_1}^{x+h_1}\int_{y_0+h_2}^{y+h_2}(x+h_1-s)^{\nu_1-1}(y+h_2-t)^{\nu_2-1}dsdt\\
&= C\|(h_1,h_2)\|_2^{\sigma}\frac{(x-x_0)^{\nu_1}(y-y_0)^{\nu_2}}{\nu_1\nu_2}\\
&\le C\|(h_1,h_2)\|_2^{\sigma}\frac{(x_N-x_0)^{\nu_1}(y_M-y_0)^{\nu_2}}{\nu_1\nu_2}.
\end{aligned}$$

Further, consider

$$|K_1|\le\int_{x_0}^{x_0+h_1}\int_{y_0}^{y_0+h_2}(x+h_1-s)^{\nu_1-1}(y+h_2-t)^{\nu_2-1}|f_L^{\alpha}(s,t)|dsdt.$$

In view of the given condition, we have $f_L^\alpha(x_0, y_0) = f(x_0, y_0) = 0$. Thus, the above estimation reduces to

$$\begin{aligned}|K_1| &\le \int_{x_0}^{x_0+h_1}\int_{y_0}^{y_0+h_2}(x+h_1-s)^{\nu_1-1}(y+h_2-t)^{\nu_2-1}|f_L^\alpha(s,t)-f_L^\alpha(x_0,y_0)|dsdt\\ &\le \int_{x_0}^{x_0+h_1}\int_{y_0}^{y_0+h_2}(x_0+h_1-s)^{\nu_1-1}(y_0+h_2-t)^{\nu_2-1}|f_L^\alpha(s,t)-f_L^\alpha(x_0,y_0)|dsdt\\ &\le C\|(h_1,h_2)\|_2^\sigma \frac{h_1^{\nu_1}h_2^{\nu_2}}{\nu_1\nu_2}.\end{aligned}$$

On the same lines for $|K_2|$ and $|K_3|$, the upper bounds are estimated as

$$\begin{aligned}|K_2| &\le C\|(h_1,h_2)\|_2^\sigma \frac{(y_M-y_0)^{\nu_2}h_1^{\nu_1}}{\nu_1\nu_2},\\ |K_3| &\le C\|(h_1,h_2)\|_2^\sigma \frac{(x_N-x_0)^{\nu_1}h_2^{\nu_2}}{\nu_1\nu_2}.\end{aligned}$$

Thus,

$$|\mathcal{I}^\nu \circ \mathcal{F}_L^\alpha(f)(x+h_1,y+h_2) - \mathcal{I}^\nu \circ \mathcal{F}_L^\alpha(f)(x,y)| \le C\|(h_1,h_2)\|_2^\sigma,$$

for some suitable constant $C$. With reference to Theorems 2.10 and 3.7, the desired result is concluded. □

Before starting the discussion of the next theorem, let us recall the following:

LEMMA 3.9 ([**37**, Lemma 3.12]). *Let $g : I \times J \to \mathbb{R}$ be a continuous function. Assume that $0 < \delta < \min\{x_N - x_0, y_M - y_0\}$, $\frac{x_N-x_0}{\delta} \le n \le 1 + \frac{x_N-x_0}{\delta}$, and $\frac{y_M-y_0}{\delta} \le m \le 1+\frac{y_M-y_0}{\delta}$ for some $n, m \in \mathbb{N}$. If $N_\delta(\mathcal{G}(g), I \times J)$ denotes the number of $\delta$-cubes that intersect with graph of $g$, then*

$$\delta^{-1}\sum_{(r,s)\in\Sigma_n\times\Sigma_m}\mathcal{R}_g[A_{rs}] \le N_\delta(\mathcal{G}(g), I\times J) \le 2nm+\delta^{-1}\sum_{(r,s)\in\Sigma_n\times\Sigma_m}\mathcal{R}_g[A_{rs}], \tag{3.1}$$

*where $A_{rs} = [x_0 + (r-1)\delta, x_0 + r\delta] \times [y_0 + (s-1)\delta, y_0 + s\delta]$ and $\mathcal{R}_g[A_{rs}] = \sup_{(x,y),(z,w)\in A_{rs}}|g(x,y) - g(z,w)|$.*

THEOREM 3.10. *Let $\alpha$, $f$, and $Lf$ be Lipschitz maps on $I \times J$. If $\|\alpha\|_\infty NM \min\{N, M\} < 1$, then*

$$\dim_{\mathcal{H}}(\mathcal{G}(f_L^\alpha)) = \dim_B(\mathcal{G}(f_L^\alpha)) = 2.$$

PROOF. For simplicity in notations, we may assume $I = [a, b]$ and $J = [c, d]$. Divide $I$ in $N$ equal sub-intervals of length $\frac{b-a}{N}$, and $J$ in $M$ equal sub-intervals of length $\frac{d-c}{M}$. Thus, the net $\Delta$ is given as

$$\Delta = \Big\{a, a+\frac{b-a}{N}, \dots, a+(N-1)\frac{b-a}{N}, b\Big\} \times \Big\{c, c+\frac{d-c}{M}, \dots, c+(M-1)\frac{d-c}{M}, d\Big\}.$$

Moreover, for any $i \in \Sigma_N$, the functions $u_i$ are obtained from equation (2.2) as follows

$$u_i(x) = \begin{cases}\frac{1}{N}x + \frac{a(N-i)+b(i-1)}{N}, & \text{if } i \text{ is odd},\\ \frac{-1}{N}x + \frac{a(N-i+1)+bi}{N}, & \text{if } i \text{ is even}.\end{cases}$$

Similarly, for any $j \in \Sigma_M$, the functions $v_j$ are estimated. Further, $f_L^\alpha$ satisfies the self-referential equation (2.3). Let $(x, y), (x_*, y_*) \in I \times J$. Notice that

$$\begin{aligned}
&\big|f_L^\alpha(u_i(x), v_j(y)) - f_L^\alpha(u_i(x_*), v_j(y_*))\big| \\
&\quad= \big|\big[f(u_i(x), v_j(y)) + \alpha(u_i(x), v_j(y))\{f_L^\alpha(x, y) - Lf(x, y)\}\big] - \\
&\qquad \big[f(u_i(x_*), v_j(y_*)) + \alpha(u_i(x_*), v_j(y_*))\{f_L^\alpha(x_*, y_*) - Lf(x_*, y_*)\}\big]\big| \\
&\quad\leq \big|f(u_i(x), v_j(y)) - f(u_i(x_*), v_j(y_*))\big| + \big|\alpha(u_i(x), v_j(y))\big| \\
&\qquad \big\{\big|f_L^\alpha(x, y) - f_L^\alpha(x_*, y_*)\big| + \big|Lf(x, y) - Lf(x_*, y_*)\big|\big\} + \\
&\qquad \big|f_L^\alpha(x_*, y_*) - Lf(x_*, y_*)\big|\big|\alpha(u_i(x), v_j(y)) - \alpha(u_i(x_*), v_j(y_*))\big|.
\end{aligned}$$

Since $\alpha$, $f$ and $Lf$ are Lipschitz maps on $I \times J$, there exist Lipschitz constants $C_\alpha, C_f$ and $C_{Lf}$, respectively. Consequently, the above inequality reduces to

$$\begin{aligned}
&\big|f_L^\alpha(u_i(x), v_j(y)) - f_L^\alpha(u_i(x_*), v_j(y_*))\big| \\
&\quad\leq C_f \|(u_i(x), v_j(y)) - (u_i(x_*), v_j(y_*))\|_2 + \|\alpha\|_\infty \big|f_L^\alpha(x, y) \\
&\qquad - f_L^\alpha(x_*, y_*)\big| + C_{Lf}\|\alpha\|_\infty \|(x, y) - (x_*, y_*)\|_2 + \\
&\qquad C_\alpha \|f_L^\alpha - Lf\|_\infty \|(u_i(x), v_j(y)) - (u_i(x_*), v_j(y_*))\|_2.
\end{aligned} \tag{3.2}$$

Notably, $I \times J = \bigcup_{(i,j) \in \Sigma_N \times \Sigma_M} u_i(I) \times v_j(J)$. This gives

$$N_\delta\big(\mathcal{G}(f_L^\alpha), I \times J\big) \leq \sum_{(i,j) \in \Sigma_N \times \Sigma_M} N_\delta\big(\mathcal{G}(f_L^\alpha), u_i(I) \times v_j(J)\big). \tag{3.3}$$

Further, assume that $0 < \delta < \min\{\frac{b-a}{N}, \frac{d-c}{M}\}$ and $n, m \in \mathbb{N}$ such that

$$\frac{b-a}{N\delta} \leq n \leq 1 + \frac{b-a}{N\delta}, \text{ and } \frac{d-c}{M\delta} \leq m \leq 1 + \frac{d-c}{M\delta}. \tag{3.4}$$

Exerting Lemma 3.9, we obtain

$$N_\delta\big(\mathcal{G}(f_L^\alpha), u_i(I) \times v_j(J)\big) \leq 2nm + \delta^{-1} \sum_{(r,s) \in \Sigma_n \times \Sigma_m} \mathcal{R}_{f_L^\alpha}[A_{rs}], \tag{3.5}$$

where $A_{rs} = \Big[a + (i-1)\frac{b-a}{N} + (r-1)\delta, a + (i-1)\frac{b-a}{N} + r\delta\Big] \times \Big[c + (j-1)\frac{d-c}{M} + (s-1)\delta, c + (j-1)\frac{d-c}{M} + s\delta\Big] \subset I_i \times J_j$ is the square of side length $\delta$. Notice that

$$
\begin{aligned}
&\mathcal{R}_{f_L^\alpha}[A_{rs}]\\
&= \sup_{\substack{(x,y),(x_*,y_*)\in I\times J\\ (u_i(x),v_j(y)),(u_i(x_*),v_j(y_*))\in A_{rs}}} \big|f_L^\alpha(u_i(x),v_j(y)) - f_L^\alpha(u_i(x_*),v_j(y_*))\big|\\
&\le C_f\sqrt{2}\delta + \|\alpha\|_\infty \sup_{\substack{(x,y),(x_*,y_*)\in I\times J\\ (u_i(x),v_j(y)),(u_i(x_*),v_j(y_*))\in A_{rs}}} \Big(\big|f_L^\alpha(x,y) - f_L^\alpha(x_*,y_*)\big| +\\
&\quad C_{Lf}\|(x,y)-(x_*,y_*)\|_2\Big) + C_\alpha\|f_L^\alpha - Lf\|_\infty\sqrt{2}\delta\\
&\le \|\alpha\|_\infty \sup_{\substack{(x,y),(x_*,y_*)\in I\times J\\ (x,y),(x_*,y_*)\in[a+(r-1)\delta N, a+r\delta N]\times[c+(s-1)\delta M, c+s\delta M]}} \big|f_L^\alpha(x,y) - f_L^\alpha(x_*,y_*)\big|\\
&\quad + C_{Lf}\|\alpha\|_\infty\sqrt{2}\delta\max\{N,M\} + C_f\sqrt{2}\delta + C_\alpha\|f_L^\alpha - Lf\|_\infty\sqrt{2}\delta\\
&= \|\alpha\|_\infty\mathcal{R}_{f_L^\alpha}([a+(r-1)\delta N, a+r\delta N]\times[c+(s-1)\delta M, c+s\delta M])\\
&\quad + C_{Lf}\|\alpha\|_\infty\sqrt{2}\delta\max\{N,M\} + C_f\sqrt{2}\delta + C_\alpha\|f_L^\alpha - Lf\|_\infty\sqrt{2}\delta.
\end{aligned}
$$

In the last estimation, without loss of generality, we assume $i, j$ to be odd. On the same lines, the remaining cases are discussed. Consequently, from equation (3.5), we obtain

$$
\begin{aligned}
&N_\delta\big(\mathcal{G}(f_L^\alpha), u_i(I)\times v_j(J)\big)\\
&\le 2nm + \delta^{-1}\sum_{(r,s)\in\Sigma_n\times\Sigma_m}\Bigg(\|\alpha\|_\infty\mathcal{R}_{f_L^\alpha}([a+(r-1)\delta N, a+r\delta N]\times[c+(s-1)\delta M,\\
&\quad c+s\delta M]) + C_{Lf}\|\alpha\|_\infty\sqrt{2}\delta\max\{N,M\} + C_f\sqrt{2}\delta + C_\alpha\|f_L^\alpha - Lf\|_\infty\sqrt{2}\delta\Bigg)\\
&= 2nm + \Big(C_{Lf}\|\alpha\|_\infty\max\{N,M\} + C_f + C_\alpha\|f_L^\alpha - Lf\|_\infty\Big)\sqrt{2}nm+\\
&\quad \delta^{-1}\|\alpha\|_\infty\sum_{(r,s)\in\Sigma_n\times\Sigma_m}\mathcal{R}_{f_L^\alpha}([a+(r-1)\delta N, a+r\delta N]\times[c+(s-1)\delta M, c+s\delta M]).
\end{aligned}
$$

Again, using equation (3.4) and Lemma 3.9, we get

$$
\begin{aligned}
&(\delta\min\{N,M\})^{-1}\sum_{(r,s)\in\Sigma_n\times\Sigma_m}\mathcal{R}_{f_L^\alpha}\Big([a+(r-1)\delta N, a+r\delta N]\times\\
&\qquad [c+(s-1)\delta M, c+s\delta M]\Big) \le N_{\delta\min\{N,M\}}\big(\mathcal{G}(f_L^\alpha), I\times J\big).
\end{aligned}
$$

Thus, using the above estimation and equation (3.3), we obtain

$$
\begin{aligned}
&N_\delta\big(\mathcal{G}(f_L^\alpha), I\times J\big)\\
&\le \sum_{(i,j)\in\Sigma_N\times\Sigma_M}\Bigg(\Big(\sqrt{2} + C_f + C_{Lf}\|\alpha\|_\infty\max\{N,M\} + C_\alpha\|f_L^\alpha - Lf\|_\infty\Big)\sqrt{2}nm+\\
&\quad \delta^{-1}\|\alpha\|_\infty\delta\min\{N,M\}N_{\delta\min\{N,M\}}\big(\mathcal{G}(f_L^\alpha), I\times J\big)\Bigg).
\end{aligned}
$$

Moreover, $\delta \leq \delta \min\{N, M\}$. This gives $N_{\delta \min\{N,M\}}\big(\mathcal{G}(f_L^\alpha), I\times J\big) \leq N_\delta\big(\mathcal{G}(f_L^\alpha), I\times J\big)$. Therefore, we get

$$\begin{aligned} &N_\delta\big(\mathcal{G}(f_L^\alpha), I \times J\big) \\ &\leq \Big(\sqrt{2} + C_f + C_{Lf}\|\alpha\|_\infty \max\{N, M\} + C_\alpha \|f_L^\alpha - Lf\|_\infty\Big)\sqrt{2}nmNM \\ &\quad + \|\alpha\|_\infty NM \min\{N, M\} N_\delta\big(\mathcal{G}(f_L^\alpha), I \times J\big). \end{aligned}$$

Consequently, using equation (3.4), above estimation reduces to

$$\begin{aligned} &N_\delta\big(\mathcal{G}(f_L^\alpha), I \times J\big) \\ &\leq \frac{\Big(\sqrt{2} + C_f + C_{Lf}\|\alpha\|_\infty \max\{N, M\} + C_\alpha \|f_L^\alpha - Lf\|_\infty\Big)\sqrt{2}nmNM}{\big(1 - \|\alpha\|_\infty NM \min\{N, M\}\big)} \\ &\leq \frac{\Big(\sqrt{2} + C_f + C_{Lf}\|\alpha\|_\infty \max\{N, M\} + C_\alpha \|f_L^\alpha - Lf\|_\infty\Big)\sqrt{2}(N\delta + b - a)(M\delta + d - c)\delta^{-2}}{\big(1 - \|\alpha\|_\infty NM \min\{N, M\}\big)}. \end{aligned}$$

Hence, from the above inequality, we obtain

$$\overline{\dim}_B(\mathcal{G}(f_L^\alpha)) = \limsup_{\delta \to 0^+} \frac{\log N_\delta(\mathcal{G}(f_L^\alpha))}{-\log \delta} \leq 2.$$

Since $f_L^\alpha$ is continuous on $I \times J$, we have

$$2 \leq \dim_{\mathcal{H}}(\mathcal{G}(f_L^\alpha)) \leq \underline{\dim}_B(\mathcal{G}(f_L^\alpha)) \leq \overline{\dim}_B(\mathcal{G}(f_L^\alpha)) \leq 2,$$

proving the result. □

### 3.1. Dimension preserving aspects with respect to fractional derivative.

LEMMA 3.11 ([**35**, Lemma 3.1]). *Let $A \subset \mathbb{R}^m$ and $f, g : A \to \mathbb{R}^n$ be continuous functions. Then*

$$\dim_{\mathcal{H}} \mathcal{G}(f + g) = \dim_{\mathcal{H}} \mathcal{G}(g),$$

*provided that $f$ is a Lipschitz function.*

LEMMA 3.12 ([**1**, Lemma 2.5]). *Let $h : I_1 \to \mathbb{R}$ be a Lipschitz map and $g : I_2 \to \mathbb{R}$ be a continuous map, where $I_1$ and $I_2$ are compact intervals of the real line. Define a map $f : I_1 \times I_2 \to \mathbb{R}$ as $f(x, y) = h(x) + g(y)$, $\forall\, (x, y) \in I_1 \times I_2$. Then*

$$\dim_{\mathcal{H}} \mathcal{G}(f) = \dim_{\mathcal{H}} \mathcal{G}(g) + 1.$$

REMARK 3.13. In view of properties of box dimension [**8**, Section 3.2], one gets that $\underline{\dim}_B$ and $\overline{\dim}_B$ are bi-Lipschitz invariant. Consequently, Lemma 3.11 and Lemma 3.12 hold for $\underline{\dim}_B$ and $\overline{\dim}_B$ also.

Now onwards intervals $I = [a, b]$ and $J = [c, d]$, for notational convenience. The following two results can be shown from [**35**], hence these are written without proof.

THEOREM 3.14. *Let $2 \leq \beta \leq 3$. The set $\mathcal{S}_\beta := \{f \in \mathcal{C}(I \times J) : \dim \mathcal{G}(f) = \beta\}$ is dense in $\mathcal{C}(I \times J)$.*

THEOREM 3.15. *Let $f \in \mathcal{C}(I \times J)$ with $f(x, y) \geq 0$, for all $(x, y) \in I \times J$. Then for any $\epsilon > 0$, there exists $h \in \mathcal{S}_\beta$ such that $h(x, y) \geq 0$, for all $(x, y) \in I \times J$, and $\|f - h\|_\infty < \epsilon$.*

PROPOSITION 3.16. *Let $2 \leq \beta < 3$ and $0 < \mu_1 < 3-\beta < 1, 0 < \mu_2 < 1$. Then the set $\mathcal{S}^{\mu}_{\beta} := \{f \in \mathcal{C}(I \times J) : \mathcal{D}^{\mu} f \in \mathcal{C}(I \times J),\ \dim_{\mathcal{H}} \mathcal{G}(f) = \beta\}$ is dense in $\mathcal{C}(I \times J)$.*

PROOF. Let $f \in \mathcal{C}(I \times J)$. Thus, by exerting Weierstrass approximation theorem, for any $\epsilon > 0$, there exists a polynomial $p \in \mathcal{C}(I \times J)$ such that

$$\|f - p\|_{\infty} < \frac{\epsilon}{2}.$$

Consider the case $\beta = 2$, then $\mathcal{P}(I \times J) \subseteq \mathcal{S}^{\mu}_{2} \subseteq \mathcal{C}(I \times J)$, where $\mathcal{P}(I \times J)$ is the set of all polynomials on $I \times J$. Since $\mathcal{P}(I \times J)$ is dense in $\mathcal{C}(I \times J)$, $\mathcal{S}^{\mu}_{2}$ is also dense in $\mathcal{C}(I \times J)$. Now, for $2 < \beta < 3$, define a sequence $\{f_n\}_{n=1}^{\infty}$ as follows

$$f_n(x, y) = p(x, y) + \frac{W_{\lambda,3-\beta}(x) + y}{n},\ \forall\, (x, y) \in I \times J, \tag{3.6}$$

where $W_{\lambda,3-\beta}$ is the Weierstrass function on $I$ specified as

$$W_{\lambda,3-\beta}(x) = \sum_{n=0}^{\infty} \lambda^{-(3-\beta)n} \sin(2\pi\lambda^n x),\ \forall\, x \in I,$$

where $\lambda \geq 2$ is an integer. Shen [**32**] computed the dimension of $W_{\lambda,3-\beta}$ as

$$\dim_{\mathcal{H}} \mathcal{G}(W_{\lambda,3-\beta}) = 2 + \frac{\log \lambda^{-(3-\beta)}}{\log \lambda} = \beta - 1.$$

Subsequently, utilizing Lemma 3.12, we obtain

$$\dim_{\mathcal{H}} \mathcal{G}(W_{\lambda,3-\beta}(x) + y) = \dim_{\mathcal{H}} \mathcal{G}(W_{\lambda,3-\beta}(x)) + 1 = \beta - 1 + 1 = \beta.$$

Again, $p \in \mathcal{C}(I \times J)$ is a polynomial on $I \times J$, so it is Lipschitz. Therefore, from Lemma 3.11, it is notable that

$$\dim_{\mathcal{H}} \mathcal{G}(f_n) = \beta.$$

It is plain to verify that $\{f_n\}_{n=1}^{\infty} \subseteq \mathcal{C}(I \times J)$. Further, with reference to the properties of fractional derivatives specified in Note 2.3, we get

$$\mathcal{D}^{\mu} f_n(x, y) = \mathcal{D}^{\mu} p(x, y) + \frac{1}{n}\Big(\mathcal{D}^{\mu} W_{\lambda,3-\beta}(x) + \mathcal{D}^{\mu} y\Big).$$

Using $0 < \mu_1 < 3 - \beta$ and [**40**, Theorem 3], we get $\mathcal{D}^{\mu} W_{\lambda,3-\beta} \in \mathcal{C}(I \times J)$. Observe that

$$|W_{\lambda,3-\beta}(x) + y| = \Big|\sum_{n=1}^{\infty} \lambda^{-(3-\beta)n} \sin(2\pi\lambda^n x) + y\Big| \leq \sum_{n=1}^{\infty} \lambda^{-(3-\beta)n} + |y| \leq C,$$

where $C$ is some constant. Therefore, from equation (3.6), we have

$$\|f_n - p\|_{\infty} \leq \frac{C}{n}.$$

Thus, for any $\epsilon > 0$, there exists positive integer $N$ such that $\|f_n - p\|_{\infty} < \frac{\epsilon}{2}$, for all $n > N$. Consequently, we obtain

$$\|f_n - f\|_{\infty} \leq \|f_n - p\|_{\infty} + \|f - p\|_{\infty} < \epsilon,\ \forall n > N.$$

Further, in view of Lemma 3.11, $\dim_{\mathcal{H}} \mathcal{G}(f_n) = \dim_{\mathcal{H}} \mathcal{G}(W_{\lambda,3-\beta}(x) + y) = \beta$, this proves the claim. □

PROPOSITION 3.17. *Let $2 \leq \beta_1 < \beta_2 < 3$ and $0 < \mu_1, \mu_2 < 3-\beta_2 < 1$. Then the set $\mathcal{S}^{\mu}_{\beta_1,\beta_2} := \{f \in \mathcal{C}(I \times J) : \mathcal{D}^{\mu} f \in \mathcal{C}(I \times J),\ \underline{\dim}_B \mathcal{G}(f) = \beta_1 < \beta_2 = \overline{\dim}_B \mathcal{G}(f)\}$ is dense in $\mathcal{C}(I \times J)$.*

PROOF. Let $f \in \mathcal{C}(I \times J)$. Thus, by exerting Weierstrass approximation theorem, for any $\epsilon > 0$, there exists a polynomial $p \in \mathcal{C}(I \times J)$ such that

$$\|f - p\|_\infty < \frac{\epsilon}{2}.$$

Now, define a sequence $\{f_n\}_{n=1}^{\infty}$ as

$$f_n(x,y) = p(x,y) + \frac{\mathcal{B}_{\beta_2,\Lambda}(x) + y}{n}, \ \forall\, (x,y) \in I \times J, \tag{3.7}$$

where $\mathcal{B}_{\beta_2,\Lambda}$ is the Besicovitch function on $I$, mentioned as

$$\mathcal{B}_{\beta_2,\Lambda}(x) = \sum_{n=1}^{\infty} (r^{\kappa^n})^{\beta_2 - 3} \sin(r^{\kappa^n} x), \ \forall\, x \in I.$$

Here, $\Lambda$ is a sequence chosen as $\{r^{\kappa^n}\}_{n=1}^{\infty}$, provided $\kappa = \frac{(\beta_1-3)(\beta_2-2)}{(\beta_1-2)(\beta_2-3)}$ and $r$ is an integer with $r \geq 2$ satisfying $r^{\kappa(\kappa-1)(\beta_2-2)} \geq 2.5$ and $r^{\kappa(\kappa-1)(3-\beta_2)} \geq 2.4$. Consequently, from [**33**, Theorem 3], we obtain

$$\underline{\dim}_B \mathcal{G}(\mathcal{B}_{\beta_2,\Lambda}) = \beta_1 - 1, \text{ and } \overline{\dim}_B \mathcal{G}(\mathcal{B}_{\beta_2,\Lambda}) = \beta_2 - 1.$$

Further, with the reference to Remark 3.13, one gets

$$\underline{\dim}_B \mathcal{G}(f_n) = \underline{\dim}_B \mathcal{G}(\mathcal{B}_{\beta_2,\Lambda}(x) + y) = \beta_1 - 1 + 1 = \beta_1, \text{ and}$$
$$\overline{\dim}_B \mathcal{G}(f_n) = \overline{\dim}_B \mathcal{G}(\mathcal{B}_{\beta_2,\Lambda}(x) + y) = \beta_2 - 1 + 1 = \beta_2.$$

Now, in view of the properties of RL fractional derivatives, we obtain

$$\mathcal{D}^\mu f_n(x,y) = \mathcal{D}^\mu p(x,y) + \frac{1}{n}\Big(\mathcal{D}^\mu \mathcal{B}_{\beta_2,\Lambda}(x) + \mathcal{D}^\mu y\Big).$$

Using the fact $0 < \mu_1, \mu_2 < 3 - \beta_2$ and [**40**, Theorem 3], we get that $\mathcal{D}^\mu \mathcal{B}_{\beta_2,\Lambda}$ is continuous on $I \times J$. As a result, $\mathcal{D}^\mu f_n \in \mathcal{C}(I \times J)$. Again, for some constant $C$, we have

$$\|f_n - p\|_\infty < \frac{C}{n}.$$

Consequently, for any $\epsilon > 0$, there exists positive integer $N$ such that $\|f_n - p\|_\infty < \frac{\epsilon}{2}$, for all $n > N$. Finally, we obtain

$$\|f_n - f\|_\infty \leq \|f_n - p\|_\infty + \|f - p\|_\infty < \epsilon, \ \forall\, n > N.$$

Thus, the proof is complete. □

LEMMA 3.18. *Let $\{f_n\}_{n=1}^{\infty}$ be a sequence of continuous functions on $I \times J$. If $f_n$ converges to $f$ uniformly on $I \times J$, then $\{\mathcal{I}^\nu f_n\}_{n=1}^{\infty}$ converges uniformly to $\mathcal{I}^\nu f$ on $I \times J$, for all $\nu = (\nu_1, \nu_2)$ with $\nu_1, \nu_2 > 0$.*

PROOF. It is observed from Remark 3.5 that $\mathcal{I}^\nu$ is the bounded linear operator on $\mathcal{C}(I \times J)$, and

$$\|\mathcal{I}^\nu f_n - \mathcal{I}^\nu f\|_\infty \leq \frac{1}{\Gamma(\nu_1+1)\Gamma(\nu_2+1)} \|f_n - f\|_\infty (b-a)^{\nu_1} (d-c)^{\nu_2}.$$

Since $\lim_{n\to\infty} \|f_n - f\|_\infty = 0$, we get $\lim_{n\to\infty} \|\mathcal{I}^\nu f_n - \mathcal{I}^\nu f\|_\infty = 0$. □

THEOREM 3.19. *Let $f \in \mathcal{C}(I \times J)$. Suppose that $\mathcal{D}^{\mu} f \in \mathcal{C}(I \times J)$ and $\dim \mathcal{G}(\mathcal{D}^{\mu} f) = \beta$, where $\beta \in [2,3]$ and $\mu = (\mu_1, \mu_2)$ with $0 < \mu_1, \mu_2 < 1$. Then there exists a sequence of continuous function $\{f_n\}_{n=1}^{\infty}$ on $I \times J$, that uniformly converges to $f$ such that $\mathcal{D}^{\mu} f_n \in \mathcal{C}(I \times J)$ and*

$$\dim \mathcal{G}(\mathcal{D}^{\mu} f_n) = \dim \mathcal{G}(\mathcal{D}^{\mu} f), \forall\, n.$$

PROOF. As $\mathcal{D}^{\mu} f \in \mathcal{C}(I \times J)$ and $\dim \mathcal{G}(\mathcal{D}^{\mu} f) = \beta$, in view of Theorem 3.14, there exists a sequence $\{g_n\}_{n=1}^{\infty} \subseteq \mathcal{S}_{\beta}$ that converges uniformly to $\mathcal{D}^{\mu} f$ on $I \times J$. Further, for $n \in \mathbb{N}$, define $f_n$ on $I \times J$ as

$$f_n(x, y) = \mathcal{I}^{\mu} g_n(x, y) + \kappa_f (x-a)^{\mu_1 - 1}(y-c)^{\mu_2 - 1}, \forall\, (x, y) \in I \times J,$$

where $\kappa_f$ is a constant depending on $f$. As a consequence of Lemma 2.4 and Remark 3.5, one gets

$$\begin{aligned}
\|f_n - f\|_{\infty} &= \sup_{(x,y)\in I\times J} |f_n(x,y) - f(x,y)| \\
&= \sup_{(x,y)\in I\times J} \left|\mathcal{I}^{\mu} g_n(x,y) + \kappa_f (x-a)^{\mu_1-1}(y-c)^{\mu_2-1} - f(x,y)\right| \\
&= \sup_{(x,y)\in I\times J} \left|\mathcal{I}^{\mu} g_n(x,y) - \mathcal{I}^{\mu}\mathcal{D}^{\mu} f(x,y)\right| \\
&= \|\mathcal{I}^{\mu} g_n - \mathcal{I}^{\mu}\mathcal{D}^{\mu} f\|_{\infty} \\
&\leq \frac{1}{\Gamma(\nu_1+1)\Gamma(\nu_2+1)} \|g_n - \mathcal{D}^{\mu} f\|_{\infty} (b-a)^{\mu_1}(d-c)^{\mu_2}.
\end{aligned}$$

As $n$ tends to $\infty$, we have $\|g_n - \mathcal{D}^{\mu} f\|_{\infty} \to 0$. This results in $\|f_n - f\|_{\infty} \to 0$. Moreover,

$$\begin{aligned}
\mathcal{D}^{\mu}(f_n)(x,y) &= \mathcal{D}^{\mu}\mathcal{I}^{\mu} g_n(x,y) + \kappa_f \mathcal{D}^{\mu}\big((x-a)^{\mu_1-1}(y-c)^{\mu_2-1}\big) \\
&= g_n(x,y) + \kappa_f \mathcal{D}^{\mu}\big((x-a)^{\mu_1-1}(y-c)^{\mu_2-1}\big).
\end{aligned}$$

Using Note 2.2, we get

$$\mathcal{D}^{\mu}\big((x-a)^{\mu_1-1}(y-c)^{\mu_2-1}\big) = 0.$$

Hence,

$$\dim \mathcal{G}(\mathcal{D}^{\mu} f_n) = \dim \mathcal{G}(g_n) = \beta = \dim \mathcal{G}(\mathcal{D}^{\mu} f), \forall\, n \in \mathbb{N},$$

this completes our proof. □

Motivated by the mean value theorem in classical calculus (Ref. [**30**]), we prove its analog in RL fractional calculus for a real-valued function of two variables.

LEMMA 3.20. *Let $\mu = (\mu_1, \mu_2)$ with $\mu_1,\ \mu_2 \in (0,1)$, and $f \in \mathcal{C}([a,b] \times [c,d])$ such that $\mathcal{D}^{\mu_1}[(x-a)^{\mu_1-1} f(x,y)] \in \mathcal{C}([a,b])$ and $\mathcal{D}^{\mu}[(y-c)^{\mu_2-1}(x-a)^{\mu_1-1} f] \in \mathcal{C}([a,b] \times [c,d])$. Then there exists $(\xi, \zeta) \in [a,x] \times [c,y]$ satisfying*

$$\begin{aligned}
&f(x,y) - f(x,c) - f(a,y) + f(a,c) \\
&\qquad = \frac{\mathcal{D}^{\mu}[(y-c)^{\mu_2-1}(x-a)^{\mu_1-1} f(x,y)]\Big|_{(\xi,\zeta)}}{\Gamma(\mu_1+1)\Gamma(\mu_2+1)} (x-a)(y-c).
\end{aligned}$$

PROOF. For any fixed $y \in [c,d]$, define

$$\Lambda(x) = f(x,y) - f(x,c), \forall\, x \in [a,b].$$

Clearly, $\Lambda$ and $\mathcal{D}^{\mu_1}[(x-a)^{\mu_1-1}\Lambda] \in \mathcal{C}([a,b])$. Using Theorem 2.5, there exists $\xi \in [a,x]$ such that

$$\Lambda(x)-\Lambda(a) = \frac{\mathcal{D}^{\mu_1}[(x-a)^{\mu_1-1}\Lambda(x)]\Big|_{x=\xi}}{\Gamma(\mu_1+1)}(x-a). \tag{3.8}$$

Again, notice that

$$\mathcal{D}^{\mu_1}[(x-a)^{\mu_1-1}\Lambda(x)] = \mathcal{D}^{\mu_1}[(x-a)^{\mu_1-1}(f(x,y)-f(x,c))].$$

By assumption, $\mathcal{D}^{\mu_1}[(x-a)^{\mu_1-1}f(x,y)]$ and $\mathcal{D}^{\mu_2}\mathcal{D}^{\mu_1}[(y-c)^{\mu_2-1}(x-a)^{\mu_1-1}f(x,y)] \in \mathcal{C}([c,d])$. Thus, using Theorem 2.5, there exists $\zeta \in [c,y]$ with

$$\begin{aligned} &\mathcal{D}^{\mu_1}[(x-a)^{\mu_1-1}f(x,y)] - \mathcal{D}^{\mu_1}[(x-a)^{\mu_1-1}f(x,c)] \\ &= \frac{\mathcal{D}^{\mu_2}\mathcal{D}^{\mu_1}[(y-c)^{\mu_2-1}(x-a)^{\mu_1-1}f(x,y)]\Big|_{y=\zeta}}{\Gamma(\mu_2+1)}(y-c). \end{aligned} \tag{3.9}$$

Combining equations (3.8) and (3.9), we get

$$\begin{aligned} &f(x,y)-f(x,c)-f(a,y)+f(a,c) \\ &= \frac{\mathcal{D}^{\mu}[(y-c)^{\mu_2-1}(x-a)^{\mu_1-1}f(x,y)]\Big|_{(\xi,\zeta)}}{\Gamma(\mu_1+1)\Gamma(\mu_2+1)}(x-a)(y-c), \end{aligned}$$

as required. □

THEOREM 3.21. *Let $\mu=(\mu_1,\mu_2)$ with $\mu_1,\ \mu_2 \in (0,1)$, and $\{f_n\}_{n=1}^{\infty} \subseteq \mathcal{C}(I\times J)$. Again, for some $(x_*,y_*) \in I\times J$, the sequences $\{f_n(x_*,y_*)\}_{n=1}^{\infty}$ are convergent, $\{f_n(x_*,.)\}_{n=1}^{\infty}$ and $\{f_n(.,y_*)\}_{n=1}^{\infty}$ converge uniformly on $J$ and $I$, respectively. Furthermore, the sequences $\mathcal{D}^{\mu_1}[(x-x_*)^{\mu_1-1}f_n] \in \mathcal{C}(I)$ and $\mathcal{D}^{\mu}[(y-y_*)^{\mu_2-1}(x-x_*)^{\mu_1-1}f_n] \in \mathcal{C}(I\times J)$, and $\{\mathcal{D}^{\mu}[(y-y_*)^{\mu_2-1}(x-x_*)^{\mu_1-1}f_n]\}_{n=1}^{\infty}$ is uniformly convergent on $I\times J$. Then the sequence $\{f_n\}_{n=1}^{\infty}$ converges uniformly to $f$ on $I\times J$, and*

$$\lim_{n\to\infty}\mathcal{D}^{\mu}[(y-y_*)^{\mu_2-1}(x-x_*)^{\mu_1-1}f_n](x,y) = \mathcal{D}^{\mu}[(y-y_*)^{\mu_2-1}(x-x_*)^{\mu_1-1}f](x,y),$$

*for all $(x,y)\in I\times J$.*

PROOF. For each $\epsilon>0$, using the hypothesis on the sequence $\left\{\mathcal{D}^{\mu}[(y-y_*)^{\mu_2-1}(x-x_*)^{\mu_1-1}f_n]\right\}_{n=1}^{\infty}$, there exists $n_0\in\mathbb{N}$ such that

$$\left|\mathcal{D}^{\mu}[(y-y_*)^{\mu_2-1}(x-x_*)^{\mu_1-1}(f_n-f_m)(x,y)]\right| < \frac{\epsilon\Gamma(\mu_1+1)\Gamma(\mu_2+1)}{4(b-a)(d-c)}, \tag{3.10}$$

for all $n,\ m\geq n_0$, and $(x,y)\in I\times J$. In view of Lemma 3.20, there exists $(\xi,\zeta)$ with $a\leq x_*\leq\xi\leq x\leq b$ and $c\leq y_*\leq\zeta\leq y\leq d$ such that

$$\begin{aligned} &(f_n-f_m)(x,y)-(f_n-f_m)(x,y_*)-(f_n-f_m)(x_*,y)+(f_n-f_m)(x_*,y_*) \\ &= \frac{(x-x_*)(y-y_*)}{\Gamma(\mu_1+1)\Gamma(\mu_2+1)}\mathcal{D}^{\mu}[(y-y_*)^{\mu_2-1}(x-x_*)^{\mu_1-1}(f_n-f_m)(x,y)]\Big|_{(\xi,\zeta)}. \end{aligned}$$

This reduces to

$$
\begin{aligned}
&\big|(f_n - f_m)(x, y)\big| \\
&\le \big|(f_n - f_m)(x, y_*)\big| + \big|(f_n - f_m)(x_*, y)\big| + \big|(f_n - f_m)(x_*, y_*)\big| + \\
&\quad \frac{|x - x_*|\,|y - y_*|}{\Gamma(\mu_1 + 1)\Gamma(\mu_2 + 1)} \left| \mathcal{D}^{\mu}[(y - y_*)^{\mu_2 - 1}(x - x_*)^{\mu_1 - 1}(f_n - f_m)(x, y)]\Big|_{(\xi,\zeta)} \right| \\
&\le \big|(f_n - f_m)(x, y_*)\big| + \big|(f_n - f_m)(x_*, y)\big| + \big|(f_n - f_m)(x_*, y_*)\big| + \\
&\quad \frac{\epsilon\,(x - x_*)(y - y_*)}{4(b - a)(d - c)}.
\end{aligned}
$$

The last line of the above inequality is obtained using equation (3.10). Moreover, by assumption of the theorem, there exist $n_1, n_2, n_3 \in \mathbb{N}$ such that

$$
\begin{aligned}
|f_n(x_*, y) - f_m(x_*, y)| &< \frac{\epsilon}{4}, \ \forall\, n, m \ge n_1, \ \text{and } y \in J, \\
|f_n(x, y_*) - f_m(x, y_*)| &< \frac{\epsilon}{4}, \ \forall\, n, m \ge n_2, \ \text{and } x \in I, \\
|f_n(x_*, y_*) - f_m(x_*, y_*)| &< \frac{\epsilon}{4}, \ \forall\, n, m \ge n_3.
\end{aligned}
$$

As a consequence of all the above estimations, we get

$$
\big|f_n(x, y) - f_m(x, y)\big| < \frac{\epsilon}{4} + \frac{\epsilon}{4} + \frac{\epsilon}{4} + \frac{\epsilon}{4} = \epsilon,
$$

for all $n$, $m \ge \max\{n_0, n_1, n_2, n_3\}$, and $(x, y) \in I \times J$. This concludes that $\{f_n\}_{n=1}^{\infty}$ is a Cauchy sequence on $I \times J$. Hence, by Cauchy's criterion of uniform convergence on $\mathcal{C}(I \times J)$, the sequence is convergent and converges to $f$ (say). Further, consider the sequence of functions $\big\{g_n : I \times J \to \mathbb{R}\big\}_{n=1}^{\infty}$ defined as

$$
g_n(x, y) = \int_a^x \int_c^y (x - s)^{-\mu_1}(y - t)^{-\mu_2}(t - y_*)^{\mu_2 - 1}(s - x_*)^{\mu_1 - 1} f_n(s, t)\,ds\,dt. \tag{3.11}
$$

Then $g_n$ is a sequence of differentiable functions on $I \times J$, and the sequences $\big\{g_n(a, .) = 0\big\}_{n=1}^{\infty}$ and $\big\{g_n(., c) = 0\big\}_{n=1}^{\infty}$ are uniformly convergent on $J$ and $I$, respectively. Also, the sequence $\mathcal{D}^{\mu}[(y - y_*)^{\mu_2 - 1}(x - x_*)^{\mu_1 - 1} f_n]$ is uniformly convergent on $I \times J$ gives $\left\{\frac{\partial^2}{\partial x \partial y} g_n\right\}$ is uniformly convergent on $I \times J$. Consequently, exerting [**1**, Theorem 2.4], we obtain

$$
\lim_{n \to \infty} \frac{\partial^2}{\partial x \partial y} g_n(x, y) = \frac{\partial^2}{\partial x \partial y} \lim_{n \to \infty} g_n(x, y), \ \forall\, (x, y) \in I \times J.
$$

Therefore,

$$
\begin{aligned}
&\lim_{n \to \infty} \frac{\partial^2}{\partial x \partial y} \int_a^x \int_c^y (x - s)^{-\mu_1}(y - t)^{-\mu_2}(t - y_*)^{\mu_2 - 1}(s - x_*)^{\mu_1 - 1} f_n(s, t)\,ds\,dt \\
&\quad = \frac{\partial^2}{\partial x \partial y} \lim_{n \to \infty} \int_a^x \int_c^y (x - s)^{-\mu_1}(y - t)^{-\mu_2}(t - y_*)^{\mu_2 - 1}(s - x_*)^{\mu_1 - 1} f_n(s, t)\,ds\,dt.
\end{aligned} \tag{3.12}
$$

Also, $\{f_n\}_{n=1}^{\infty} \subseteq \mathcal{C}(I \times J)$ and converges uniformly to $f$ on $I \times J$. This results in

$$
\begin{aligned}
&\frac{\partial^2}{\partial x \partial y} \lim_{n \to \infty} \int_a^x \int_c^y (x - s)^{-\mu_1}(y - t)^{-\mu_2}(t - y_*)^{\mu_2 - 1}(s - x_*)^{\mu_1 - 1} f_n(s, t)\,ds\,dt \\
&\quad = \frac{\partial^2}{\partial x \partial y} \int_a^x \int_c^y (x - s)^{-\mu_1}(y - t)^{-\mu_2}(t - y_*)^{\mu_2 - 1}(s - x_*)^{\mu_1 - 1} f(s, t)\,ds\,dt.
\end{aligned} \tag{3.13}
$$

Combining equations (3.12) and (3.13), we get the required result. □

DEFINITION 3.22. A function $f : I \times J \to \mathbb{R}$ is defined as $\mu$-fractional convex function if

$$\mathcal{D}^{\mu} f(x, y) \geq 0, \, \forall \, (x, y) \in I \times J,$$

where $\mu = (\mu_1, \mu_2)$ with $0 < \mu_1, \mu_2 < 1$.

THEOREM 3.23. *Let $f : I \times J \to \mathbb{R}$ be a $\mu$-fractional convex function with $\mathcal{D}^{\mu} f \in \mathcal{C}(I \times J)$. Then for any $\epsilon > 0$, there exists $\mu$-fractional convex function $g$ such that $\mathcal{D}^{\mu} g \in \mathcal{S}_{\beta}$ and $\|f - g\|_{\infty} < \epsilon$.*

PROOF. Let $\epsilon > 0$. Since $f$ is $\mu$-fractional convex function and $\mathcal{D}^{\mu} f \in \mathcal{C}(I \times J)$, Theorem 3.15 gives a function $h \in \mathcal{S}_{\beta}$ such that $h(x, y) \geq 0, \, \forall \, (x, y) \in I \times J$, and

$$\|\mathcal{D}^{\mu} f - h\|_{\infty} < \frac{\Gamma(\mu_1 + 1)\Gamma(\mu_2 + 1)\epsilon}{(b - a)^{\mu_1}(d - c)^{\mu_2}}.$$

Define $g(x, y) = \mathcal{I}^{\mu} h(x, y) + \kappa_f (x - a)^{\mu_1 - 1}(y - b)^{\mu_2 - 1}, \, \forall \, (x, y) \in I \times J$, and $\kappa_f \in \mathbb{R}$ depends on $f$. Then

$$\begin{aligned} \|f - g\|_{\infty} &= \sup_{(x,y) \in I \times J} \left| f(x, y) - g(x, y) \right| \\ &= \sup_{(x,y) \in I \times J} \left| f(x, y) - \mathcal{I}^{\mu} h(x, y) - \kappa_f (x - a)^{\mu_1 - 1}(y - b)^{\mu_2 - 1} \right|. \end{aligned}$$

Exerting Theorem 2.4, we get

$$\|f - g\|_{\infty} = \sup_{(x,y) \in I \times J} \left| \mathcal{I}^{\mu} \mathcal{D}^{\mu} f(x, y) - \mathcal{I}^{\mu} h(x, y) \right| = \left\| \mathcal{I}^{\mu} \mathcal{D}^{\mu} f - \mathcal{I}^{\mu} h \right\|_{\infty}.$$

Now, Lemma 3.18 yields

$$\|f - g\|_{\infty} \leq \frac{(b - a)^{\mu_1}(d - c)^{\mu_2}}{\Gamma(\mu_1 + 1)\Gamma(\mu_2 + 1)} \|\mathcal{D}^{\mu} f - h\|_{\infty} < \epsilon.$$

Further, from Notes 2.2 and 2.3, we have

$$\begin{aligned} \mathcal{D}^{\mu} g(x, y) &= \mathcal{D}^{\mu} \mathcal{I}^{\mu} h(x, y) + \kappa_f \mathcal{D}^{\mu} (x - a)^{\mu_1 - 1}(y - b)^{\mu_2 - 1} \\ &= h(x, y) \geq 0, \, \forall \, (x, y) \in I \times J. \end{aligned}$$

Therefore, $\mathcal{D}^{\mu} g \in \mathcal{S}_{\beta}$. This completes the proof. □

## Statements and Declarations

- The authors have no competing interests to declare that are relevant to the content of this article.
- There is no data associated with this paper.

## Acknowledgments

Both authors thank IIIT Allahabad (Ministry of Education, India) for giving a good environment for research.

## References


[1] V. Agrawal, T. Som, and S. Verma, *On bivariate fractal approximation*, J. Anal. **30** (2022), no. 4, 1765–1783, DOI 10.1007/s41478-022-00430-0. MR4498869

[2] E. Agrawal and S. Verma, *Fractal surfaces in Hölder and Sobolev spaces*, J. Anal. **32** (2024), no. 2, 1161–1179, DOI 10.1007/s41478-023-00672-6. MR4716296

[3] M. F. Barnsley, *Fractal functions and interpolation*, Constr. Approx. **2** (1986), no. 4, 303–329, DOI 10.1007/BF01893434. MR892158

[4] P. Bouboulis and L. Dalla, *A general construction of fractal interpolation functions on grids of* $\mathbb{R}^n$, European J. Appl. Math. **18** (2007), no. 4, 449–476, DOI 10.1017/S0956792507007024. MR2344315

[5] S. Chandra and S. Abbas, *Analysis of fractal dimension of mixed Riemann-Liouville integral*, Numer. Algorithms **91** (2022), no. 3, 1021–1046, DOI 10.1007/s11075-022-01290-2. MR4493354

[6] S. Chandra and S. Abbas, *On fractal dimensions of fractal functions using function spaces*, Bull. Aust. Math. Soc. **106** (2022), no. 3, 470–480, DOI 10.1017/S0004972722000685. MR4510139

[7] J. B. Conway, *A course in functional analysis*, 96 Springer, 2019.

[8] K. Falconer, *Fractal geometry*, 2nd ed., John Wiley & Sons, Inc., Hoboken, NJ, 2003. Mathematical foundations and applications, DOI 10.1002/0470013850. MR2118797

[9] J. S. Geronimo and D. Hardin, *Fractal interpolation surfaces and a related 2-D multiresolution analysis*, J. Math. Anal. Appl. **176** (1993), no. 2, 561–586, DOI 10.1006/jmaa.1993.1232. MR1224164

[10] P. Guo, C. Li, and G. Chen, *On the fractional mean-value theorem*, Internat. J. Bifur. Chaos Appl. Sci. Engrg. **22** (2012), no. 5, 1250104, 6, DOI 10.1142/S0218127412501040. MR2947001

[11] Gurubachan, V. V. M. S. Chandramouli, and S. Verma, *Fractal dimension of $\alpha$-fractal functions without endpoint conditions*, Mediterr. J. Math. **21** (2024), no. 3, Paper No. 71, 23, DOI 10.1007/s00009-024-02610-7. MR4719420

[12] B. R. Hunt, *The Hausdorff dimension of graphs of Weierstrass functions*, Proc. Amer. Math. Soc. **126** (1998), no. 3, 791–800, DOI 10.1090/S0002-9939-98-04387-1. MR1452806

[13] J. E. Hutchinson, *Fractals and self-similarity*, Indiana Univ. Math. J. **30** (1981), no. 5, 713–747, DOI 10.1512/iumj.1981.30.30055. MR625600

[14] M. Jahn and P. Massopust, *An integral RB operator*, J. Fixed Point Theory Appl. **26** (2024), no. 4, Paper No. 37, 19 pp., DOI 10.1007/s11784-024-01125-8. MR4795079

[15] Y. S. Liang, *Fractal dimension of Riemann-Liouville fractional integral of 1-dimensional continuous functions*, Fract. Calc. Appl. Anal. **21** (2018), no. 6, 1651–1658, DOI 10.1515/fca-2018-0087. MR3912448

[16] Y.-S. Liang, *Progress on estimation of fractal dimensions of fractional calculus of continuous functions*, Fractals **27** (2019), no. 5, 1950084, 15 pp., DOI 10.1142/S0218348X19500841. MR3999132

[17] Y. S. Liang, Q. Zhang, and K. Yao, *Fractal dimension of fractional calculus of certain interpolation functions* (Chinese, with English and Chinese summaries), Chinese Ann. Math. Ser. A **38** (2017), no. 1, 117–124; English transl., Chinese J. Contemp. Math. **38** (2017), no. 1, 93–100. MR3676963

[18] Z. Liang and H.-J. Ruan, *Construction and box dimension of recurrent fractal interpolation surfaces*, J. Fractal Geom. **8** (2021), no. 3, 261–288, DOI 10.4171/jfg/105. MR4321218

[19] Y. Luchko, *Fractional derivatives and the fundamental theorem of fractional calculus*, Fract. Calc. Appl. Anal. **23** (2020), no. 4, 939–966, DOI 10.1515/fca-2020-0049. MR4151092

[20] R. Małysz, *The Minkowski dimension of the bivariate fractal interpolation surfaces*, Chaos Solitons Fractals **27** (2006), no. 5, 1147–1156, DOI 10.1016/j.chaos.2005.05.007. MR2164847

[21] P. R. Massopust, *Fractal surfaces*, J. Math. Anal. Appl. **151** (1990), no. 1, 275–290, DOI 10.1016/0022-247X(90)90257-G. MR1069462

[22] P. R. Massopust, *Fractal interpolation over nonlinear partitions*, Chaos Solitons Fractals **162** (2022), Paper No. 112503, 8 pp., DOI 10.1016/j.chaos.2022.112503. MR4467626

[23] R. D. Mauldin and S. C. Williams, *On the Hausdorff dimension of some graphs*, Trans. Amer. Math. Soc. **298** (1986), no. 2, 793–803, DOI 10.2307/2000650. MR860394

[24] W. Metzler and C. H. Yun, *Construction of fractal interpolation surfaces on rectangular grids*, Internat. J. Bifur. Chaos Appl. Sci. Engrg. **20** (2010), no. 12, 4079–4086, DOI 10.1142/S0218127410027933. MR2775262

[25] M. A. Navascués, *Fractal approximation*, Complex Anal. Oper. Theory **4** (2010), no. 4, 953–974, DOI 10.1007/s11785-009-0033-1. MR2735314

[26] M. A. Navascués, *Fractal polynomial interpolation*, Z. Anal. Anwendungen **24** (2005), no. 2, 401–418, DOI 10.4171/ZAA/1248. MR2174271

[27] K. B. Oldham and J. Spanier, *The fractional calculus*, Mathematics in Science and Engineering, Vol. 111, Academic Press [Harcourt Brace Jovanovich, Publishers], New York-London, 1974. Theory and applications of differentiation and integration to arbitrary order; With an annotated chronological bibliography by Bertram Ross. MR361633

[28] I. Podlubny, *Geometric and physical interpretation of fractional integration and fractional differentiation*, Fract. Calc. Appl. Anal. **5** (2002), no. 4, 367–386. Dedicated to the 60th anniversary of Prof. Francesco Mainardi. MR1967839

[29] H.-J. Ruan and Q. Xu, *Fractal interpolation surfaces on rectangular grids*, Bull. Aust. Math. Soc. **91** (2015), no. 3, 435–446, DOI 10.1017/S0004972715000064. MR3338968

[30] W. Rudin, *Principles of mathematical analysis*, 3rd ed., International Series in Pure and Applied Mathematics, McGraw-Hill Book Co., New York-Auckland-Düsseldorf, 1976. MR385023

[31] S. G. Samko, A. A. Kilbas, and O. I. Marichev, *Fractional integrals and derivatives: Theory and applications*, Gordon and Breach Science Publishers, Yverdon, 1993. MR1347689

[32] W. Shen, *Hausdorff dimension of the graphs of the classical Weierstrass functions*, Math. Z. **289** (2018), no. 1-2, 223–266, DOI 10.1007/s00209-017-1949-1. MR3803788

[33] S. Zhou and G. He, *On a class of Besicovitch functions to have exact box dimension: a necessary and sufficient condition*, Anal. Theory Appl. **20** (2004), no. 2, 175–181, DOI 10.1007/BF02901441. MR2095460

[34] S. Verma, *Some Results on Fractal Functions, Fractal Dimensions and Fractional Calculus*, Ph.D. thesis, Indian Institute of Technology Delhi, India, 2020.

[35] S. Verma and P. R. Massopust, *Dimension preserving approximation*, Aequationes Math. **96** (2022), no. 6, 1233–1247, DOI 10.1007/s00010-022-00893-3. MR4514041

[36] S. Verma and P. Viswanathan, *A fractal operator associated with bivariate fractal interpolation functions on rectangular grids*, Results Math. **75** (2020), no. 1, Paper No. 28, 26 pp., DOI 10.1007/s00025-019-1152-2. MR4053993

[37] S. Verma and P. Viswanathan, *Bivariate functions of bounded variation: fractal dimension and fractional integral*, Indag. Math. (N.S.) **31** (2020), no. 2, 294–309, DOI 10.1016/j.indag.2020.01.006. MR4068578

[38] S. Verma and P. Viswanathan, *Parameter identification for a class of bivariate fractal interpolation functions and constrained approximation*, Numer. Funct. Anal. Optim. **41** (2020), no. 9, 1109–1148, DOI 10.1080/01630563.2020.1738458. MR4116000

[39] J. Wang, K. Yao, and Y. Liang, *On the connection between the order of Riemann-Liouvile fractional calculus and Hausdorff dimension of a fractal function*, Anal. Theory Appl. **32** (2016), no. 3, 283–290, DOI 10.4208/ata.2016.v32.n3.6. MR3605523

[40] B. Yu and Y. Liang, *Approximation with continuous functions preserving fractal dimensions of the Riemann-Liouville operators of fractional calculus*, Fract. Calc. Appl. Anal. **26** (2023), no. 6, 2805–2836, DOI 10.1007/s13540-023-00215-7. MR4673084

[41] N. Zhao, *Construction and application of fractal interpolation surfaces*, The Visual Computer **12** (1996), 132–146.

Department of Applied Sciences, IIIT Allahabad, Prayagraj, India 211015
*Email address*: ekta.agrawal5346@gmail.com

Department of Applied Sciences, IIIT Allahabad, Prayagraj, India 211015
*Email address*: saurabhverma@iiita.ac.in

Contemporary Mathematics
Volume **825**, 2025
https://doi.org/10.1090/conm/825/16506

# Optimal quantization for nonuniform discrete distributions

Russel Cabasag, Samir Huq, Eric Mendoza,
and Mrinal Kanti Roychowdhury

ABSTRACT. This paper explores the process of optimal quantization for several types of discrete probability distributions. Quantization is a technique used to approximate a complex distribution with a smaller set of representative points, which is important in fields such as data compression and signal processing. We begin by examining two specific nonuniform distributions over a finite set of values and identify the best representative points for different levels of approximation. We then extend our analysis to two infinite discrete distributions: one supported on the reciprocals of natural numbers and another on the natural numbers themselves. For these distributions, we compute the optimal sets of representatives and assess how well they approximate the original distributions. Finally, we address the reverse problem—determining the underlying distribution when the optimal sets are known. Our results provide both theoretical insights and computational techniques useful in information theory and data analysis.

## 1. Introduction

Quantization is the process of converting a continuous analog signal into a digital signal of $k$ discrete levels, or converting a digital signal of $n$ levels into another digital signal of $k$ levels, where $k < n$. It is essential when analog quantities are represented, processed, stored, or transmitted by a digital system, or when data compression is required. It is a classic and still very active research topic in source coding and information theory. It has broad applications in engineering and technology. In fact, it is useful in data compression and cluster analysis. For some details one can see [**1**–**7**]. Recently, the quantization theory has been extended to constrained quantization and conditional quantization (see [**8**–**10**]). Following the introduction of constrained quantization, the theory of quantization is now categorized into two main branches: constrained quantization and unconstrained quantization. To gain a quick understanding and practical grasp of research in constrained and unconstrained quantization, the paper [**8**] serves as a helpful starting point. For a more in-depth treatment of unconstrained quantization, one may refer to [**11**].

Let $\mathbb{R}^d$ denote the $d$-dimensional Euclidean space equipped with a metric $\|\cdot\|$ compatible with the Euclidean topology. Let $P$ be a Borel probability measure

---

2020 *Mathematics Subject Classification.* Primary 60E05, 94A34.
*Key words and phrases.* Discrete distribution, optimal sets of $n$-means, quantization error.

on $\mathbb{R}^d$ and $\alpha$ be a locally finite subset of $\mathbb{R}^d$, i.e., intersection of $\alpha$ with any bounded subset of $\mathbb{R}^d$ is finite. This implies that $\alpha$ is countable and closed. Then, $\int \min_{a\in\alpha} \|x-a\|^2 dP(x)$ is often referred to as the *cost*, or *distortion error* for $\alpha$ with respect to the probability measure $P$, and is denoted by $V(P;\alpha)$. Write $\mathcal{D}_n := \{\alpha \subset \mathbb{R}^d : 1 \leq \text{card}(\alpha) \leq n\}$. Then, $\inf\{V(P;\alpha) : \alpha \in \mathcal{D}_n\}$ is called the *nth quantization error* for the probability measure $P$, and is denoted by $V_n := V_n(P)$. A set $\alpha$ for which the infimum occurs and contains no more than $n$ elements is called an *optimal set of n-points*. It is known that for a Borel probability measure $P$, if its support contains at least $n$ elements and $\int \|x\|^2 dP(x)$ is finite, then an optimal set of $n$-points always has exactly $n$-elements (see [**3**, **11**–**13**]). Recently, optimal quantization for different uniform distributions have been investigated by several authors, for example, see [**14**–**18**].

Given a finite set $\alpha \subset \mathbb{R}^d$, the Voronoi region generated by $a \in \alpha$ is defined by

$$M(a|\alpha) = \{x \in \mathbb{R}^d : \|x-a\| = \min_{b\in\alpha} \|x-b\|\},$$

i.e., the Voronoi region generated by $a \in \alpha$ is the set of all elements in $\mathbb{R}^d$ which are nearest to $a$, and the set $\{M(a|\alpha) : a \in \alpha\}$ is called the *Voronoi diagram* or *Voronoi tessellation* of $\mathbb{R}^d$ with respect to $\alpha$.

Let us now state the following proposition (see [**2**, **11**]).

PROPOSITION 1.1. *Let $\alpha$ be an optimal set of n-points for $P$, and $a \in \alpha$. Then, (i) $P(M(a|\alpha)) > 0$, (ii) $P(\partial M(a|\alpha)) = 0$, (iii) $a = E(X : X \in M(a|\alpha))$, where $M(a|\alpha)$ is the Voronoi region of $a \in \alpha$, i.e., $M(a|\alpha)$ is the set of all elements $x$ in $\mathbb{R}^d$ which are closest to $a$ among all the elements in $\alpha$, and $\partial M(a|\alpha)$ represents the boundary of the Voronoi region $M(a|\alpha)$.*

By the above proposition, we see that in unconstrained quantization, the elements in an optimal set of $n$-points are the conditional expectations in their own Voronoi regions. Because of this fact, in unconstrained quantization, an optimal set of $n$-points is termed an optimal set of $n$-means.

**1.2. Delineation.** In this paper, we investigate the optimal quantization for finite, and infinite discrete distributions. Section 2 contains the basic preliminaries. In Section 3, we calculate the optimal sets of $n$-means and the $n$th quantization errors for all $1 \leq n \leq 6$ for two nonuniform discrete distributions with support $\{1,2,3,4,5,6\}$ associated with two different probability vectors. In Section 4 first, for a probability distribution $P$ with support $\{\frac{1}{n} : n \in \mathbb{N}\}$ associated with a mass function $f$, given by $f(x) = \frac{1}{2^k}$ if $x = \frac{1}{k}$ for $k \in \mathbb{N}$, and zero otherwise, we determine the optimal sets of $n$-means and the $n$th quantization errors for all positive integers up to $n = 300$. Then, for a probability distribution $P$ with support the set $\mathbb{N}$ of natural number associated with a mass function $f$, given by $f(x) = \frac{1}{2^k}$ if $x = k$ for $k \in \mathbb{N}$, and zero otherwise, we determine the optimal sets of $n$-means and the $n$th quantization errors for all positive integers $n$. In Section 5, we discuss for a discrete distribution, if the optimal sets are given, how to obtain the probability distributions.

## 2. Basic Preliminaries

Let $P$ be a Borel probability measure on $\mathbb{R}^d$, and $U$ be the largest open subset of $\mathbb{R}^d$ for which $P(U) = 0$. Then, $\mathbb{R}^d \setminus U$ is called the support of $P$, and is denoted by supp$(P)$. Let $P$ be a uniform distribution defined on the set $\{1,2,3,4,5,6\}$. Then, the random variable $X$ associated with the probability distribution is a discrete random variable with probability mass function $f$ given by

$$f(x) = P(X = x) = \frac{1}{6}, \text{ where } x \in \{1, 2, 3, 4, 5, 6\}.$$

It is not difficult to show that if $\alpha_n$ is an optimal set of $n$-means for $P$, then

$$\alpha_1 = \{3.5\},\ \alpha_2 = \{2, 5\},\ \alpha_3 = \{1.5, 3.5, 5.5\},\ \alpha_4 = \{1.5, 3.5, 5, 6\},$$
$$\alpha_5 = \{1.5, 3, 4, 5, 6\},\ \text{and } \alpha_6 = \text{supp}(P).$$

REMARK 2.1. Optimal sets are not unique. For example, in the above, the set $\alpha_5$ can be any one of the following sets:

$$\{1.5, 3, 4, 5, 6\},\ \{1, 2.5, 4, 5, 6\},\ \{1, 2, 3.5, 5, 6\},\ \{1, 2, 3, 4.5, 6\},\ \{1, 2, 3, 4, 5.5\}.$$

In the following sections we give our main results.

## 3. Optimal quantization for nonuniform discrete distributions

Let $P$ be a nonuniform distribution defined on the set $\{1, 2, 3, 4, 5, 6\}$ associated with a probability vector of the form $(x, (1-x)x, (1-x)^2x, (1-x)^3x, (1-x)^4x, (1-x)^5)$, i.e., the probability mass function $f$ is given by

$$f(j) = P(X = j) = \begin{cases} x & \text{if } j = 1, \\ (1-x)^{j-1}x & \text{if } j \in \{2, 3, 4, 5\}, \\ (1-x)^5 & \text{if } j = 6, \\ 0 & \text{otherwise,} \end{cases}$$

where $0 < x < 1$. Let $X$ be a random variable with probability distribution $P$. Then,

$$E(X) = 1x + \sum_{j=2}^{5} jx(1-x)^{j-1} + 6(1-x)^5 = -x^5 + 6x^4 - 15x^3 + 20x^2 - 15x + 6,$$

and so, the optimal set of one-mean is the set $\{-x^5 + 6x^4 - 15x^3 + 20x^2 - 15x + 6\}$ with quantization error the variance $V$ of the random variable $X$, where

$$\begin{aligned} V = V_1 = E\|X - E(X)\|^2 &= \sum_{j=1}^{6} f(j)\Big(j - E(X)\Big)^2 \\ &= -x(-55 + 275x - 627x^2 + 858x^3 - 781x^4 + 495x^5 - 220x^6 + 66x^7 - 12x^8 + x^9). \end{aligned}$$

Notice that $\text{supp}(P) = \{1, 2, 3, 4, 5, 6\}$, and it forms the optimal set of six-means for the probability distribution $P$ for any $0 < x < 1$. For $i, j \in \{1, 2, \cdots, 6\}$ with $i \le j$, write

$$a[i, j] := E(X : X \in \{i, i+1, \cdots, j\}) = \frac{\sum_{x=i}^{j} xf(x)}{\sum_{x=i}^{j} f(x)}.$$

In the following two subsections, we determine the optimal sets of $n$-means and the $n$th quantization errors for all $2 \le n \le 5$ for the probability distribution $P$: one for $x = \frac{1}{2}$, and one for $x = \frac{7}{10}$.

**3.1. Optimal quantization for the probability distribution $P$ with $x = \frac{1}{2}$.** In this case, the probability mass function $f$ is given by

$$f(j) = P(X = j) = \begin{cases} \frac{1}{2^j} & \text{if } j \in \{1, 2, 3, 4, 5\}, \\ \frac{1}{2^5} & \text{if } j = 6, \\ 0 & \text{otherwise.} \end{cases}$$

We now give the following propositions.

PROPOSITION 3.2. *The optimal set of two-means is given by* $\{a[1, 2], a[3, 6]\}$ *with quantization error* $V_2 = \frac{341}{768}$.

PROOF. Notice that $a[1, 2] = \frac{4}{3}$, and $a[3, 6] = \frac{31}{8}$. Let us consider the set $\beta := \{\frac{4}{3}, \frac{31}{8}\}$. Since $2 < \frac{1}{2}(\frac{4}{3} + \frac{31}{8}) = 2.60417 < 3$, the distortion error due to the set $\beta$ is given by

$$\sum_{j=1}^{6} f(j) \min_{a\in\beta}(j - a)^2 = \sum_{j=1}^{2} f(j)\Big(j - \frac{4}{3}\Big)^2 + \sum_{j=3}^{6} f(j)\Big(j - \frac{31}{8}\Big)^2 = \frac{341}{768}.$$

Since $V_2$ is the quantization error for two-means, we have $V_2 \le \frac{341}{768} = 0.44401$. Let $\alpha := \{a_1, a_2\}$ be an optimal set of two-means. Without any loss of generality, we can assume that $1 \le a_1 < a_2 \le 6$. Notice that the Voronoi region of $a_1$ must contain 1. Suppose that the Voronoi region of $a_1$ contains the set $\{1, 2, 3\}$. Then, as $a[1, 3] = \frac{11}{7}$, we have

$$V_2 \ge \sum_{j=1}^{3} f(j)\Big(j - \frac{11}{7}\Big)^2 = \frac{13}{28} = 0.464286 > V_2,$$

which gives a contradiction. Hence, we can assume that the Voronoi region of $a_1$ does not contain $\{1, 2, 3\}$. Next, suppose that the Voronoi region of $a_1$ contains only the element 1. Then, the Voronoi region of $a_2$ contains all the remaining elements, and so

$$a_2 = a[2, 6] = \frac{47}{16}$$

implying

$$V_2 = \sum_{j=2}^{6} f(j)\Big(j - \frac{47}{16}\Big)^2 = \frac{367}{512} = 0.716797 > V_2,$$

which yields a contradiction. Hence, we can assume that the Voronoi region of $a_1$ contains only the elements 1 and 2, and the remaining elements are contained in the Voronoi region of $a_2$ implying

$$a_1 = a[1, 2] = \frac{4}{3}, \text{ and } a_2 = a[3, 6] = \frac{31}{8}$$

with quantization error $V_2 = \frac{341}{768}$. Thus, the proof of the proposition is complete. □

PROPOSITION 3.3. *The optimal set of three-means is given by* $\{1, a[2, 3], a[4, 6]\}$ *with quantization error* $V_3 = \frac{65}{384}$.

PROOF. Notice that $a[2,3] = \frac{7}{3}$, and $a[4,6] = \frac{19}{4}$. The distortion error due to the set $\beta := \{1, \frac{7}{3}, \frac{19}{4}\}$ is given by

$$\sum_{j=1}^{6} f(j) \min_{a\in\beta}(j-a)^2 = \sum_{j=2}^{3} f(j)\Big(j-\frac{7}{3}\Big)^2 + \sum_{j=4}^{6} f(j)\Big(j-\frac{19}{4}\Big)^2 = \frac{65}{384}.$$

Since $V_3$ is the quantization error for three-means, we have $V_3 \leq \frac{65}{384} = 0.169271$. Let $\alpha := \{a_1, a_2, a_3\}$ be an optimal set of three-means such that $1 \leq a_1 < a_2 < a_3 \leq 6$. Notice that the Voronoi region of $a_1$ must contain 1. Suppose that the Voronoi region of $a_1$ also contains 3. Then,

$$V_3 \geq \sum_{j=1}^{3} f(j)(j-a[1,3])^2 = \frac{13}{28} > V_3,$$

which yields a contradiction. Thus, we can assume that the Voronoi region of $a_1$ does not contain 3. Suppose that the Voronoi region of $a_1$ contains only the two elements 1 and 2. Then, the Voronoi region of $a_2$ must contain 3. The following two cases can arise:

CASE 1. *The Voronoi region of $a_2$ does not contain 4.*
Then, we must have $a_2 = 3$, and $a_4 = a[4,6]$ yielding

$$V_3 \geq \sum_{j=1}^{2} f(j)(j-a[1,2])^2 + \sum_{j=4}^{6} f(j)(j-a[4,6])^2 = \frac{97}{384} = 0.252604 > V_3,$$

which is a contradiction.

CASE 2. *The Voronoi region of $a_2$ contains 4.*
Then,

$$V_3 \geq \sum_{j=1}^{2} f(j)(j-a[1,2])^2 + \sum_{j=3}^{4} f(j)(j-a[3,4])^2 = \frac{5}{24} = 0.208333 > V_3,$$

which leads to a contradiction.

Hence, by Case 1 and Case 2, we can assume that the Voronoi region of $a_1$ contains only the element 1, i.e., $a_1 = 1$. Then, the Voronoi region of $a_2$ must contain 2. Suppose that the Voronoi region of $a_2$ also contains the set $\{2,3,4\}$. Then,

$$V_3 \geq \sum_{j=2}^{4} f(j)(j-a[2,4])^2 = \frac{13}{56} = 0.232143 > V_3,$$

which yields a contradiction. Thus, we can assume that the Voronoi region of $a_2$ does not contain 4. Suppose that the Voronoi region of $a_2$ contains only the element 2. Then, the Voronoi region of $a_3$ must contain the remaining elements, which yields

$$V_3 \geq \sum_{j=3}^{6} f(j)(j-a[3,6])^2 = \frac{71}{256} = 0.277344 > V_3,$$

which is a contradiction. Hence, we can assume that the Voronoi region of $a_2$ contains only the two elements 2 and 3 implying the fact that the Voronoi region of $a_3$ contains the elements 4, 5, and 6. Thus, we have

$$a_1 = 1,\ a_2 = a[2,3] = \frac{7}{3},\ \text{and } a_3 = a[4,6] = \frac{19}{4},$$

with quantization error $V_3 = \frac{65}{384}$, which yields the proposition. □

PROPOSITION 3.4. *The optimal set of four-means is* $\{1, 2, a[3,4], a[5,6]\}$ *with quantization error* $V_4 = \frac{11}{192}$.

PROOF. The distortion error due to the set $\beta := \{1, 2, a[3,4], a[5,6]\}$ is given by

$$\sum_{j=1}^{6} f(j) \min_{a\in\beta}(j-a)^2 = \sum_{j=3}^{4} f(j)(j - a[3,4])^2 + \sum_{j=5}^{6} f(j)(j - a[5,6])^2 = \frac{11}{192}.$$

Since $V_4$ is the quantization error for four-means, we have $V_4 \leq \frac{11}{192} = 0.0572917$. Let $\alpha := \{a_1, a_2, a_3, a_4\}$ be an optimal set of four-means. Without any loss of generality, we can assume that $1 \leq a_1 < a_2 < a_3 < a_4 \leq 6$. The Voronoi region of $a_1$ must contain 1. Suppose that the Voronoi region of $a_1$ contains 2 as well. Then,

$$V_4 \geq \sum_{j=1}^{2} f(j)(j - a[1,2])^2 = \frac{1}{6} > V_4,$$

which gives a contradiction. Hence, we can assume that the Voronoi region of $a_1$ contains only the element 1, i.e., $a_1 = 1$. Then, the Voronoi region of $a_2$ must contain 2. Suppose that the Voronoi region of $a_2$ also contains 3. Then,

$$V_4 \geq \sum_{j=2}^{3} f(j)(j - a[2,3])^2 = \frac{1}{12} = 0.0833333 > V_4,$$

which leads to a contradiction. Hence, the Voronoi region of $a_2$ does not contain 3, i.e., $a_2 = 2$. Then, the Voronoi region of $a_3$ must contain 3. Suppose that the Voronoi region of $a_3$ contains the set $\{3, 4, 5\}$. Then, we have

$$V_4 \geq \sum_{j=3}^{5} f(j)(j - a[3,5])^2 = \frac{13}{112} = 0.116071 > V_4,$$

which yields a contradiction. Thus, we can assume that the Voronoi region of $a_3$ does not contain 5. Suppose that the Voronoi region of $a_3$ contains 3 only. Then, the Voronoi region of $a_5$ contains $4, 5, 6$, which implies

$$V_4 = \sum_{j=4}^{6} f(j)(j - a[4,6])^2 = \frac{11}{128} = 0.0859375 > V_4,$$

which gives a contradiction. Hence, the Voronoi region of $a_3$ contains 3 and 4 yielding $a_3 = a[3,4]$, and $a_4 = a[5,6]$. Thus, the optimal set of four-means is $\{1, 2, a[3,4], a[5,6]\}$ with quantization error $V_4 = \frac{11}{192}$, which is the proposition. □

Using the similar technique as the previous proposition, the following proposition can be proved.

PROPOSITION 3.5. *The optimal set of five-means is* $\{1, 2, 3, 4, a[5,6]\}$ *with quantization error* $V_5 = \frac{1}{64}$.

**3.6. Optimal quantization for the probability distribution $P$ with $x = \frac{7}{10}$.** In this case, the probability mass function $f$ is given by

$$f(j) = P(X = j) = \begin{cases} \frac{7}{10} & \text{if } j = 1, \\ (\frac{3}{10})^{j-1}\frac{7}{10} & \text{if } j \in \{2,3,4,5\}, \\ (\frac{3}{10})^5 & \text{if } j = 6, \\ 0 & \text{otherwise.} \end{cases}$$

We now give the following propositions.

PROPOSITION 3.7. *The optimal set of two-means is given by* $\{1, a[2,6]\}$ *with quantization error* $V_2 = \frac{174296997}{1000000000}$.

PROOF. The distortion error due to the set $\beta := \{1, a[2,6]\}$ is given by

$$\sum_{j=1}^{6} f(j) \min_{a\in\beta}(j-a)^2 = \sum_{j=2}^{6} f(j)(j - a[2,6])^2 = \frac{174296997}{1000000000}.$$

Since $V_2$ is the quantization error for two-means, we have $V_2 \leq \frac{174296997}{1000000000} = 0.174296997$. Let $\alpha := \{a_1, a_2\}$ be an optimal set of two-means. Without any loss of generality, we can assume that $1 \leq a_1 < a_2 \leq 6$. Notice that the Voronoi region of $a_1$ must contain 1. Suppose that the Voronoi region of $a_1$ contains the set $\{1, 2, 3\}$. Then,

$$V_2 \geq \sum_{j=1}^{3} f(j)(j - a[1,3])^2 = \frac{4809}{13900} = 0.345971 > V_2,$$

which gives a contradiction. Hence, we can assume that the Voronoi region of $a_1$ does not contain 3. Next, suppose that the Voronoi region of $a_1$ contains the set $\{1, 2\}$. Then, the Voronoi region of $a_2$ contains all the remaining elements, and so

$$V_2 = \sum_{j=1}^{2} f(j)(j - a[1,2])^2 + \sum_{j=3}^{6} f(j)(j - a[3,6])^2 = \frac{272139987}{1300000000} = 0.209338 > V_2,$$

which yields a contradiction. Hence, we can assume that the Voronoi region of $a_1$ contains only the element 1, and the remaining elements are contained in the Voronoi region of $a_2$ implying

$$a_1 = 1, \text{ and } a_2 = a[2,6]$$

with quantization error $V_2 = \frac{174296997}{1000000000}$. Thus, the proof of the proposition is complete. □

PROPOSITION 3.8. *The optimal set of three-means is given by* $\{1, 2, a[3,6]\}$ *with quantization error* $V_3 = \frac{4779999}{100000000}$.

PROOF. The distortion error due to the set $\beta := \{1, 2, a[3,6]\}$ is given by

$$\sum_{j=3}^{6} f(j) \min_{a\in\beta}(j-a)^2 = \sum_{j=3}^{6} f(j)(j - a[3,6])^2 = \frac{4779999}{100000000} = 0.04779999.$$

Since $V_3$ is the quantization error for three-means, we have $V_3 \leq 0.04779999$. Let $\alpha := \{a_1, a_2, a_3\}$ be an optimal set of three-means such that $1 \leq a_1 < a_2 < a_3 \leq 6$.

Notice that the Voronoi region of $a_1$ must contain 1. Suppose that the Voronoi region of $a_1$ also contains 2. Then,

$$V_3 \geq \sum_{j=1}^{2} f(j)(j - a[1,2])^2 = \frac{21}{130} = 0.161538 > V_3,$$

which yields a contradiction. Thus, we can assume that the Voronoi region of $a_1$ contains only the element 1, i.e., $a_1 = 1$. The Voronoi region of $a_2$ contains 2. Suppose that the Voronoi region of $a_2$ also contains the set $\{2, 3\}$. Then,

$$V_3 \geq \sum_{j=2}^{3} f(j)(j - a[2,3])^2 = \frac{63}{1300} = 0.0484615 > V_3,$$

which is a contradiction. Hence, the Voronoi region of $a_2$ contains only the element 2, which yields $a_2 = 2$, and $a_3 = a[3, 6]$, with quantization error $V_3 = \frac{4779999}{100000000}$. Thus, the proof of the proposition is complete. □

Following the similar techniques as given in Proposition 3.8, we can prove the following two propositions.

PROPOSITION 3.9. *The optimal set of four-means is given by* $\{1, 2, 3, a[4, 6]\}$ *with quantization error* $V_4 = \frac{112833}{10000000}$.

PROPOSITION 3.10. *The optimal set of five-means is given by* $\{1, 2, 3, 4, a[5, 6]\}$ *with quantization error* $V_5 = \frac{1701}{1000000}$.

## 4. Optimal quantization for infinite discrete distributions

In this section, for $n \in \mathbb{N}$, we investigate the optimal sets of $n$-means for two different infinite discrete distributions. We give them in the following two subsections.

**4.1. Optimal quantization for an infinite discrete distribution with support $\{\frac{1}{n} : n \in \mathbb{N}\}$.** Let $\mathbb{N} := \{1, 2, 3, \cdots\}$ be the set of natural numbers, and let $P$ be a Borel probability measure on the set $\{\frac{1}{n} : n \in \mathbb{N}\}$ with probability mass function $f$ given by

$$f(x) = \begin{cases} \frac{1}{2^k} & \text{if } x = \frac{1}{k} \text{ for } k \in \mathbb{N}, \\ 0 & \text{otherwise.} \end{cases}$$

Then, $P$ is a Borel probability measure on $\mathbb{R}$, and the support of $P$ is given by $\text{supp}(P) = \{\frac{1}{n} : n \in \mathbb{N}\}$. In this section, for the probability measure $P$, we investigate the optimal sets of $n$-means and the $n$th quantization errors for $n \in \mathbb{N}$. For $k, \ell \in \mathbb{N}$, where $k \leq \ell$, write

$$[k, \ell] := \{\frac{1}{n} : n \in \mathbb{N} \text{ and } k \leq n \leq \ell\}, \text{ and } [k, \infty) := \{\frac{1}{n} : n \in \mathbb{N} \text{ and } n \geq k\}.$$

Further, write

$$Av[k, \ell] := E\Big(X : X \in [k, \ell]\Big) = \frac{\sum_{n=k}^{\ell} \frac{1}{2^n} \frac{1}{n}}{\sum_{n=k}^{\ell} \frac{1}{2^n}}, \; Av[k, \infty) := E\Big(X : X \in [k, \infty)\Big) = \frac{\sum_{n=k}^{\infty} \frac{1}{2^n} \frac{1}{n}}{\sum_{n=k}^{\infty} \frac{1}{2^n}},$$

$$Er[k, \ell] := \sum_{n=k}^{\ell} \frac{1}{2^n} \Big(\frac{1}{n} - Av[k, \ell]\Big)^2, \text{ and } Er[k, \infty) := \sum_{n=k}^{\infty} \frac{1}{2^n} \Big(\frac{1}{n} - Av[k, \infty)\Big)^2.$$

Notice that $E(X) := E(X : X \in \text{supp}(P)) = \sum_{n=1}^{\infty} \frac{1}{2^n}\frac{1}{n} = Av[1,\infty) = \log(2)$, and so the optimal set of one-mean is the set $\{\log(2)\}$ with quantization error

$$V(P) = \sum_{n=1}^{\infty} \frac{1}{2^n}\Big(\frac{1}{n} - \log(2)\Big)^2 = Er[1,\infty) = \frac{1}{12}\left(\pi^2 - 18\log^2(2)\right) = 0.101788.$$

PROPOSITION 4.2. *The set $\{Av[2,\infty), 1\}$ forms the optimal set of two-means for the probability measure $P$ with quantization error*

$$V_2(P) = Er[2,\infty) = \frac{1}{12}\left(\pi^2 - 12 - 30\log^2(2) + 24\log(2)\right) = 0.0076288597.$$

PROOF. Consider the set $\beta := \{Av[2,\infty), 1\}$. Since $\frac{1}{3} < \frac{1}{2}(Av[2,\infty) + 1) < 1$, the Voronoi region of 1 contains only the element 1, and the Voronoi region of $Av[2,\infty)$ contains the set $\{\frac{1}{n} : n \geq 2\}$. Hence, the distortion error due to the set $\beta$ is given by

$$V(P;\beta) = Er[2,\infty) = \frac{1}{12}\left(\pi^2 - 12 - 30\log^2(2) + 24\log(2)\right) = 0.0076288597.$$

Since $V_2(P)$ is the quantization error for two-means, we have $V_2(P) \leq 0.0076288597$. Let $\alpha := \{a_2, a_1\}$ be an optimal set of two-means. Due to Proposition 1.1, we can assume that $0 \leq a_2 < a_1 \leq 1$. The Voronoi region of $a_1$ must contain 1. Suppose that the Voronoi region of $a_1$ also contains $\frac{1}{2}$. Then,

$$V_2(P) \geq Er[1,2] = \frac{1}{24} = 0.0416667 > V_2(P),$$

which leads to a contradiction. Hence, we can assume that the Voronoi region of $a_1$ does not contain $\frac{1}{2}$. Again, by Proposition 1.1, the Voronoi region of $a_2$ cannot contain the element 1. Thus, we have $a_2 = Av[2,\infty)$, and $a_1 = 1$, and the corresponding quantization error is $V_2(P) = Er[2,\infty) = 0.0076288597$. Thus, the proof of the proposition is complete. □

PROPOSITION 4.3. *The set $\{Av[3,\infty), \frac{1}{2}, 1\}$ forms the optimal set of three-means for the probability measure $P$ with quantization error $V_3(P) = Er[3,\infty) =$ 0.00116437359.*

PROOF. Consider the set $\beta := \{Av[3,\infty), \frac{1}{2}, 1\}$. Since, $\frac{1}{3} < \frac{1}{2}(Av[3,\infty) + \frac{1}{2}) < \frac{1}{2}$, and $\frac{1}{2} < \frac{1}{2}(\frac{1}{2} + 1) < 1$, the distortion error due to the set $\beta := \{Av[3,\infty), \frac{1}{2}, 1\}$ is given by

$$V(P;\beta) = Er[3,\infty) = \frac{1}{24}\left(2\pi^2 - 51 - 108\log^2(2) + 120\log(2)\right) = 0.00116437359.$$

Since $V_3(P)$ is the quantization error for three-means, we have $V_3(P) \leq 0.00116437359$. Let $\alpha := \{a_3, a_2, a_1\}$ be an optimal set of three-means such that $0 \leq a_3 < a_2 < a_1 \leq 1$. Proceeding as Proposition 4.2, we can show that $a_1 = 1$. Suppose that the Voronoi region of $a_2$ contains $\frac{1}{2}$ and $\frac{1}{3}$. Then,

$$V_3(P) \geq Er[2,3] = 0.002314814815 > V_3(P),$$

which is a contradiction. Hence, the Voronoi region of $a_2$ cannot contain $\frac{1}{3}$. Thus, we have

$$a_3 = Av[3,\infty),\ a_2 = \frac{1}{2},\ \text{and } a_1 = 1,$$

with quantization error $V_3(P) = Er[3,\infty) = 0.00116437359$. Thus, the proof of the proposition is complete. □

PROPOSITION 4.4. *The set* $\{Av[4,\infty), \frac{1}{3}, \frac{1}{2}, 1\}$ *forms the optimal set of four-means for the probability measure* $P$ *with quantization error* $V_4(P) = Er[4,\infty) = 0.0002418966477$.

PROOF. The proof of this proposition is similar to the proof of Proposition 4.3. □

PROPOSITION 4.5. *The set* $\{Av[5,\infty), \frac{1}{4}, \frac{1}{3}, \frac{1}{2}, 1\}$ *forms the optimal set of five-means for the probability measure* $P$ *with quantization error* $V_5(P) = Er[5,\infty) = 0.00005991266593$.

PROOF. The distortion error due to the set $\beta := \{Av[5,\infty), \frac{1}{4}, \frac{1}{3}, \frac{1}{2}, 1\}$ is given by

$$V(P;\beta) := Er[5,\infty) = Er[5,\infty) = 0.00005991266593.$$

Since $V_5(P)$ is the quantization error for five-means, we have

$$V_5(P) \leq 0.00005991266593.$$

Let $\alpha := \{a_5, a_4, a_3, a_2, a_1\}$ be an optimal set of five-means such that $0 \leq a_5 < a_4 < a_3 < a_2 < a_1 \leq 1$. Proceeding as Proposition 4.3, we can show that $a_1 = 1$, $a_2 = \frac{1}{2}$, and $a_3 = \frac{1}{3}$. We now show that $a_4 = \frac{1}{4}$. Suppose that the Voronoi region of $a_4$ contains $\frac{1}{4}$, $\frac{1}{5}$, and $\frac{1}{6}$. Then,

$$V_5(P) \geq Er[4,6] = 0.0001116071429 > V_5(P),$$

which is a contradiction. Assume that the Voronoi region of $a_4$ contains only the elements $\frac{1}{4}$, and $\frac{1}{5}$. Then, the Voronoi region of $a_5$ contains the set $[6,\infty)$, and so we have

$$V_5(P) = Er[6,\infty) + Er[4,5] = 0.00006872664638 > V_5(P),$$

which leads to a contradiction. Hence, we can assume that the Voronoi region of $a_4$ contains only the element $\frac{1}{4}$. Thus, we have $a_5 = Av[5,\infty)$, $a_4 = \frac{1}{4}$, $a_3 = \frac{1}{3}$, $a_2 = \frac{1}{2}$, and $a_1 = 1$ with quantization error $V_5(P) = Er[5,\infty) = 0.00005991266593$. Thus, the proof of the proposition is complete. □

PROPOSITION 4.6. *The set* $\{Av[7,\infty), Av[5,6], \frac{1}{4}, \frac{1}{3}, \frac{1}{2}, 1\}$ *forms the optimal set of six-means for the probability measure* $P$ *with quantization error*

$$V_6(P) = Er[7,\infty) + Er[5,6] = 0.00001658886625.$$

PROOF. Notice that $\frac{1}{7} = 0.142857 < \frac{1}{2}(Av[7,\infty) + Av[5,6]) = 0.158488 < 0.166667 = \frac{1}{6}$, and $\frac{1}{5} < \frac{1}{2}(Av[5,6] + \frac{1}{4}) < \frac{1}{4}$. Hence, the distortion error due to the set $\beta := \{Av[7,\infty), Av[5,6], \frac{1}{4}, \frac{1}{3}, \frac{1}{2}, 1\}$ is given by

$$V(P;\beta) = Er[7,\infty) + Er[5,6] = 0.00001658886625.$$

Since $V_6(P)$ is the distortion error for six-means, we have $V_6(P) \leq 0.00001658886625$. Let $\alpha := \{a_6, a_5, a_4, a_3, a_2, a_1\}$ be an optimal set of six-means such that $0 \leq a_6 < a_5 < \cdots < a_1 \leq 1$. Proceeding in the similar way as in the proof of Proposition 4.3, we can show that $a_3 = \frac{1}{3}$, $a_2 = \frac{1}{2}$, and $a_1 = 1$. Proceeding in the similar way as in the proof of Proposition 4.5, we can show that $a_4 = \frac{1}{4}$. We now show that $a_5 = Av[5,6]$. Notice that the Voronoi region of $a_5$ must contain $\frac{1}{5}$. Suppose that the Voronoi region of $a_5$ contains $\frac{1}{7}$ and $\frac{1}{6}$ as well. Then,

$$V_6(P) \geq Er[5,7] = 0.00002576328150 > V_6(P),$$

which leads to a contradiction. Suppose that the Voronoi region of $a_5$ contains only the element $\frac{1}{5}$, i.e., $a_5 = \frac{1}{5}$. Then,

$$V_6(P) = Er[6, \infty) = 0.00001664331305 > V_6(P),$$

which yields a contradiction. Hence, we can assume that the Voronoi region of $a_5$ contains only the two elements $\frac{1}{6}$ and $\frac{1}{5}$. Thus, we have

$$a_6 = Av[7, \infty),\ a_5 = Av[5, 6],\ a_4 = \frac{1}{4},\ a_3 = \frac{1}{3},\ a_2 = \frac{1}{2},\ \text{and } a_1 = 1,$$

and the quantization error is $V_6(P) = Er[7, \infty) + Er[5, 6] = 0.00001658886625$. Thus, the proof of the proposition is complete. □

In the following proposition, we calculate the optimal set of $n$-means and the $n$th quantization error for $n = 300$.

PROPOSITION 4.7. *The set* $\{Av[301, \infty), Av[299, 300], \frac{1}{298}, \frac{1}{297}, \cdots, \frac{1}{3}, \frac{1}{2}, 1\}$ *forms the optimal set of* 300*-means for the probability measure* $P$ *with quantization error* $V_{300}(P) = Er[301, \infty) + Er[299, 300] = 1.56431764258240960617412 8 \times 10^{-100}$.

PROOF. Notice that $\frac{1}{301} = 0.003322259136 < \frac{1}{2}(Av[301, \infty) + Av[299, 300]) = 0.003326047849 < 0.003333333333 = \frac{1}{300}$, and

$$\frac{1}{299} = 0.003344481605 < \frac{1}{2}(Av[299, 300] + \frac{1}{298}) = 0.003348235106 < 0.003355704698 = \frac{1}{298}.$$

Hence, the distortion error due to the set

$$\beta := \{Av[301, \infty), Av[299, 300], \frac{1}{198}, \frac{1}{197}, \cdots, \frac{1}{3}, \frac{1}{2}, 1\}$$

is given by

$$V(P; \beta) = Er[301, \infty) + Er[299, 300] = 1.564317642582409606174128 \times 10^{-100}.$$

Since $V_{300}(P)$ is the distortion error for 300-means, we have

$$V_{300}(P) \leq 1.564317642582409606174128 \times 10^{-100}.$$

Let $\alpha := \{a_{300}, a_{299}, \cdots, a_3, a_2, a_1\}$ be an optimal set of 300-means such that $0 \leq a_{300} < a_{299} < \cdots < a_1 \leq 1$. Proceeding in the similar way as in the proof of Proposition 4.3, we can show that $a_{297} = \frac{1}{297}$, $a_{296} = \frac{1}{296}$, $\cdots$, $a_3 = \frac{1}{3}$, $a_2 = \frac{1}{2}$, and $a_1 = 1$. Proceeding in the similar way as in the proof of Proposition 4.5, we can show that $a_{298} = \frac{1}{298}$. We now show that $a_{299} = Av[299, 300]$. The Voronoi region of $a_{299}$ must contain $\frac{1}{299}$. Suppose that the Voronoi region of $a_{299}$ contains $\frac{1}{i}$ for $i = 299, 300, 301, 302$. Then,

$$V_{300}(P) \geq Er[299, 302] = 1.953916208081117722202350 \times 10^{-100} > V_{300}(P),$$

which leads to a contradiction. Assume that the Voronoi region of $a_{299}$ contains only the elements $\frac{1}{i}$ for $i = 299, 300, 301$. Then,

$$V_{300}(P) = Er[302, \infty) + Er[299, 301] = 1.698521119259119376459397 \times 10^{-100} > V_{300}(P),$$

which yields a contradiction. Assume that the Voronoi region of $a_{299}$ contains only the element $\frac{1}{299}$. Then,

$$V_{300}(P) = Er[300, \infty) = 2.345910694878821203973953 \times 10^{-100} > V_{300}(P),$$

which gives a contradiction. Hence, we can assume that the Voronoi region of $a_{299}$ contains only the two elements $\frac{1}{299}$ and $\frac{1}{300}$. Thus, we have

$$a_{300} = Av[301, \infty),\ a_{299} = Av[299, 300],\ a_{298} = \frac{1}{298}, \cdots, a_4 = \frac{1}{4},\ a_3 = \frac{1}{3},\ a_2 = \frac{1}{2}, \text{ and } a_1 = 1,$$

and the quantization error is given by

$$V_{300}(P) = Er[301, \infty) + Er[299, 300] = 1.5643176425824096061741 28 \times 10^{-100}.$$

Thus, the proof of the proposition is complete. ☐

We now give the following theorem.

THEOREM 4.8. *For any positive integer $n$, the sets $\{Av[n, \infty), \frac{1}{n-1}, \cdots, \frac{1}{3}, \frac{1}{2}, 1\}$, where $1 \leq n \leq 5$, form the optimal sets of $n$-means for the probability measure $P$ with quantization errors $V_n(P) := Er[n, \infty)$. For the positive integers $n$, where $6 \leq n \leq 300$, the sets $\{Av[n+1, \infty), Av[n-1, n], \frac{1}{n-2}, \cdots, \frac{1}{3}, \frac{1}{2}, 1\}$ form the optimal sets of $n$-means for the probability measure $P$ with quantization errors*

$$V_n(P) = Er[n + 1, \infty) + Er[n - 1, n].$$

PROOF. As a consequence of Proposition 4.2 through Proposition 4.5, it follows that for $1 \leq n \leq 5$, the sets

$$\left\{\text{Av}[n, \infty), \frac{1}{n-1}, \cdots, \frac{1}{3}, \frac{1}{2}, 1\right\}$$

form the optimal sets of $n$-means for the probability measure $P$ with corresponding quantization errors

$$V_n(P) = Er[n, \infty).$$

Proceeding in the similar way as Proposition 4.6 and Proposition 4.7, we can show that for any positive integer $n$, where $6 \leq n \leq 300$, the sets $\{Av[n + 1, \infty), Av[n - 1, n], \frac{1}{n-2}, \cdots, \frac{1}{3}, \frac{1}{2}, 1\}$ form the optimal sets of $n$-means for the probability measure $P$ with quantization errors

$$V_n(P) = Er[n + 1, \infty) + Er[n - 1, n].$$

Thus, we complete the proof of the theorem. ☐

The following problem remains open.

OPEN 4.9. Proceeding in the similar way as Proposition 4.6 and Proposition 4.7, it can be shown that the set $\{Av[n+1, \infty), Av[n-1, n], \frac{1}{n-2}, \cdots, \frac{1}{3}, \frac{1}{2}, 1\}$ also forms an optimal set of $n$-means for $n = 301$. It is still not known whether the sets $\{Av[n + 1, \infty), Av[n - 1, n], \frac{1}{n-2}, \cdots, \frac{1}{3}, \frac{1}{2}, 1\}$ give the optimal sets of $n$-means for all positive integers $n \geq 6$. If not, then the least upper bound of $n \in \mathbb{N}$ for which such sets give the optimal sets of $n$-means for the probability measure $P$ is not known yet.

**4.10. Optimal quantization for an infinite discrete distribution with support $\{n : n \in \mathbb{N}\}$.** Let $\mathbb{N} := \{1, 2, 3, \cdots\}$ be the set of natural numbers, and let $P$ be a Borel probability measure on the set $\{n : n \in \mathbb{N}\}$ with probability density function $f$ given by

$$f(x) = \begin{cases} \frac{1}{2^n} & \text{if } x = n \text{ for } n \in \mathbb{N}, \\ 0 & \text{otherwise.} \end{cases}$$

Then, $P$ is a Borel probability measure on $\mathbb{R}$, and the support of $P$ is the set $\mathbb{N}$ of natural numbers. In this section, our goal is to determine the optimal sets of $n$-means and the $n$th quantization errors for all positive integers $n$ for the probability measure $P$. For $k, \ell \in \mathbb{N}$, where $k \leq \ell$, write

$$[k, \ell] := \{n : n \in \mathbb{N} \text{ and } k \leq n \leq \ell\}, \text{ and } [k, \infty) := \{n : n \in \mathbb{N} \text{ and } n \geq k\}.$$

Further, write

$$Av[k, \ell] := E\Big(X : X \in [k, \ell]\Big) = \frac{\sum_{n=k}^{\ell} \frac{n}{2^n}}{\sum_{n=k}^{\ell} \frac{1}{2^n}}, \ Av[k, \infty) := E\Big(X : X \in [k, \infty)\Big) = \frac{\sum_{n=k}^{\infty} \frac{n}{2^n}}{\sum_{n=k}^{\infty} \frac{1}{2^n}},$$

$$Er[k, \ell] := \sum_{n=k}^{\ell} \frac{1}{2^n}\Big(n - Av[k, \ell]\Big)^2, \text{ and } Er[k, \infty) := \sum_{n=k}^{\infty} \frac{1}{2^n}\Big(n - Av[k, \infty)\Big)^2.$$

Notice that $E(P) := E(X : X \in \text{supp}(P)) = \sum_{n=1}^{\infty} \frac{n}{2^n} = Av[1, \infty) = 2$, and so the optimal set of one-mean is the set $\{2\}$ with quantization error

$$V(P) = \sum_{n=1}^{\infty} \frac{1}{2^n}(n - 2)^2 = Er[1, \infty) = 2.$$

PROPOSITION 4.11. *The optimal set of two-means is given by* $\{Av[1, 2], Av[3, \infty)\}$ *with quantization error* $V_2 = \frac{2}{3}$.

PROOF. We see that $Av[1, 2] = \frac{4}{3}$, and $Av[3, \infty) = 4$. Since $\frac{4}{3} < \frac{1}{2}(\frac{4}{3} + 4) < 4$, the distortion error due to the set $\beta := \{\frac{4}{3}, 4\}$ is given by

$$V(P; \beta) = Er[1, 2] + Er[3, \infty) = \frac{2}{3}.$$

Since $V_2$ is the quantization error for two-means, we have $V_2 \leq \frac{2}{3}$. Let $\alpha := \{a_1, a_2\}$, where $1 \leq a_1 < a_2 < \infty$, be an optimal set of two-means. Notice that the Voronoi region of $a_1$ must contain 1. Suppose that the Voronoi region of $a_1$ contains the set $\{1, 2, 3, 4\}$. Then,

$$V_2 \geq \sum_{j=1}^{4} \frac{1}{2^j}(j - Av[1, 4])^2 = Er[1, 4] = \frac{97}{120} = 0.808333 > V_2,$$

which yields a contradiction. Hence, we can assume that the Voronoi region of $a_1$ contains only the set $\{1, 2, 3\}$, and so the Voronoi region of $a_2$ contains the set $\{n : n \geq 4\}$. Then, we have

$$V_2 = Er[1, 3] + Er[4, \infty) = \frac{5}{7} = 0.714286 > V_2,$$

which is a contradiction. Next, suppose that the Voronoi region of $a_1$ contains only the element 1, and so the Voronoi region of $a_2$ contains the set $\{n : n \geq 2\}$. Then, we have

$$V_2 = Er[2, \infty) = 1 > V_2,$$

which leads to a contradiction. Hence, we can assume that the Voronoi region of $a_1$ contains the set $\{1, 2\}$, and so the Voronoi region of $a_2$ contains $\{3, 4, 5, \cdots\}$ yielding $a_1 = Av[1, 2]$, and $a_2 = Av[3, \infty)$, and the corresponding quantization error is $V_2 = \frac{2}{3}$. Thus, the proof of the proposition is complete. □

PROPOSITION 4.12. *The sets* $\{1, Av[2, 3], Av[4, \infty)\}$, *and* $\{Av[1, 2], Av[3, 4], Av[5, \infty)\}$ *form two different optimal sets of three-means with quantization error* $V_3 = \frac{1}{3}$.

PROOF. The distortion error due to set $\beta := \{1, Av[2, 3], Av[4, \infty)\}$ is given by

$$V(P; \beta) = Er[2, 3] + Er[4, \infty) = \frac{1}{3}.$$

Notice that the distortion error due to the set $\{Av[1, 2], Av[3, 4], Av[5, \infty)\}$ is also $\frac{1}{3}$. Since $V_3$ is the quantization error for three-means, we have $V_3 \leq \frac{1}{3}$. Let $\alpha := \{a_1, a_2, a_3\}$ be an optimal set of three-means, where $1 \leq a_1 < a_2 < a_3 < \infty$. Suppose that the Voronoi region of $a_1$ contains the set $\{1, 2, 3\}$. Then,

$$V_3 \geq \sum_{j=1}^{3} \frac{1}{2^j}(j - Av[1, 3])^2 = \frac{13}{28} > \frac{1}{3} > V_3,$$

which leads to a contradiction. Hence, we can assume that the Voronoi region of $a_1$ contains either the set $\{1\}$, or the set $\{1, 2\}$. Consider the following two cases:

CASE 1. *The Voronoi region of* $a_1$ *contains only the set* $\{1\}$.

In this case, the Voronoi region of $a_2$ must contain the element 2. Suppose that the Voronoi region of $a_2$ contains the set $\{2, 3, 4, 5\}$. Then,

$$V_3 \geq \sum_{j=2}^{5} \frac{1}{2^j}(j - Av[2, 5])^2 = \frac{97}{240} = 0.404167 > V_3,$$

which yields a contradiction. Assume that the Voronoi region of $a_2$ contains only the set $\{2, 3, 4\}$, and so the Voronoi region of $a_3$ contains the set $\{n : n \geq 5\}$. Then, the distortion error is

$$V_3 = Er[2, 4] + Er[5, \infty) = \frac{5}{14} = 0.357143 > V_3,$$

which gives a contradiction. Next, assume that the Voronoi region of $a_2$ contains only the element 2, and so the Voronoi region of $a_3$ contains the set $\{n : n \geq 3\}$. Then, the distortion error is

$$V_3 = Er[3, \infty) = \frac{1}{2} > V_3,$$

which is a contradiction. Hence, in this case, we can conclude that the Voronoi region of $a_2$ contains only the set $\{2, 3\}$ yielding $a_1 = 1$, $a_2 = Av[2, 3]$, and $a_3 = Av[4, \infty)$ with quantization error $V_3 = \frac{1}{3}$.

CASE 2. *The Voronoi region of* $a_1$ *contains only the set* $\{1, 2\}$.

In this case, the Voronoi region of $a_2$ must contain the element 3. Suppose that the Voronoi region of $a_2$ contains the set $\{3, 4, 5, 6\}$. Then,

$$V_3 \geq \sum_{j=1}^{2} \frac{1}{2^j}(j - Av[1, 2])^2 + \sum_{j=3}^{6} \frac{1}{2^j}(j - Av[3, 6])^2 = \frac{59}{160} = 0.36875 > V_3,$$

which yields a contradiction. Assume that the Voronoi region of $a_2$ contains only the set $\{3, 4, 5\}$, and so the Voronoi region of $a_3$ contains the set $\{n : n \geq 6\}$. Then, the distortion error is

$$V_3 = Er[1, 2] + Er[3, 5] + Er[6, \infty) = \frac{29}{84} = 0.345238 > V_3,$$

which gives a contradiction. Next, assume that the Voronoi region of $a_2$ contains only the element 3, and so the Voronoi region of $a_3$ contains the set $\{n : n \geq 4\}$. Then, the distortion error is

$$V_3 = Er[1, 2] + Er[4, \infty) = \frac{5}{12} = 0.416667 > V_3,$$

which yields a contradiction. Hence, in this case, we can conclude that the Voronoi region of $a_2$ contains only the set $\{3, 4\}$ yielding $a_1 = Av[1, 2]$, $a_2 = Av[3, 4]$, and $a_3 = Av[5, \infty)$ with quantization error $V_3 = \frac{1}{3}$.

By Case 1 and Case 2, the proof of the proposition is complete. □

We need the following lemma.

LEMMA 4.13. *Let $n \geq 4$, and let $\alpha_n$ be an optimal set of $n$-means. Then, $\alpha_n$ must contain the set $\{1, 2, \cdots, (n-3)\}$.*

PROOF. The distortion error due to the set $\beta := \{1, 2, \cdots, (n-3), (n-2), Av[n-1, n], Av[n+1, \infty)\}$ is given by

$$V(P; \beta) = Er[n-1, n] + Er[n+1, \infty) = \frac{2^{3-n}}{3}.$$

Since $V_n$ is the quantization error for $n$-means, we have $V_n \leq \frac{2^{3-n}}{3}$. Let $\alpha_n := \{a_1, a_2, \cdots, a_n\}$ be an optimal set of $n$-means such that $1 \leq a_1 < a_2 < \cdots < a_n < \infty$. We show that $a_1 = 1$, $a_2 = 2$, $\cdots$, $a_{n-3} = n-3$. Notice that the Voronoi region of $a_1$ must contain the element 1. Suppose that the Voronoi region of $a_1$ also contains the element 2. Then,

$$V_n > \sum_{j=1}^{2} \frac{1}{2^j}(j - Av[1, 2])^2 = \frac{1}{6} \geq \frac{2^{3-n}}{3} \geq V_n,$$

which is a contradiction. Hence, we can conclude that the Voronoi region of $a_1$ contains only the element 1 yielding $a_1 = 1$. Thus, we can deduce that there exists a positive integer $k$, where $1 \leq k < n-3$, such that $a_1 = 1$, $a_2 = 2$, $\cdots$, $a_k = k$. We now show that $a_{k+1} = k+1$. Notice that the Voronoi region of $a_{k+1}$ must contain $k+1$. Suppose that the Voronoi region of $a_{k+1}$ also contains the element $k+2$. Then, as $k < n-3$, we have

$$V_n > \sum_{j=k+1}^{k+2} \frac{1}{2^j}(j - Av[k+1, k+2])^2 = \frac{2^{-k-1}}{3} \geq \frac{2^{3-n}}{3} \geq V_n,$$

which is a contradiction. Hence, we can conclude that the Voronoi region of $a_{k+1}$ contains only the element $k+1$ yielding $a_{k+1} = k+1$. Thus, by the Principle of Mathematical Induction, we deduce that $a_1 = 1$, $a_2 = 2$, $\cdots$, $a_{n-3} = n-3$. Thus, the proof of the lemma is complete. □

THEOREM 4.14. *Let $n \geq 4$, and let $\alpha_n$ be an optimal set of $n$-means. Then, either $\alpha_n = \{1, 2, 3, \cdots, n-3, n-2, Av[n-1, n], Av[n+1, \infty)\}$, or $\alpha_n = \{1, 2, 3, \cdots, n-3, Av[n-2, n-1], Av[n, n+1], Av[n+2, \infty)\}$ with quantization error $V_n = \frac{2^{3-n}}{3}$.*

PROOF. As shown in the proof of Lemma 4.13, we have $V_n \leq \frac{2^{3-n}}{3}$. Let $\alpha_n := \{a_1, a_2, \cdots, a_n\}$ be an optimal set of $n$-means such that $1 \leq a_1 < a_2 < \cdots < a_n < \infty$. By Lemma 4.13, we have $a_1 = 1$, $a_2 = 2$, $\cdots$, $a_{n-3} = n - 3$. Recall that $n \geq 4$. Suppose that the Voronoi region of $a_{n-2}$ contains the set $\{n-2, n-1, n\}$. Then,

$$V_n \geq \sum_{j=n-2}^{n} \frac{1}{2^j}(j - Av[n-2, n])^2 = \frac{13}{7}2^{1-n} > \frac{2^{3-n}}{3} \geq V_n,$$

which leads to a contradiction. Hence, we can assume that the Voronoi region of $a_{n-2}$ contains either the set $\{n-2\}$, or the set $\{n-2, n-1\}$. Consider the following two cases:

CASE 1. *The Voronoi region of $a_{n-2}$ contains only the set $\{n-2\}$.*

Proceeding along the similar lines as Case 1 in the proof of Proposition 4.12, we can show that the Voronoi region of $a_{n-1}$ contains only the set $\{n-1, n\}$ yielding $a_{n-2} = n - 2$, $a_{n-1} = Av[n-1, n]$, and $a_n = Av[n+1, \infty)$ with quantization error $V_n = \frac{2^{3-n}}{3}$.

CASE 2. *The Voronoi region of $a_{n-2}$ contains only the set $\{n-2, n-1\}$.*

Proceeding along the similar lines as Case 2 in the proof of Proposition 4.12, we can show that the Voronoi region of $a_{n-1}$ contains only the set $\{n, n+1\}$ yielding $a_{n-2} = Av[n-2, n-1]$, $a_{n-1} = Av[n, n+1]$, and $a_n = Av[n+2, \infty)$ with quantization error $V_n = \frac{2^{3-n}}{3}$.

By Case 1 and Case 2, the proof of the theorem is complete. □

## 5. Probability distributions when the optimal sets are given

Let $P$ be a discrete probability measure on $\mathbb{R}$ with support a finite or an infinite set $\{1, 2, 3, \cdots\}$. Let $(p_1, p_2, p_3, \cdots)$ be a probability vector associated with $\{1, 2, 3, \cdots\}$ such that the probability mass function $f$ of $P$ is given by $f(k) = p_k$ if $k \in \{1, 2, 3, \cdots\}$, and zero otherwise. For $k, \ell \in \{1, 2, 3, \cdots\}$ with $k \leq \ell$, write

$$[k, \ell] := \{n : k \leq n \leq \ell\}, \text{ and } [k, \infty) := \{k, k+1, \cdots\}.$$

For a random variable $X$ with distribution $P$, let $Av[k, \ell]$ represent the conditional expectation of $X$ given that $X$ takes values on the set $\{k, k+1, k+2, \cdots, \ell\}$, i.e.,

$$Av[k, \ell] = E(X : X \in [k, \ell]),$$

where $k, \ell \in \{1, 2, 3, \cdots\}$ with $k \leq \ell$. On the other hand, by $Av[k, \infty)$ it is meant $Av[k, \infty) = E(X : X \in [k, \infty))$, where $k \in \{1, 2, 3, \cdots\}$. Let $\alpha_n$ be an optimal set of $n$-means for $P$, where $n \in \mathbb{N}$. In this section, our goal is to find a set of probability vectors $(p_1, p_2, p_3, \cdots)$ such that for all $n \in \mathbb{N}$, the optimal sets of $n$-means are given by $\alpha_n = \{1, 2, 3, \cdots, n-1, Av[n, \infty)\}$.

Consider the following two cases:

CASE 1. *$\{1, 2, 3, \cdots\}$ is a finite set.*

In this case, there exists a positive integer $m$, such that the support of $P$ is given by $\{1, 2, 3, \cdots, m\}$. Notice that for any $k \in \{1, 2, \cdots, m\}$, in this case by

$[k, \infty)$ it meant the set $[k, m]$. If $m = 1$, then $\alpha_1 = \{1\}$; and if $m = 2$, then $\alpha_1 = \{Av[1, \infty)\}$, and $\alpha_2 = \{1, Av[2, \infty)\}$, i.e., there is nothing to prove as in the cases of $m = 1$ and $m = 2$, they are true for any associated probability vector. So, we can assume that $m \geq 3$. Define the probability vector $(p_1, p_2, \cdots, p_m)$ as follows:

$$p_j = \begin{cases} x & \text{if } j = 1, \\ (1-x)^{j-1}x & \text{if } 2 \leq j \leq m-1, \\ (1-x)^{j-1} & \text{if } j = m. \end{cases} \tag{1}$$

For the sets $\alpha_n$ to form the optimal sets of $n$-means for all $1 \leq n \leq m$, we must have

$$(n-1) \leq \frac{1}{2}(n - 1 + Av[n, \infty)) \leq n \tag{2}$$

and

$$V(P; \{n-1, Av[n, \infty)\}) \leq V(P; \{Av[n-1, n], Av[n+1, \infty)\}) \tag{3}$$

for $2 \leq n \leq m$. In this regard we give two examples: Example 5.1 and Example 5.2.

CASE 2. $\{1, 2, 3, \cdots\}$ *is an infinite set.*
Define the probability vector $(p_1, p_2, p_3 \cdots)$ as follows:

$$p_j = \begin{cases} x & \text{if } j = 1, \\ (1-x)^{j-1}x & \text{if } 2 \leq j. \end{cases} \tag{4}$$

For the sets $\alpha_n$ to form the optimal sets of $n$-means for all $1 \leq n$, we must have

$$(n-1) \leq \frac{1}{2}(n - 1 + Av[n, \infty)) \leq n \tag{5}$$

and

$$V(P; \{n-1, Av[n, \infty)\}) \leq V(P; \{Av[n-1, n], Av[n+1, \infty)\}) \tag{6}$$

for $2 \leq n$. After some calculation, we see that there exists a real number $y$ for which the inequalities given by (5) and (6) are satisfied if $y \leq x < 1$. The ten-digit rational approximation of such a $y$ is 0.6666666667. Hence, a set of probability vectors $(p_1, p_2, p_3, \cdots)$ for which the given sets $\alpha_n$ form the optimal sets of $n$ means for $1 \leq n$ is given by

$$\{(x, (1-x)x, (1-x)^2x, (1-x)^3x, (1-x)^4x, \cdots) : 0.6666666667 \leq x < 1\}.$$

EXAMPLE 5.1. Let $m = 6$ in Case 1. Then, for $0 < x < 1$ we have

$$p_1 = x,\ p_2 = (1-x)x,\ p_3 = (1-x)^2x,\ p_4 = (1-x)^3x,\ p_5 = (1-x)^4x, \text{ and } p_6 = (1-x)^5.$$

After solving the inequalities given by (2), we have $0.4812099363 < x < 1$. Again, solving the inequality (3), we have $0.6628057756 \leq x < 1$. Notice that 0.4812099363 and 0.6628057756 are the ten-digit rational approximations of two real numbers. Thus, the inequalities given by (2) and (3) are true if $0.6628057756 \leq x < 1$. Hence, a set of probability vectors $(p_1, p_2, \cdots, p_6)$ for which the given sets $\alpha_n$ form the optimal sets of $n$ means for $1 \leq n \leq 6$ is given by

$$\{(x, (1-x)x, (1-x)^2x, (1-x)^3x, (1-x)^4x, (1-x)^5) : 0.6628057756 \leq x < 1\}.$$

EXAMPLE 5.2. Let $m = 7$ in Case 1. Then, proceeding as Example 5.1, we see that (2) and (3) are true if $0.6654212000 \leq x < 1$. Hence, a set of probability vectors $(p_1, p_2, \cdots, p_7)$ for which the given sets $\alpha_n$ form the optimal sets of $n$ means for $1 \leq n \leq 7$ is given by

$$\{x, (1-x)x, (1-x)^2x, (1-x)^3x, (1-x)^4x, (1-x)^5x, (1-x)^6\}$$

where $0.6654212000 \leq x < 1$.

## 6. Conclusion

In this paper, we investigated the problem of optimal quantization for several discrete probability distributions, both finite and infinite. Starting with specific nonuniform discrete distributions supported on a finite set, we computed the optimal sets of $n$-means and the corresponding quantization errors for various values of $n$. We then extended our analysis to infinite discrete distributions, including those supported on the set of reciprocals of natural numbers and on the natural numbers themselves. For each distribution, we established explicit constructions for optimal sets and determined the associated quantization errors, sometimes for values of $n$ as large as 300.

Moreover, we addressed the inverse problem of recovering probability distributions from known optimal sets. Our study included identifying the conditions under which given sets form optimal quantizers and proposing conjectures supported by computational verification. These results contribute to the broader understanding of quantization in discrete settings, offering both theoretical insights and practical tools applicable in areas such as data compression and signal processing.

Future work may explore whether the conjectured optimal sets continue to hold beyond the computed range and further generalize the reverse problem to more complex classes of distributions.

## Declaration

**Authors' contributions:** Each author contributed equally to this manuscript. All authors have read and agreed to the published version of the manuscript.

**Funding:** This research received no external funding.

**Data availability:** No data were used to support this study.

**Conflicts of interest.** The authors declare no conflict of interest.

## References

[1] Q. Du, V. Faber, and M. Gunzburger, *Centroidal Voronoi tessellations: applications and algorithms*, SIAM Rev. **41** (1999), no. 4, 637–676, DOI 10.1137/S0036144599352836. MR1722997

[2] A. Gersho and R. M. Gray, *Vector quantization and signal compression*, Kluwer Academy publishers: Boston, 1992.

[3] A. György and T. Linder, *On the structure of optimal entropy-constrained scalar quantizers*, IEEE Trans. Inform. Theory **48** (2002), no. 2, 416–427, DOI 10.1109/18.978755. MR1891255

[4] R. M. Gray and D. L. Neuhoff, *Quantization*, IEEE Trans. Inform. Theory **44** (1998), no. 6, 2325–2383, DOI 10.1109/18.720541. Information theory: 1948–1998. MR1658787

[5] D. Pollard, *Quantization and the method of $k$-means*, IEEE Trans. Inform. Theory **28** (1982), no. 2, 199–205, DOI 10.1109/TIT.1982.1056481. MR651814

[6] P. L. Zador, *Asymptotic quantization error of continuous signals and the quantization dimension*, IEEE Trans. Inform. Theory **28** (1982), no. 2, 139–149, DOI 10.1109/TIT.1982.1056490. MR651809

[7] R. Zam, *Lattice Coding for Signals and Networks: A Structured Coding Approach to Quantization, Modulation, and Multiuser Information Theory*, Cambridge University Press, 2014.

[8] M. Pandey and M.K. Roychowdhury, *Constrained quantization for probability distributions*, to appear, J. Fractal Geom., 2025.

[9] M. Pandey and M. K. Roychowdhury, *Constrained quantization for the Cantor distribution*, J. Fractal Geom. **11** (2024), no. 3-4, 319–341, DOI 10.4171/jfg/147. MR4812406

[10] M. Pandey and M. K. Roychowdhury, *Conditional constrained and unconstrained quantization for probability distributions*, `arXiv:2312.02965 [math.PR]`.

[11] S. Graf and H. Luschgy, *Foundations of quantization for probability distributions*, Lecture Notes in Mathematics, vol. 1730, Springer-Verlag, Berlin, 2000, DOI 10.1007/BFb0103945. MR1764176

[12] R. M. Gray, J. C. Kieffer, and Y. Linde, *Locally optimal block quantizer design*, Inform. and Control **45** (1980), no. 2, 178–198, DOI 10.1016/S0019-9958(80)90313-7. MR584832

[13] E. F. Abaya and G. L. Wise, *Some remarks on the existence of optimal quantizers*, Statistics & Probability Letters, Volume 2, Issue 6, December 1984, Pages 349-351.

[14] C. P. Dettmann and M. K. Roychowdhury, *Quantization for uniform distributions on equilateral triangles*, Real Anal. Exchange **42** (2017), no. 1, 149–166. MR3702559

[15] G. Pena, H. Rodrigo, M. K. Roychowdhury, J. Sifuentes, and E. Suazo, *Quantization for uniform distributions on hexagonal, semicircular, and elliptical curves*, J. Optim. Theory Appl. **188** (2021), no. 1, 113–142, DOI 10.1007/s10957-020-01771-1. MR4200945

[16] M. K. Roychowdhury, *Optimal quantizers for some absolutely continuous probability measures*, Real Anal. Exchange **43** (2018), no. 1, 105–136, DOI 10.14321/realanalexch.43.1.0105. MR3816435

[17] J. Rosenblatt and M. K. Roychowdhury, *Optimal quantization for piecewise uniform distributions*, Unif. Distrib. Theory **13** (2018), no. 2, 23–55. MR3883505

[18] M. K. Roychowdhury and W. Salinas, *Quantization for a mixture of uniform distributions associated with probability vectors*, Unif. Distrib. Theory **15** (2020), no. 1, 105–142. MR4221193

School of Mathematical and Statistical Sciences, University of Texas Rio Grande Valley, 1201 West University Drive, Edinburg, Texas 78539-2999

*Email address*: `rljcabasag07@gmail.com`

*Email address*: `samir.huq01@gmail.com`

*Email address*: `ericmendoza128@gmail.com`

*Email address*: `mrinal.roychowdhury@utrgv.edu`

Contemporary Mathematics
Volume **825**, 2025
https://doi.org/10.1090/conm/825/16507

# Coherent upper conditional previsions based on Hausdorff measures and its applications in Artificial Intelligence

Serena Doria

ABSTRACT. In a metric space $(\Omega, d)$ the model of coherent upper conditional prevision based on Hausdorff outer measures is applied to represent uncertainty in neural networks when the distribution over data is unknown. Let $G$ be the group of all bi-Lipschitz self functions on $\Omega$. The standard Borel $G$ group is introduced and coherent conditional prevision is defined with respect to a quasi-invariant measure $\nu$ on the $G$ space $\Omega$ if the conditioning event has positive and finite $\nu$ measure. Otherwise it is defined with respect to 0-1 valued finitely additive, but not countably additive, probability.

## 1. Introduction

Coherent conditional previsions, as described by Bruno de Finetti [**6**],[**7**], are a fundamental concept in the framework of subjective probability theory. De Finetti's approach focuses on the idea that probabilities reflect an individual's personal beliefs or degrees of uncertainty rather than objective frequencies.

In this context, coherent conditional previsions represent a consistent way to update beliefs or make predictions in the face of new information. De Finetti's theory starts with the notion that probabilities are subjective assessments of uncertainty. Each individual may have their own probabilities for events based on their knowledge, information, and beliefs. A conditional prevision is essentially a probability assessment made under certain conditions or given certain information. It represents what an individual expects the probability of an event to be, given some specified background information or condition. Coherence refers to the consistency of an individual's beliefs or probability assessments. In the context of conditional previsions, coherence implies that the individual's beliefs should satisfy certain logical constraints when updated with new information. These constraints ensure that the individual's beliefs do not lead to inconsistencies or contradictions. When new information becomes available, individuals update their conditional previsions to reflect this new knowledge. De Finetti proposed rules for updating beliefs in a coherent manner, ensuring that the resulting beliefs remain consistent with the principles of probability theory. In the subjective approach to probability advocated by de Finetti, countable additivity is not considered a defining property because it does not allow for extensions beyond a $\sigma$-field that is properly contained

---

2020 *Mathematics Subject Classification.* Primary 60A05, 60A10, 60B05.

in the class of all subsets of the sample space $\Omega$. The arbitrary extension of probability is one of the desired features of subjective probability [**6**],[**7**],[**16**],[**19**],[**20**],[**23**]. However, when unbounded random variables are taken into account, coherent previsions can induce an ordering of random variables that violate the Monotone Convergence Theorem, as highlighted by Seidenfeld et al. [**22**], which hinders the possibility of indifference between random variables. To address this issue, a model of coherent upper conditional previsions based on Hausdorff outer measures has been proposed in [**14**], which is defined within a linear space comprising Choquet integrable bounded and unbounded random variables. If the conditioning event possesses a positive and finite Hausdorff outer measure in its Hausdorff dimension [**17**] coherent upper conditional previsions are established using the Choquet integral [**5**], defined with respect to Hausdorff outer measures, which are particular monotone set functions or capacities [**4**]. However, if the conditioning event does not satisfy this criterion, the previsions are defined using a finitely additive probability that takes values of either 0 or 1 but is not countably additive. This new model of coherent upper conditional previsions has been introduced for bounded random variables in [**8**–**12**]. In cases where the conditioning event has a positive and finite Hausdorff outer measure in its Hausdorff dimension, the provided coherent upper conditional prevision adheres to the Monotone Convergence Theorem due to the continuity from below exhibited by Hausdorff outer measures. This ensures that coherent upper conditional previsions, defined by Hausdorff outer measures when the conditioning event has a positive and finite Hausdorff outer measure, maintain the principle of indifference between unbounded random variables that share the same distribution. In this paper the model of coherent upper conditional prevision based on Hausdorff outer measures is applied to represent uncertainty in neural networks when the distribution over data is unknown. Moreover denote by $G$ the group of all bi-Lipschitz self functions on $\Omega$, the standard Borel $G$ group is introduced and coherent conditional prevision is defined with respect to a quasi-invariant measure $\nu$ on the $G$ space $\Omega$ if the conditioning event has positive and finite $\nu$ measure.

## 2. Coherent upper conditional previsions

Let $\Omega$ be a non-empty set, let $\mathbf{B}$ be a partition of $\Omega$ and denote by $\wp(\Omega)$, the family of all subsets of $\Omega$. A random variable is a function $X : \Omega \to \widehat{\Re} = \Re \cup \{-\infty, +\infty\}$, let $\mathcal{U}(\Omega)$ be the class of all random variables. $\mathcal{U}(\Omega)$ is not a linear space in fact, if random variables take values $-\infty, +\infty$ then the sum between two of them can be not defined (when for the same $\omega$ one takes value $+\infty$ and the others $-\infty$). Let $L(\Omega) \subset \mathcal{U}(\Omega)$ be the linear space of all bounded random variables defined on $\Omega$; for every $B \in \mathbf{B}$ denote by $X|B$ the restriction of $X$ to $B$ and by $\sup(X|B)$ the supremum of values that $X$ assumes on $B$. Let $L(B)$ be the linear space of all bounded random variables $X|B$. Denote by $I_A$ the indicator function of any event $A \in \wp(B)$, i.e. $I_A(\omega) = 1$ if $\omega \in A$ and $I_A(\omega) = 0$ if $\omega \in A^c$. For every $B \in \mathbf{B}$ let $\mathcal{K}(B)$ be a linear space of random variables $X|B$ with $X \in \mathcal{U}(\Omega)$. Coherent upper conditional previsions $\overline{P}(\cdot|B)$ are real valued functionals defined on a linear space $\mathcal{K}(B)$.

DEFINITION 1. Coherent upper conditional previsions are functionals $\overline{P}(\cdot|B)$ defined on a linear space $\mathcal{K}(B)$ with values in the real number, such that the following axioms of coherence hold for every $X$ and $Y$ in $\mathcal{K}(B)$ and every strictly positive constant $\lambda$:

1) $\overline{P}(X|B) \leq \sup(X|B)$;
2) $\overline{P}(\lambda X|B) = \lambda \overline{P}(X|B)$ (positive homogeneity);
3) $\overline{P}(X+Y|B) \leq \overline{P}(X|B) + \overline{P}(Y|B)$ (subadditivity).

If $\mathcal{K}(B)$ coincides with $L(B)$ the previous definition is the definition of coherent upper conditional prevision given in Walley [**23**]. Suppose that $\overline{P}(X|B)$ is a coherent upper conditional prevision on $\mathcal{K}$. Then its conjugate coherent lower conditional prevision is defined by the conjugacy property $\underline{P}(X|B) = -\overline{P}(-X|B)$. If for every $X$ belonging to $\mathcal{K}(B)$ we have $P(X|B) = \underline{P}(X|B) = \overline{P}(X|B)$ then $P(X|B)$ is called a coherent *linear* conditional prevision and if $\mathcal{K} = L(B)$ it is a coherent linear, positive and positively homogenous functional [**6**],[**7**], [**19**], [**20**] and [**23**, Corollary 2.8.5].

From axioms 1)-3) and by the conjugacy property we have that

$$1 \leq \underline{P}(I_B|B) \leq \overline{P}(I_B|B) \leq 1$$

so that

$$\underline{P}(I_B|B) = \overline{P}(I_B|B) = 1$$

The unconditional coherent upper prevision $\overline{P} = \overline{P}(\cdot|\Omega)$ is obtained as a particular case when the conditioning event is $\Omega$. Coherent upper conditional probabilities are obtained when only 0-1 valued random variables are considered.

DEFINITION 2. Given a partition $\mathbf{B}$ and a random variable $X \in L(\Omega)$, a coherent upper conditional prevision $\overline{P}(X|\mathbf{B})$ is a random variable on $\Omega$ equal to $\overline{P}(X|B)$ if $\omega \in B$.

DEFINITION 3. A bounded random variable $X \in L(\Omega)$ is called $\mathbf{B}$-*measurable* or measurable with respect to a partition $\mathbf{B}$ of $\Omega$ if it is constant on the atoms of the partition.

The following necessary condition for coherence holds [**23**, p. 292]:

PROPOSITION 1. *If $P(X|B)$ is a coherent linear previsions for every $B$ that belongs to a partition $\mathbf{B}$ of $\Omega$ then $P(X|\mathbf{B}) = X$ for all random variables $X \in L(\Omega)$ that are $\mathbf{B}$-measurable.*

This property may be not satisfied by the axiomatic definition of conditional expectation, where conditional expectation is defined with respect to a $\sigma$-field of conditioning events by the Radon-Nikodym derivative[**2**, Section 34] . In [**12**] it has been proven that every time the $\sigma$-field $\mathbf{G}$ is properly contained in $\mathbf{F}$ and it contains all singletons of $[0,1]$ then the conditional prevision defined by the Radon-Nikodym derivative is not coherent. It occurs because one of the defining properties of the Radon-Nikodym derivative is to be measurable with respect to the $\sigma$-field of the conditioning events and this requirement contradicts the necessary condition for the coherence of a linear conditional prevision recalled in Proposition 1.

**2.1. Coherent upper conditional previsions defined by Hausdorff outer measures.** Given a metric space $(\Omega, d)$ the diameter of a non empty set $U$ of $\Omega$ is defined as $|U| = \sup\left\{d(x,y) : x, y \in U\right\}$ and if a subset $A$ of $\Omega$ is such that $A \subset \bigcup_i U_i$ and $0 < |U_i| < \delta$ for each $i$, the class $\left\{U_i\right\}$ is called a $\delta$-cover of $A$.

Let $s$ be a non-negative real number. For $\delta > 0$ we define $h^s_\delta(A) = \inf \sum_{i=1}^{\infty} |U_i|^s$, where the infimum is overall (countable) $\delta$-covers $\left\{U_i\right\}$.

The *s-dimensional Hausdorff measure* of a set $A$ ([**17**],[**21**]), that denoted by $h^s(A)$, is defined as follows

$$h^s(A) = \lim_{\delta \to 0} h^s_\delta(A).$$

This limit exists but may be infinite, since $h^s_\delta(A)$ increases as $\delta$ decreases. The *Hausdorff dimension* of a set $A$, $\dim_H(A)$, is defined as the unique value, such that

$$h^s(A) = \infty \text{ if } 0 \leq s < \dim_H(A),$$
$$h^s(A) = 0 \text{ if } \dim_H(A) < s < \infty.$$

REMARK 2.1. If $\Omega$ is not separable, it implies that for sufficiently small $\delta > 0$, there does not exist a countable cover of $\Omega$ using sets with a diameter less than $\delta$. As a result, the infimum in the definition of Hausdorff outer measure is taken over the empty set, resulting in a value of $+\infty$. Consequently, as $\delta$ approaches zero, the limit of the outer measure also tends to $+\infty$. This means that for a non-separable set $\Omega$, the Hausdorff measure $h^s(\Omega)$ is equal to $+\infty$ for any $s \in [0, +\infty)$, indicating that the Hausdorff dimension of $\Omega$ is also $+\infty$.

THEOREM 1. *Let $\Omega = [0,1]^n$ and let $\mathbf{F}$ and $\mathbf{G}$ be two $\sigma$-field of subsets of $\Omega$ such that $\mathbf{G}$ is properly contained in $\mathbf{F}$ and it contains all singletons of $\Omega$. Let $\mathbf{B}$ be the partition of singletons and let $X$ be the indicator function of an event $A$ belonging to $\mathbf{F} - \mathbf{G}$. If we define the conditional prevision $P(X|\mathbf{B})$ equal to the Radon-Nikodym derivative with probability 1, that is*

$$P(X|\mathbf{B}) = E[X|\mathbf{G}]$$

*except on a subset $N$ of $[0,1]^n$ of $P$-measure zero, then the conditional prevision $P(X|\mathbf{B})$ is not coherent.*

A new model of coherent upper conditional previsions based on Hausdorff outer measures on a metric space has been introduced for bounded and unbounded random variables [**14**] where the following theorem has been proven. The Choquet integral of $X$ with respect to $\mu$ is defined if $\mu(\Omega) < +\infty$ through

$$\int X d\mu = \int_{-\infty}^{0} (\mu\left\{\omega \in \Omega : X(\omega) > x\right\} - \mu(\Omega))dx + \int_{0}^{+\infty} \mu\left\{\omega \in \Omega : X(\omega) > x\right\}dx$$

where the integrals in the second member of the formula are Riemann integrals.

DEFINITION 4. An random variable $X$ is *Choquet integrable* with respect to a monotone set function $\mu^*_B$ if the Choquet integral is finite, that is

$$-\infty < \int X d\mu^*_B < +\infty$$

Let $L^*(B)$ be the linear space of random variables which are Choquet integrable with respect to $\mu_B^*$ [**13**, Theorem 11].

THEOREM 2. *Let $(\Omega, d)$ be a metric space and let $\mathbf{B}$ be a partition of $\Omega$. For $B \in \mathbf{B}$ denote by $s$ the Hausdorff dimension of the conditioning event $B$ and by $h^s$ the Hausdorff $s$-dimensional outer measure. Let $m_B$ be a 0-1 valued finitely additive, but not countably additive, probability on $\wp(B)$. Then for each $B \in \mathbf{B}$ the functional $\overline{P}(X|B)$ defined on the linear space $L^*(B)$ by*

$$\overline{P}(X|B) = \begin{cases} \frac{1}{h^s(B)} \int_B X dh^s & \text{if } 0 < h^s(B) < +\infty \\ \int_B X dm_B & \text{if } h^s(B) \in \{0, +\infty\} \end{cases}$$

*is a coherent upper conditional prevision if $B$ has positive and finite Hausdorff measure in its Hausdorff dimension and it is a linear prevision whose restriction to events assumes only the values 0-1 if $B$ has Hausdorff outer measure in its Hausdorff dimension equal to zero or infinity.*

THEOREM 3. *Consider a metric space $(\Omega, d)$ and let $\mathbf{B}$ be a partition of $\Omega$. For each $B \in \mathbf{B}$, let $s$ denote the Hausdorff dimension of the conditioning event $B$, and let $h^s$ represent the $s$-dimensional Hausdorff outer measure. Suppose $m_B$ is a finitely additive probability defined on $\wp(B)$, which takes values of either 0 or 1, but is not countably additive. For every $B \in \mathbf{B}$, we define a function on $\wp(B)$ as follows*

$$\overline{P}(A|B) = \begin{cases} \frac{h^s(A \cap B)}{h^s(B)} & if \quad 0 < h^s(B) < +\infty, \\ \\ m_B & if \quad h^s(B) \in \{0, +\infty\} \end{cases}$$

*is a coherent upper conditional probability.*

Let us consider a set $B \in \mathbf{B}$ with a positive and finite Hausdorff outer measure corresponding to its Hausdorff dimension $s$. In this scenario, we define the monotone set function $\mu_B^*$ for every $A \in \wp(B)$ as follows: $\mu_B^*(A) = \frac{h^s(AB)}{h^s(B)}$. This function serves as a coherent upper conditional probability and exhibits properties such as submodularity and continuity from below. Furthermore, when we restrict it to the $\sigma$-field of all $\mu_B^*$-measurable sets, it becomes a Borel regular countably additive probability.

EXAMPLE 2.2. Let $\Omega = [0,1]$ and let be $\mathbf{B}$ be a partition of $\Omega$ containing the Cantor set $B$; the conditioning event $B$ has zero probability with respect to the coherent upper unconditional probability since $\overline{P}(B|\Omega)\frac{h^1(B)}{h^1(\Omega)} = 0$ since it is a fractal set with a non-integer Hausdorff dimension $s$ with positive and finite Hausdorff outer measure in its Hausdorff dimension. By Theorem 3 we have for every $A \in \wp(B)$:

$$\overline{P}(X|B) = \frac{h^s(A \cap B)}{h^s(B)} \text{ if } 0 < h^s(B) < +\infty.$$

When considering $B \in \mathbf{B}$ such that $h^s(B)$ falls within the set $\{0, +\infty\}$, the consistent upper conditional probability is established through a finitely additive probability $m_B$ on $\wp(B)$, which takes values of either 0 or 1 and is not countably additive. Notably, the realm of 0-1 valued finitely additive probabilities is in direct bijective correspondence with ultrafilters, as outlined in the context of $\mathcal{A}$.

DEFINITION 5. An ultrafilter $\mathcal{A}$ is a class of subsets of $\wp(B)$ such that

a) $\oslash \notin \mathcal{A}$
b) $A, B \in \mathcal{A} \Rightarrow A \cap B \in \mathcal{A}$
c) $A \in \mathcal{A}; A \subset B \subset \Omega \Rightarrow B \in \mathcal{A}$
d) $\forall A \in \wp(\Omega)$ either $A \in \mathcal{A}$ or $A^c \in \mathcal{A}$

Given an ultrafilter $\mathcal{A} \subset \wp(\Omega)$ a 0-1 valued finitely additive probability $m_B$ on $\wp(\Omega)$ can be defined by $m_B(A) = 1$ if $A \in \mathcal{A}$ and $m_B(A) = 0$ if $A^c \in \mathcal{A}$.

EXAMPLE 2.3. Consider the ultrafilter $\mathcal{A}$ over the set $\Omega$, comprising sets whose complements are finite. In this context, if $A$ represents any finite set, then the value of $m_\Omega(A)$ is 0.

EXAMPLE 2.4. Consider the scenario where $\Omega$ is synonymous with $\mathcal{N}$. Let $\mathcal{A}$ represent the ultrafilter of $\Omega$, which pertains to sets where the complement is finite. Take $A$ to be the set $\{2n : n \in \mathcal{N}\}$. According to property d) as defined in Definition 5, we can infer that if $A \in \mathcal{A}$, then it follows that $m_\Omega(A) = 1$, and in cases where $A^c \in \mathcal{A}$, it holds that $m_\Omega(A) = 0$, or alternatively, $m_\Omega(A^c) = 1$.

REMARK 2.5. Theorem 2 establishes that when considering each conditioning set $B$ with Hausdorff outer measure equal to zero or infinity, conditional probability is defined as a finitely additive probability that takes values 0 or 1. This definition ensures that, under these conditions, the conditional probability's restriction to the Borel $\sigma$-field complies with Dubins' concept of full conditional probability. Specifically, it adheres to the general compound rule for any Borelian sets $A$, $B$, and $C$.

$$P(A \cap B|C) = P(A|B \cap C)P(B|C).$$

Note that coherent conditional expected values can always be defined by axioms 1)-3) of Definition 1 regardless of whether the density function $f(x)$ of a random variable exists or whether the integral defining the moments of order $k$ is finite, i.e., whether the integral extended over the real axis of the function $|x^k|f(x)$ is finite.

EXAMPLE 2.6. The Cauchy distribution of parameters $(x_0, y_0)$ is the probability distribution of a random variable $T$ defined as the distance, in the Euclidean Cartesian plane, from the origin to which the x-axis is intersected by a line passing through the point $(x_0, y_0)$ and inclined at an angle $\Theta$ with uniform distribution on the interval $[0, \pi]$. The distribution function of a random variable with Cauchy distribution can be obtained as inverse of quantile function, which outputs the value of a random variable such that its probability is less than or equal to an input probability value. Since for the Cauchy distribution with parameters $(x_0, y_0)$ the lines that form angles with the x-axis less than $\Theta$ correspond to values less than $x = x_0 - y_0 \tan(\pi(\theta - 1/2))$ then the quantile function is $Q(p; x_0, y_0) = x_0 + y_0 \tan\left[\pi\left(\Theta - \frac{1}{2}\right)\right]$ and the distribution function is

$$F(x) = \frac{1}{\pi} arctg\Big(\frac{x - x_0}{y_0}\Big) + \frac{1}{2}$$

The density function is

$$f(x) = F^{'}(x) = \frac{1}{\pi}\Big(\frac{y_0}{y_0^2 + (x - x_0)^2}\Big)$$

Expected value and variance of the Cauchy distribution are undefined since The Cauchy distribution does not have finite moments of order greater than or equal to one because $\int_{-infty}^{+infty} x^k f(x)$ do not exist for $k \geq 1$; The Cauchy distribution has no moment generating function. For the Cauchy random variable $T$ coherent conditional prevision can be defined by

$$\overline{P}(X|B) = \int_B X dm_B$$

where $m_B$ is a 0-1 valued finitely additive, but not countably additive, probability on $\wp(B)$

## 3. Merging of coherent conditional probability

In [**15**] the distance between coherent conditional probabilities, defined by Hausdorff measures with respect to metrics that are bi-Lipschitz equivalent is proven to converge to zero as the amount of information increases. these coherent conditional probabilities, symbolize opinions that converge or merge with the accumulation of information [**3**]. The results hold since Hausdorff outer measures are metric outer measures and so all Borelian sets are measurable with respect to Hausdorff outer measures. So when the

Two distinct notions of equivalence for metrics can be applied : bi-Lipschitz equivalence and topological equivalence.

Definition 6. Given a metric space $(\Omega, d)$, a metric $d'$ on $\Omega$ is bi-Lipschitz equivalent to the metric $d$ if there exist two positive real constants $\alpha, \beta$ such that

$$\alpha d'(x,y) \leq d(x,y) \leq \beta d'(x,y)$$

Definition 7. Given a metric space $(\Omega, d)$ and a metric $d'$ on $\Omega$ then $d$ and $d'$ are topological equivalent if they induce the same topology.

Proposition 2. *Let $(\Omega, d)$ be a metric space and $d'$ be a metric on $\Omega$ bi-Lipschitz equivalent to $d$. Then $d$ and $d'$ are topological equivalent.*

Definition 8. Let $\mu$ and $\nu$ be two coherent upper probabilities measures on the same $\sigma$-field $\mathcal{F}$. Then $\nu$ is absolutely continuous with respect to $\mu$, ($\mu << \nu$, for short) if $\mu(A) = 0 \Rightarrow \nu(A) = 0$ for every $A \in \mathcal{F}$.

In Theorem 5 of [**15**] it has been proven that for every $B \in \mathbf{B}$ with positive and finite Hausdorff outer measures in its dimensions in both metric spaces, the coherent upper conditional probabilities $\mu_B$ and $\nu_B$, defined respectively in $(\Omega, d)$ and $(\Omega, d')$ as in Theorem 1, are mutually absolutely continuous.

The following theorem [**15**, Theorem 9] has been proved

Theorem 4. *Let $(\Omega, d)$ be a metric space where $\Omega$ is a set with positive and finite Hausdorff outer measure in its Hausdorff dimension $s$ and let $d_i$ be a metric bi-Lipschitz equivalent to $d$. Let $\mathcal{B}$ the Borel $\sigma$-field of the metric spaces $(\Omega, d)$ and $(\Omega, d_i)$. Let $H \in \mathbf{F}_\infty$ with Hausdorff dimension equal to $s$. Denote by $h^s$ and by $h_i^s$ respectively the $s$-dimensional Hausdorff measure with respect to the metric $d$ and $d_i$. Let $H$ be an atom of positive and finite $s$-dimensional Hausdorff outer measure in the two metric spaces and $\mu_H(A)$ and $\mu_H^i(A)$ the coherent upper conditional probabilities defined by*

$$\mu_H(A) = \frac{h^s(A)}{h^s(H)} \text{ and } \mu_H^i(A) = \frac{h_i^s(A)}{h_i^s(H)},$$

*then* $\mu_H^i$ *merges to* $\mu_H$ *for all* $A \in \wp(\Omega)$.

## 4. Coherent conditional previsions on standard Borel G-space

In this section, denoted by $G$, the group of all bi-Lipschizt self-maps on $\Omega$, coherent conditional previsions are defined on the standard Borel G-space when the conditioning event has positive $\nu$ measure, where $\nu$ is a quasi-invariant measure on the Borel $G$-space $(\Omega, \mathcal{F})$.

DEFINITION 9. A standard Borel space is a set $\Omega$ equipped with the Borel $\sigma$-field $\mathcal{F}$ generated by a topology that makes $\Omega$ homeomorphic to a complete separable metric space (a Polish space). So a standard Borel space is any set that can be given a structure such that it has the same measurable structure as a Polish space.

The definition of standard Borel space assures that $\Omega$ is homeomorphic to a complete separable metric space; if $\Omega$ is a compact metric separable metric space then all open sets of the topological space are Borelian sets.

DEFINITION 10. A group action of a group $G$ on a set $\Omega$ is a function : $G \times \Omega \Rightarrow \Omega$ that satisfies two properties:
Identity $e \cdot \omega = \omega$ for all $\omega \in \Omega$ and the identity element $e$ of $G$.
Compatibility $g(h \cdot \omega) = (gh)\omega$ for all $g, h \in G$ and $\omega \in \Omega$

DEFINITION 11. A standard Borel $G$-space is a standard Borel space $\Omega$ equipped with a Borel measurable action of a group $G$. This means: $\Omega$ is a standard Borel space, the function $G \times \Omega \Rightarrow \Omega$ is measurable with respect to the Borel $\sigma$-algebra on $G$, $\Omega$ and $G \times \Omega$. This measurability condition means that for every subset $B \subseteq \Omega$ and any element $g \in G$ the set $\{\omega : g \cdot \omega \in B\}$ must be also a Borel subset of $\Omega$.

DEFINITION 12. A Borel probability measure $\mu$ on a $G$-space $(\Omega, \mathcal{F})$ is called quasi-invariant [**24**] if, for each $g \in G$, $g(\mu)$ is equivalent to $\mu$, meaning that $\mu(B) = 0$ if and only if $\mu(g^{-1}(B)) = 0$ for any $B \in \mathcal{B}$. The class of all quasi-invariant measures on a $G$-space $(\Omega, \mathcal{F})$ is denoted as $\mathcal{M}_{q.i.}$

Let $(\Omega, d)$ be a metric space.
A function $f : \Omega \to \Omega$ is bi-Lipschitz if $\forall x, y \in \Omega$ there exist two positive constants $\alpha$ and $\beta$ such that

$$\alpha d(x,y) \leq d(f(x), f(y)) \leq \beta d(x,y)$$

If $\Omega \subseteq \Re$ and $d$ is the Euclidean distance the previous condition can be written as:

$$\alpha|x-y| \leq |f(x) - f(y)| \leq \beta|x-y|$$

PROPOSITION 3. *The class of all bi-Lipschitz self-maps on* $\Omega$ *forms a group under the operation of function composition. It is denoted by* $BiLip(\Omega, d)$.

PROOF. $BiLip(\Omega)$ has the following group properties:
i) Identity Element, in fact the identity map $id : \Omega \to \Omega$ defined by $id(\omega) = \omega$ is the identity element of the group. It is trivially bi-Lipschitz with constants $\alpha = \beta = 1$.
ii) Inverse: For any $f \in BiLip(\Omega)$ its inverse $f^{-1} \in BiLip(\Omega)$. If $f$ is bi-Lipschitz with constants $\alpha$ and $\beta$, then its inverse $f^{-1}$ is bi-Lipschitz with constants $\frac{1}{\beta}$ and $\frac{1}{\alpha}$

iii) Closure Under Composition: If $f, g \in BiLip(\Omega)$ then their composition $f \circ g \in BiLip(\Omega)$. If $f$ is bi-Lipschitz with constants $\alpha_f, \beta_f$ and $g$ is bi-Lipschitz with constants $\alpha_g, \beta_g$ then $f \circ g$ is bi-Lipschitz with constants $\alpha_f \cdot \alpha_g, \beta_f \cdot \beta_g$ □

THEOREM 5. *Let $\Omega$ be a standard Borel space, let $\mathbf{B}$ be a partition of $\Omega$, let $G = BiLip(\Omega)$ and let $\mathcal{M}_{q.i.}$ be the class of all quasi-invariant measures on the $G$-space $(\Omega, \mathcal{F})$. Suppose $m_B$ is a finitely additive probability defined on $\wp(B)$, which takes values of either 0 or 1, but is not countably additive. For every $B \in \mathbf{B}$, for any $\nu \in \mathcal{M}_{q.i.}$ and for each Borel measurable random variable $X \in L^*(\Omega)$ then*

$$\overline{P}(X|B) = \begin{cases} \frac{1}{\nu(B)} \int_B X d\nu & \text{if } 0 < \nu(B) < +\infty \\ \int_B X dm_B & \text{if } \nu(B) \in \{0, +\infty\} \end{cases}$$

*is a coherent conditional prevision.*

## 5. Conditional expectation in Deep Neural Networks

In this section we analyse how the model of coherent upper conditional prevision defined in Theorem 2 can be applied in a neural network. A deep neural network (DNN) is a type of artificial neural network (ANN) with multiple layers between the input and output layers. These layers are known as hidden layers, and they enable the network to learn and model complex patterns in data. Here, we provide a brief overview of how a DNN operates. The first layer of the network, named input layer, receives the input data. Each neuron in this layer represents a feature of the input. The intermediate layers between the input and output layers are called hidden layers. A DNN can have one or more hidden layers, each consisting of a number of neurons. These layers perform nonlinear transformations of the inputs and progressively extract higher-level features. The final layer of the network, is the output layer that produces the output predictions. The number of neurons in this layer corresponds to the number of desired output classes or values. The process of forward propagation involves passing the input data through the network to obtain the output. Each neuron in the network performs the following operations:

Each neuron computes a weighted sum of its inputs. For neuron $j$ in layer $l$, the weighted sum is given by:

$$z_j^l = \sum_i w_{ji}^l a_i^{l-1} + b_j^l$$

where $w_{ji}^l$ are the weights, $a_i^{l-1}$ are the activations from the previous layer, and $b_j^l$ is the bias term. The weighted sum is then passed through an activation function to introduce nonlinearity. Common activation functions include the sigmoid, hyperbolic tangent (tanh), and rectified linear unit (ReLU):

$$a_j^l = \sigma(z_j^l)$$

where $\sigma$ represents the activation function. Training a DNN involves adjusting the weights and biases to minimize the error in the network's predictions. A loss function $\mathcal{L}$ that measures the difference between the network's predictions and the true values, is defined. Common loss functions include mean squared error (MSE) for regression tasks and cross-entropy loss for classification tasks. Compute the gradients of the loss function with respect to each weight and bias in the network using the chain rule of calculus. This process is known as backpropagation. Update the weights and biases using an optimization algorithm such as stochastic gradient descent (SGD) or Adam. The update rule for a weight $w$ using SGD is:

$$w \leftarrow w - \eta \frac{\partial \mathcal{L}}{\partial w}$$

where $\eta$ is the learning rate.

Conditional expectation is a concept from probability theory that has useful applications in deep learning. It involves the expected value of a random variable given that certain conditions are met or certain information is known. In the context of deep learning ([**1**], [**18**]) conditional expectation can be used in various ways to enhance model performance and interpretability . The conditional expectation of a random variable $Y$ given another random variable $X$ is denoted as $\mathbb{E}[Y \mid X]$ and is defined in the axiomati approach as:

$$\mathbb{E}[Y \mid X] = \int y\, f_{Y|X}(y \mid x)\, dy$$

where $f_{Y|X}(y \mid x)$ is the conditional probability density function of $Y$ given $X$. This expectation gives the average value of $Y$ when $X$ is known.

Conditional expectation is particularly useful in deep learning for several reasons, including handling uncertainty, generating data, and enhancing model interpretability. Here are some specific applications:

Conditional expectation is integrated into deep learning models as occurs in probabilistic neural networks. Suppose we have input features $X$ and target variable $Y$. The network learns a mapping $f_\theta$ parameterized by $\theta$, which models the conditional distribution $P(Y \mid X)$. The goal is to find the parameters $\theta$ that maximize the likelihood of the observed data:

$$\theta^* = \arg\max_\theta \sum_{i=1}^{N} \log P(y_i \mid x_i; \theta)$$

By optimizing this likelihood, the network effectively learns the conditional expectation $\mathbb{E}[Y \mid X]$, providing predictions that account for the uncertainty and variability in the data. The problem is that the probability distribution or the density function of the population from which the data for the training of the neural network is not always known. In the next subsection we will propose to use coherent conditional previsions as defined in Theorem 2 the determine if the neural network have learnt from the data.

**5.1. Application of the coherent upper conditional previsions defined by Hausdorff outer measures in artificial intelligence.** Let $t \in T = [0, \tau]$, and let $(T, d)$ be a metric space, let $\mathcal{Z} = \mathcal{X} \times \mathcal{Y}$ be the space of data where $\mathcal{X}$ and $\mathcal{Y}$ are respectively an input and an output space that are continuous sets with positive and finite Hausdorf measure in their respective Hausdorff dimensions; a training set of $n$ elements is a subclass $\mathcal{S}$ of $n$ subsets of $\mathcal{Z}$, denoted by $\mathcal{S} = \left\{ z_1(t), \ldots, z_n(t) \right\}$ where $z_i(t) = (x_i(t), y_i(t))$. An iterative training algorithm $\mathcal{A}(\mathcal{S})$ returns the entire evolution of the parameters, i.e. for each $z \in \mathcal{S}$ we have $\mathcal{A}(z(t)) = \omega$ so that we can defined

$$\mathcal{W}_{\mathcal{S}} = \left\{ \omega \in \Re^d \exists t \in [0, T] : z(t) \in \mathcal{S} : \mathcal{A}(z(t)) = \omega . \right\} \tag{5.1}$$

The loss function $l(w,z) : \Re^d \times \mathcal{Z}$ is defined as the measure induced by a single data set $Z_i$ for a particular values $w$ of the random variables $W$, which represent the weights in the neural network.

The population risk $\mathcal{R}(w)$ is defined by the conditional prevision with respect to the $r$-dimensional Hausdorff measure where $r$ is the Hausdorff dimension of the set $\mathcal{Z}$:

$$\mathcal{R}(w) = P[l(w,z)] = \int_{\mathcal{Z}} l(w,z(t))dh^r \tag{5.2}$$

Denoted by $s$ the Hausdorff dimension of $\mathcal{W}_{\mathcal{S}}$ the empirical risk is defined as

$$\tilde{\mathcal{R}}(w,S) = P[l(w,z_i)] = \int_{\mathcal{W}_{\mathcal{S}}} l(w,z_i)dh^s \tag{5.3}$$

The generalization error is a random variable depending on the training set $S$ and it is defined as

$$\sup_{w\in\mathcal{W}_{\mathcal{S}}} \left|\tilde{\mathcal{R}}(w,S) - \mathcal{R}(w,z)\right| = \left|\int_{\mathcal{W}_{\mathcal{S}}} l(w,z_i)dh^s - \int_{\mathcal{Z}} l(w,z(t))dh^r\right|.$$

## References


[1] M. Abdar et al., A review of uncertainty quantification in deep learning: Techniques, applications and challenges, Information Fusion, 76, 243-297, (2021).

[2] P. Billingsley, *Probability and measure*. Wiley, (1986).

[3] D. Blackwell and L. Dubins, *Merging of opinions with increasing information*, Ann. Math. Statist. **33** (1962), 882–886, DOI 10.1214/aoms/1177704456. MR149577

[4] G. Choquet, *Theory of capacity*. Ann. Inst. Fourier, 5, 131-295, (1953).

[5] D. Denneberg, *Non-additive measure and integral*. Kluwer Academic Publishers, (1994).

[6] B. de Finetti, *Probability, induction and statistics. The art of guessing*, Wiley Series in Probability and Mathematical Statistics, John Wiley & Sons, London-New York-Sydney, 1972. MR440638

[7] B. de Finetti, *Theory of probability: a critical introductory treatment. Vol. 1*, Wiley Series in Probability and Mathematical Statistics, John Wiley & Sons, London-New York-Sydney, 1974. MR440640

[8] S. Doria, *Probabilistic independence with respect to upper and lower conditional probabilities assigned by Hausdorff outer and inner measures*, Internat. J. Approx. Reason. **46** (2007), no. 3, 617–635, DOI 10.1016/j.ijar.2007.02.006. MR2373748

[9] S. Doria, Characterization of a coherent upper conditional prevision as the Choquet integral with respect to its associated Hausdorff outer measure. Ann. Oper. Res., **195**, 33-48 (2012).

[10] S. Doria, Symmetric coherent upper conditional prevision by the Choquet integral with respect to Hausdorff outer measure. Ann. Oper. Res., **229**, 377-396 (2014).

[11] S. Doria, On the disintegration property of a coherent upper conditional prevision by the Choquet integral with respect to its associated Hausdorff outer measure. Ann. Oper. Res., **216**, 253-269 (2017).

[12] S. Doria, B. Dutta, and R. Mesiar, *Integral representation of coherent upper conditional prevision with respect to its associated Hausdorff outer measure: a comparison among the Choquet integral, the pan-integral and the concave integral*, Int. J. Gen. Syst. **47** (2018), no. 6, 569–592, DOI 10.1080/03081079.2018.1473392. MR3817444

[13] S. Doria, *Coherent upper conditional previsions defined by Hausdorff outer measures for unbounded random variables*, International Symposium on Imprecise Probabilities: Theories and Applications, Proc. Mach. Learn. Res. (PMLR), vol. 103, Proceedings of Machine Learning Research PMLR, [place of publication not identified], 2019, pp. 159–166. MR4020916

[14] S. Doria, *Disintegration property of coherent upper conditional previsions with respect to Hausdorff outer measures for unbounded random variables*, Int. J. Gen. Syst. **50** (2021), no. 3, 262–280, DOI 10.1080/03081079.2021.1872561. MR4254993

[15] S. Doria, Merging of coherent upper conditional probabilities defined by Hausdorff outer measures. Chaos, Solitons and Fractals, 181,147715, (2024).

[16] L. E. Dubins, *Finitely additive conditional probabilities, conglomerability and disintegrations*, Ann. Probability **3** (1975), 89–99, DOI 10.1214/aop/1176996451. MR358891
[17] K. J. Falconer, *Fractal geometry: mathematical foundations and applications*. Chichester, Wiley, (1990).
[18] J. Gawlikowski, C. R. N. Tassi, M. Ali et al., A survey of uncertainty in deep neural networks. Artif Intell Rev 56 (Suppl 1), 1513–1589 (2023).
[19] E. Regazzini, *Finitely additive conditional probabilities*. Rend. Sem. Mat. Fis., Milano, 5 5,69-89, (1985).
[20] E. Regazzini, *de Finetti's coherence and statistical inference*, Ann. Statist. **15** (1987), no. 2, 845–864, DOI 10.1214/aos/1176350379. MR888444
[21] C. A. Rogers, *Hausdorff measures*. Cambridge University Press, (1970).
[22] T. Seidenfeld, M. J. Schervish, and J. B. Kadane, *Preference for equivalent random variables: a price for unbounded utilities*, J. Math. Econom. **45** (2009), no. 5-6, 329–340, DOI 10.1016/j.jmateco.2008.12.002. MR2522457
[23] P. Walley, *Statistical Reasoning with Imprecise Probabilities*. Chapman and Hall, London, (1991).
[24] R. J. Zimmer, Ergodic Theory and Semisimple Groups, Birkauser Verlag, (1984).

Contemporary Mathematics
Volume **825**, 2025
https://doi.org/10.1090/conm/825/16508

# Fractal interpolation surfaces generated by Rakotch type contraction mappings

Vasileios Drakopoulos and Song-Il Ri

ABSTRACT. In the development of fractal interpolation theory, a primary goal is to discover new families of contractive mappings capable of generating fractal interpolation functions. In this article, we propose Rakotch-type contraction mappings to construct novel fractal interpolation surfaces with broad applicability and meaningful theoretical significance.

## 1. Introduction

Fractal interpolation theory describes the laws governing many disordered and irregular phenomena in nature. It is a powerful driving force for the development of multimedia technologies, including computer-aided cinema production, and has found wide application in solving scientific and technological problems such as data compression, speech and fingerprint analysis, medical diagnostic programs, image processing, signal processing and function approximation.

Fractal interpolation functions, or FIFs for short, are widely used in various applications because they can describe many natural geometries and analyze time series data with fractal characteristics relatively accurately. The concept of fractal interpolation functions, their construction method using the Banach fixed point theorem and the existence theorem of the attractor of an iterated function system, or IFS for short, was intoduced in [**1**], [**2**]. Fractal interpolation theory employs IFSs defined over a wide range of sets (e.g., interval, rectangular domain, Sierpiński gasket, Koch curve) to generate, model, and analyze highly elegant and complicated natural images.

The concept of IFS is introduced as a natural generalisation of the well-known Banach fixed point theorem, and a FIF is the attractor of an appropriately chosen IFS. Since fractal interpolation offers advantages over traditional polynomial interpolation in modelling the self-similarity of natural systems, fractal interpolation curves, or FICs for short, have become a powerful tool for modeling numerous natural curves, such as coastlines, mountain ranges, cloud profiles, and horizons. Likewise, fractal interpolation surfaces, or FISs for short, have been widely used to generate and describe natural surfaces such as rock formations, metal surfaces, topographies, and forests (see [**6**]). As a result, many scholars have studied fractal interpolation theory, which is not only fundamental and theoretically important, but also widely applied in various fields, including computer graphics (see [**7**]).

2020 *Mathematics Subject Classification.* Primary 28A80, 37L30; Secondary 54H25.

However, compared to the diverse and complicated curves, surfaces, and phenomena in nature, the coverage of FICs and FISs studied to date is limited. Generating new FICs and FISs and analysing their complexity is one of the most important challenges in fractal interpolation theory and a core research topic. The construction of new IFSs and the generation of new FIFs face several issues, including extending the domain of the functions that constitute IFSs and applying well-known fixed point theorems.

Few studies have addressed the problem of better modeling curves and surfaces of various natural systems, generalizing fixed point theorems (such as the Banach fixed point theorem) to provide a wider application of FIC and FIS generation methods, establishing the uniqueness of FIFs on typical fractal curves, and exploring the complexity of FICs and FISs. In [**12**] and [**11**], we first demonstrated that by using the Rakotch fixed point theorem, a generalization of the Banach fixed point theorem, it is possible to generate FICs and FISs with broader coverage. However, in [**12**] and [**11**], only a few non-Banach contractions restricted to generating FICs and FISs were used. In [**17**], the authors presented notable counterexamples showing that, in general, attractors of IFSs cannot be generated using Kannan (or Reich) contraction mappings.

The results of [**17**] and [**12**] indicate that fixed point theorems applicable to the generation of attractors of IFSs, and, more importantly, those applicable to the generation of FIFs, are extremely limited. In particular, these results highlight that finding IFSs capable of generating fractals, especially contractive mappings capable of generating FIFs, is one of the primary challenges in fractal theory. Although [**12**] shows that using Rakotch contraction mappings, which do not satisfy the Banach contraction condition, can generate more generalized FIFs than Barnsley's FIFs, finding such Rakotch contraction mappings is not straightforward. The existence of the bivariable fractal interpolation function using the Rakotch contraction theory and iterated function system for a countable data set is proved in [**18**]. Hence, the challenge of finding a new family of Rakotch contractions capable of generating FICs or FISs remains one of the most important problems in fractal interpolation theory. To address this fundamental and critical problem in fractal interpolation theory, we propose a new family of Rakotch contractions and apply it to the generation of FISs.

This article is organised as follows. In Section 2, we review previous studies on fractal interpolation using Rakotch contractions, and in Section 3, we present a new family of Rakotch contractions and the resulting FISs. The new family of Rakotch contractions proposed in this article will contribute to the further development of fractal interpolation theory.

## 2. Preliminaries

In this section, we recall fractal interpolation principles using Rakotch contraction mappings.

**2.1. On contraction mappings.** Firstly, we recall the definition of a Rakotch contraction mapping; see also [**9**], p.463, Corollary,[**10**], p.257, see [**16**], see [**14**], [**15**], p.100, [**12**].

DEFINITION 2.1. Let $(X, d)$ be a metric space and $f\colon X \to X$ a transformation. Let $\alpha\colon (0, +\infty) \to (0, +\infty)$ be a function satisfying for all $x, y \in X$,

$$d(f(x), f(y)) \leq \alpha(d(x, y)),$$

where $\alpha$ is an increasing function and for every $t > 0$,

$$\lim_{n\to\infty} \alpha^n(t) = 0.$$

Then $f$ is called *Matkowski contraction mapping* (cf. [**8**]), and in particular, if for every $t > 0$, $\alpha(t) < t$ and the function $\frac{\alpha(t)}{t}$ is a decreasing function, then $f$ is called *Rakotch contraction mapping* (cf. [**9**]).

Obviously, any mapping that satisfies Banach contraction condition also satisfies Rakotch (or Matkowski) contraction condition, but any mapping that satisfies Rakotch (or Matkowski) contraction condition does not always satisfy Banach contraction condition. Also, any mapping that satisfies Rakotch contraction condition also satisfies Matkowski contraction condition, but any mapping that satisfies Matkowski contraction condition does not always satisfy Rakotch contraction condition; for examples see [**11**].

**2.2. Assumptions.** Let $(X, d)$ be a complete metric space and $(\mathcal{H}(X), h_d)$ be a set generated by $(X, d)$, where

$$\mathcal{H}(X) - \{A \subset X \mid A \text{ is compact}$$

and $h_d$ expresses the Hausdorff metric on $\mathcal{H}(X)$.

Let $\{(x_i, y_j, z_{i,j}) \in \mathbb{R}^3 : i \in \{0, 1, 2, \dots, M\}, j \in \{0, 1, 2, \dots, N\}\}$ be a data set, where $M \geq 2$, $N \geq 2$, $-\infty < x_0 < x_1 < x_2 < \cdots < x_M < +\infty$, $-\infty < y_0 < y_1 < y_2 < \cdots < y_N < +\infty$ and $z_{i,j} \in \mathbb{R}$.

Let for $(x, y, z) \in [x_0, x_M] \times [y_0, y_N] \times \mathbb{R}$,

$$w_{m,n}\begin{pmatrix} x \\ y \\ z \end{pmatrix} := \begin{pmatrix} u_m x + v_m \\ p_n y + q_n \\ F_{m,n}(x, y, z) \end{pmatrix},$$

where

$$u_m := \frac{x_m - x_{m-1}}{x_N - x_0}, \qquad v_m := x_{m-1} - \frac{x_m - x_{m-1}}{x_N - x_0} x_0,$$
$$p_n := \frac{y_n - y_{n-1}}{y_M - y_0}, \qquad q_n := y_{n-1} - \frac{y_n - y_{n-1}}{y_M - y_0} y_0,$$

$F_{m,n}$ is a continuous function such that $F_{m,n}\colon [x_0, x_M] \times [y_0, y_N] \times \mathbb{R} \to \mathbb{R}$ and every $w_{m,n}$ satisfies the following conditions

$$w_{m,n}\begin{pmatrix} x_0 \\ y_0 \\ z_{0,0} \end{pmatrix} = \begin{pmatrix} x_{m-1} \\ y_{n-1} \\ z_{m-1,n-1} \end{pmatrix}, \quad w_{i,j}\begin{pmatrix} x_0 \\ y_N \\ z_{0,N} \end{pmatrix} = \begin{pmatrix} x_{m-1} \\ y_n \\ z_{m-1,n} \end{pmatrix},$$
$$w_{i,j}\begin{pmatrix} x_M \\ y_0 \\ z_{M,0} \end{pmatrix} = \begin{pmatrix} x_m \\ y_{n-1} \\ z_{m,n-1} \end{pmatrix}, \quad w_{i,j}\begin{pmatrix} x_M \\ y_N \\ z_{M,N} \end{pmatrix} = \begin{pmatrix} x_m \\ y_n \\ z_{m,n} \end{pmatrix}.$$

**2.3. Bounded bivariable FIFs generated by Rakotch contraction mappings.** The following theorem (see [**12**]) summarises the generative principle of bivariable FIFs (not necessarily continuous) by using Rakotch contraction mappings.

THEOREM 2.2.
1) *If for every* $m \in \{1, 2, \ldots, M\}$ *and* $n \in \{1, 2, \ldots, N\}$*, a mapping* $F_{m,n}(\cdot, \cdot, z)$ *satisfies Matkowski contraction condition, then there is a bounded function that passes through the given data.*
2) *If for every* $m \in \{1, 2, \ldots, M\}$ *and* $n \in \{1, 2, \ldots, N\}$*, a mapping* $F_{m,n}(\cdot, \cdot, z)$ *satisfies Rakotch contraction condition, and* $F_{m,n}(x, \cdot, \cdot)$ *and* $F_{m,n}(\cdot, y, \cdot)$ *satisfy Lipschitz condition, then the graph of a bounded function that passes through the given data is an attractor of the IFS defined above.*

REMARK 2.3. Unlike a univariable FIF that passes through the given data, a bivariable FIF that passes through the given data does not always satisfy continuous condition (see [**19**], p.630-631).

In order to obtain bivariable FIFs where the continuity are guaranteed, one can add more assumptions for the data set or the vertical scaling factors; see [**3**], [**5**] or [**12**].

THEOREM 2.4.
1) *If for every* $m \in \{1, 2, \ldots, M\}$ *and* $n \in \{1, 2, \ldots, N\}$*, a mapping* $F_{m,n}(\cdot, \cdot, z)$ *satisfies Banach contraction condition, and data points of every given boundary are collinear, then there exists a continuous function that passes through the given data.*
2) *If for every* $m \in \{1, 2, \ldots, M\}$ *and* $n \in \{1, 2, \ldots, N\}$*, a mapping* $F_{m,n}(\cdot, \cdot, z)$ *satisfies Rakotch contraction condition, and data points of all the boundaries are coplanar, then there exists a continuous function that passes through the given data.*
3) *If for every* $m \in \{1, 2, \ldots, M\}$ *and* $n \in \{1, 2, \ldots, N\}$*, a mapping* $F_{m,n}(\cdot, \cdot, z)$ *satisfies Banach (or Rakotch) contraction condition, and the values of vertical scaling functions that are not constants in all the boundaries are equal to zero, then there exists a continuous function that passes through the given data.*

REMARK 2.5.
1) In [**11**], the author used only Rakotch contraction mappings $f(x) = \frac{1}{1+x}$ and $g(x) = \frac{x}{1+x}$ $(x \geq 0)$ that do not satisfy Banach contraction condition. By using these Rakotch contraction mappings, the author of [**11**] gave a novel idea and an original way for new FIF generation.
2) In [**13**], the author introduced the Rakotch contraction mappings

$$\left\{ g_n(x) := \frac{x}{1+nx} | n \in \mathbb{N} \right\}$$

that do not satisfy Banach contraction condition. This Rakotch contraction family has an essential and important significance on the solution of new FIS generation problem. In fact, since

$$\frac{x}{1+x} \geq \frac{x}{1+2x} \geq \frac{x}{1+3x} \geq \frac{x}{1+4x} \geq \cdots \geq \frac{x}{1+nx} \geq \cdots$$

and for all $n \in \mathbb{N}$ and for $\delta(t) := \frac{t}{1+t}$ $(t \geq 0)$,

$$|g_n(x) - g_n(x)| \leq \delta(|x - y|),$$

we can obtain a new Rakotch contraction family that does not satisfy Banach contraction condition defined for $m, n \in \mathbb{N}$ as

$$g_{m,n}(x) := \begin{cases} \frac{x}{1+mx} & \text{if } x \in [0, \frac{n-1}{m}] \\ \frac{1}{n}x & \text{if } x > \frac{n-1}{m}. \end{cases}$$

3) In [**4**], in order to present new FIFs, the authors used a trigonometric function $\sin x$ $(x > 0)$ that is a Rakotch contraction mapping which does not satisfy Banach contraction condition.

Indeed, $\sin x$ is not a Banach contraction mapping on $(0, +\infty)$. In fact, if there is $0 < c < 1$ such that for all $x \neq y$ $(x, y \in (0, +\infty))$,

$$\left| \frac{\sin x - \sin y}{x - y} \right| \leq c,$$

then for all $x \in (0, +\infty)$,

$$\left| \lim_{y \to x} \frac{\sin x - \sin y}{x - y} \right| = |\cos x| \leq c$$

a contradiction, and so a function $\sin x$ is not a Banach contraction mapping on $(0, +\infty)$. On the other hand, for all $x, y \in (0, +\infty)$,

$$|\sin x - \sin y| = 2 \left| \cos \frac{x+y}{2} \sin \frac{x-y}{2} \right| \leq 2 \left| \sin \frac{x-y}{2} \right| \leq 2 \left| \sin \frac{|x-y|}{2} \right|.$$

For each $t > 0$, let

$$\alpha(t) := \begin{cases} 2\sin\frac{t}{2} & \text{if } t \in [0, \pi] \\ \frac{2}{\pi}t & \text{if } t > \pi. \end{cases}$$

Then for all $t \in (0, +\infty)$, we have $\alpha(t) < t$. Note that $\alpha\colon [0, +\infty) \to [0, +\infty)$ is an increasing function, and a mapping $t \to \frac{\alpha(t)}{t}$ is a decreasing function, and for all $x, y \in (0, +\infty)$,

$$|\sin x - \sin y| \leq \alpha\, |x - y|.$$

Therefore, a trigonometric function $\sin x$ is a Rakotch contraction mapping that does not satisfy Banach contraction condition.

## 3. FISs generated by new Rakotch contraction mappings

In this section, we present new Rakotch contraction mappings that do not satisfy Banach contraction condition by giving a bivariable FIF and a FIS generated by these Rakotch contraction mappings.

**3.1. New Rakotch contraction mappings.** Let $\alpha(t) := \frac{t}{\sqrt{1+t^2}}$ for all $t \geq 0$. Then for all $n \in \mathbb{N}$ and for all $t \geq 0$, $\alpha^n(t) = \frac{t}{\sqrt{1+nt^2}}$ . Therefore, for all $t \geq 0$, $\lim_{n\to\infty} \alpha^n(t) = 0$. Furthermore, if $t > 0$, then $\alpha(t) = \frac{t}{\sqrt{1+t^2}} < t$ because $\sqrt{1+t^2} > 1$. Additionally, $\alpha$ is an increasing function because $\frac{d\alpha(t)}{dt} = \frac{1}{(1+t^2)^{\frac{3}{2}}} > 0$ for all $t \geq 0$. Therefore if $t > 0$, $\alpha(t) < t$, and for all $t > 0$, $\lim_{n\to+\infty} \alpha^n(x) = 0$, and a function $\alpha(t)$ is an increasing function and the function $\frac{\alpha(t)}{t} = \frac{1}{\sqrt{1+t^2}}$ is decreasing function.

For each $x > 0$, let $f(x) := \frac{x}{\sqrt{1+x^2}}$. Then, for all $x, y > 0$,

$$\begin{aligned} |f(x) - f(y)| = |\frac{x}{\sqrt{1+x^2}} - \frac{y}{\sqrt{1+y^2}}| &= \frac{|\sqrt{x^2+x^2y^2} - \sqrt{y^2+x^2y^2}|}{\sqrt{1+x^2}\sqrt{1+y^2}} \\ &= \frac{|x^2 - y^2|}{\sqrt{1+x^2}\sqrt{1+y^2}(x\sqrt{1+y^2} + y\sqrt{1+x^2})} \\ &\leq \frac{|x^2-y^2|}{\sqrt{1+x^2}\sqrt{1+y^2}(x+y)} \leq \frac{|x-y|}{\sqrt{1+x^2}\sqrt{1+y^2}} \\ &\leq \frac{|x-y|}{\sqrt{1+|x-y|^2}} = \alpha(|x-y|). \end{aligned}$$

Therefore, for all $x > 0$, $f(x)$ is a new Rakotch contraction mapping. Then, obviously, $f(x)$ does not satisfy Banach contraction condition. In fact, the mapping $f(x) = \frac{x}{\sqrt{1+x^2}}$ $(x \geq 0)$ has a unique fixed point $x_0 = 0$. Fix a real number $c \in (0,1)$. Then for all $x \in (0, \sqrt{\frac{1}{c^2} - 1})$,

$$d(f(0), f(x)) = \frac{x}{\sqrt{1+x^2}} > \frac{x}{\sqrt{1 + (\sqrt{\frac{1}{c^2}-1})^2}} = cx = cd(0,x).$$

Therefore, the mapping $f(x) = \frac{x}{\sqrt{1+x^2}}$ $(x \geq 0)$ does not satisfy Banach contraction condition.

Now, we give another Rakotch contraction mapping. For each $t \geq 0$, let $\beta(t) := \frac{t}{\sqrt[3]{1+t^3}}$. Then for all $n \in \mathbb{N}$ and for all $t \geq 0$, $\beta^n(t) = \frac{t}{\sqrt[3]{1+nt^3}}$ . Therefore, for all $t \geq 0$, $\lim_{n\to\infty} \beta^n(t) = 0$. Also, if $t > 0$, then $\beta(t) = \frac{t}{\sqrt[3]{1+t^3}} < t$ because $\sqrt[3]{1+t^3} > 1$. Moreover, $\beta$ is an increasing function because for all $t \geq 0$, $\frac{d\beta(t)}{dt} = \frac{1}{(1+t^3)^{\frac{4}{3}}} > 0$. Hence for all $t > 0$, $\beta(t) < t$, and for all $t > 0$, $\lim_{n\to\infty} \beta^n(x) = 0$, and the function $\beta(t)$ is an increasing function and the function $\frac{\beta(t)}{t} = \frac{1}{\sqrt[3]{1+t^3}}$ is a decreasing function.

For each $x > 0$, let $g(x) := \frac{x}{\sqrt[3]{1+x^3}}$. Then, obviously, for all $x \in \mathbb{R}$, $f(x) = \frac{x}{\sqrt[2]{1+x^2}} \leq \frac{x}{\sqrt[3]{1+x^3}} = g(x)$. While, for all $x, y > 0$,

$$\begin{aligned} |g(x) - g(y)| = |\frac{x}{\sqrt[3]{1+x^3}} - \frac{y}{\sqrt[3]{1+y^3}}| &= \frac{|\sqrt[3]{x^3+x^3y^3} - \sqrt[3]{y^3+x^3y^3}|}{\sqrt[3]{1+x^3}\sqrt[3]{1+y^3}} \\ &= \frac{|x^3-y^3|}{\sqrt[3]{1+x^3}\sqrt[3]{1+y^3}\{(x^3+x^3y^3)^{\frac{2}{3}} + [(x^3+x^3y^3)(y^3+x^3y^3)]^{\frac{1}{3}} + (y^3+x^3y^3)^{\frac{2}{3}}\}} \\ &= \frac{|x-y|(x^2+xy+y^2)}{\sqrt[3]{1+x^3}\sqrt[3]{1+y^3}\{x^2(1+y^3)^{\frac{2}{3}} + xy[(1+y^3)(1+x^3)]^{\frac{1}{3}} + y^2(1+x^3)^{\frac{2}{3}}\}} \\ &\leq \frac{|x-y|}{\sqrt[3]{1+x^3}\sqrt[3]{1+y^3}} \leq \frac{|x-y|}{\sqrt[3]{1+|x-y|^3}} = \beta(|x-y|). \end{aligned}$$

Therefore, for all $x > 0$, $g(x)$ is a new Rakotch contraction mapping that does not satisfy Banach contraction condition.

By using induction, we can see that for every $n \in \mathbb{N}$, the mapping $\frac{x}{\sqrt[n]{1+x^n}}$ is a Rakotch contraction mapping that does not satisfy Banach contraction condition, and

$$\frac{x}{1+x} \leq \frac{x}{\sqrt[2]{1+x^2}} \leq \frac{x}{\sqrt[3]{(1+x^3)}} \leq \frac{x}{\sqrt[4]{(1+x^4)}} \leq \cdots \leq \frac{x}{\sqrt[n]{(1+x^n)}} \leq \cdots.$$

In case of $n = 1$, our function is equal to the function of [**9**] (or [**12**]).

We present new Rakotch contraction mappings in this manner that do not satisfy Banach contraction condition such that

$$\left\{ f_n(x) := \frac{x}{\sqrt[n]{1+x^n}} \mid n \in \mathbb{N} \right\}.$$

REMARK 3.1. Obviously, a Rakotch contraction is a generalized Banach contraction, but it is not an easy matter to find a new Rakotch contraction mappings that does not satisfy Banach contraction condition because it is difficult to make a mapping $f(x)$ and a function $\alpha(t)$ such that for all $x, y \in X$,

$$d(f(x), f(y)) \leq \alpha(d(x, y)),$$

where $\alpha\colon (0, +\infty) \to (0, +\infty)$ is a function satisfying for all $t > 0$, $\alpha(t) < t$, and for all $t > 0$, $\lim_{n\to\infty} \alpha^n(x) = 0$, and also where a function $\alpha(t)$ is an increasing function and the function $\frac{\alpha(t)}{t}$ is a decreasing function.

Even if it is difficult to define a mapping $f(x)$ and a function $\alpha(t)$, one can always generate a new FIF by finding a new Rakotch contraction mapping $f(x)$ that does not satisfy Banach contraction condition; see **Subsection 3.2** and **Subsection 3.3**.

**3.2. A bivariable FIF generated by new Rakotch contraction mappings.** By using a new Rakotch contraction mapping, we give a bounded bivariable FIF (not necessarily continuous).

EXAMPLE 3.2. Let $x_0 = 0$, $0 < x_1 < 1$, $x_2 = 1$, $y_0 = 0$, $0 < y_1 < 1$, $y_2 = 1$. Let $(x_0, y_0, z_{0,0})$, $(x_1, y_0, z_{1,0})$, $(x_2, y_0, z_{2,0})$, $(x_0, y_1, z_{0,1})$, $(x_1, y_1, z_{1,1})$, $(x_2, y_1, z_{2,1})$, $(x_0, y_2, z_{0,2})$, $(x_1, y_2, z_{1,2})$, $(x_2, y_2, z_{2,2})$ be a data set, where $z_{0,0}, z_{1,0}, z_{2,0}, z_{0,1}, z_{1,1}, z_{2,1}, z_{0,2}, z_{1,2}, z_{2,2} \in \mathbb{R}$.

Let $\alpha_{1,1}(z) := \alpha$, where $0 \leq \alpha < 1$. Then $\alpha_{1,1}(z)$ is a Banach contraction (see [**3**], p.54).

Let $\alpha_{1,2}(z) := d(x, y)z$, where $d(x, y) := \lambda xy(1 - x)(1 - y)$ and $0 < \lambda < 16$. Then $\alpha_{1,2}(z)$ is a Banach contraction (see [**5**], p.1897).

Let $\alpha_{2,1}(z) := \frac{z}{1+z}$. Then $\alpha_{2,1}(z)$ is a Rakotch contraction which does not satisfy Banach contraction condition (see [**12**]).

Let $\alpha_{2,2}(z) := \frac{z}{\sqrt{1+z^2}}$. Then $\alpha_{2,2}(z)$ is a new Rakotch contraction which does not satisfy Banach contraction condition (see **Subsection 3.1**).

For each $x \in [0, 1]$, $y \in [0, 1]$, $z \in \mathbb{R}$, $m \in \{1, 2\}$ and $n \in \{1, 2\}$, let

$$w_{m,n} \begin{pmatrix} x \\ y \\ z \end{pmatrix} := \begin{pmatrix} u_m x + v_m \\ p_n y + q_n \\ a_{m,n}x + b_{m,n}y + c_{m,n}xy + \alpha_{m,n}(z) + d_{m,n} \end{pmatrix},$$

where we assume that

$$w_{m,n} \begin{pmatrix} x_0 \\ y_0 \\ z_{0,0} \end{pmatrix} = \begin{pmatrix} x_{m-1} \\ y_{n-1} \\ z_{m-1,n-1} \end{pmatrix}, \quad w_{m,n} \begin{pmatrix} x_0 \\ y_2 \\ z_{0,2} \end{pmatrix} = \begin{pmatrix} x_{m-1} \\ y_n \\ z_{m-1,n} \end{pmatrix},$$

$$w_{m,n} \begin{pmatrix} x_2 \\ y_0 \\ z_{2,0} \end{pmatrix} = \begin{pmatrix} x_m \\ y_{n-1} \\ z_{m,n-1} \end{pmatrix}, \quad w_{m,n} \begin{pmatrix} x_2 \\ y_2 \\ z_{2,2} \end{pmatrix} = \begin{pmatrix} x_m \\ y_n \\ z_{m,n} \end{pmatrix}.$$

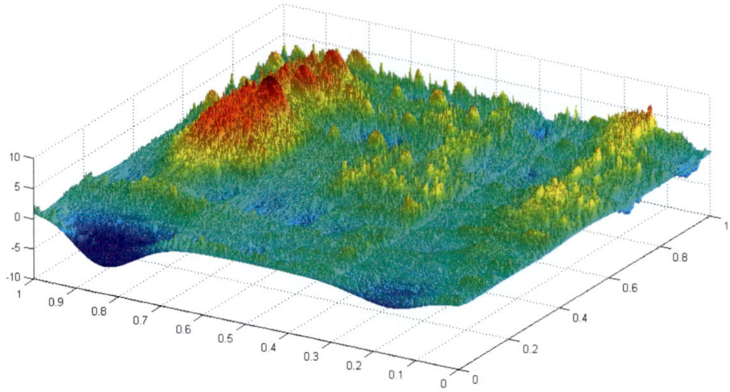


FIGURE 1. The graph of a FIF generated by Rakotch contraction mappings.

For each $t \geq 0$, let

$$\beta(t) := \max\left\{\alpha t, \max_{(x,y)\in[0,1]\times[0,1]d(x,y)} t, \frac{t}{\sqrt{1+t^2}}\right\}.$$

Since for all $t \geq 0$, $0 \leq \frac{t}{1+t} \leq \frac{t}{\sqrt{1+t^2}}$ by **Subsection 3.1**, we can see that $\alpha_{1,1}(z)$, $\alpha_{1,2}(z)$, $\alpha_{2,1}(z)$ and $\alpha_{2,2}(z)$ are Rakotch contraction mappings, that is, for all $x \in [0,1]$ and $y \in [0,1]$,

$$|\alpha_{i,j}(x) - \alpha_{i,j}(y)| \leq \beta(|x-y|).$$

Therefore, by Theorem 2.2, there is a bounded bivariable FIF (not necessarily continuous) which passes through the data points $(x_0, y_0, z_{0,0})$, $(x_1, y_0, z_{1,0})$, $(x_2, y_0, z_{2,0})$, $(x_0, y_1, z_{0,1})$, $(x_1, y_1, z_{1,1})$, $(x_2, y_1, z_{2,1})$, $(x_0, y_2, z_{0,2})$, $(x_1, y_2, z_{1,2})$, $(x_2, y_2, z_{2,2})$.

**3.3. A FIS generated by new Rakotch contraction mappings.** By using a new Rakotch contraction mapping and the coplanarity of data points of all the boundaries, we give a new FIS that the continuity of a bivariable FIF is guaranteed.

EXAMPLE 3.3. In Example 3.2, let

$$z_{0,0} = z_{0,1} = z_{0,2} = z_{1,0} = z_{1,2} = z_{2,0} = z_{2,1} = z_{2,2},$$

and $z_{1,1}$ be any fixed real value. Then, by Theorem 2.4, there is a FIS which passes through the data points $(x_0, y_0, z_{0,0})$, $(x_1, y_0, z_{1,0})$, $(x_2, y_0, z_{2,0})$, $(x_0, y_1, z_{0,1})$, $(x_1, y_1, z_{1,1})$, $(x_2, y_1, z_{2,1})$, $(x_0, y_2, z_{0,2})$, $(x_1, y_2, z_{1,2})$, $(x_2, y_2, z_{2,2})$.

A FIS generated by a new Rakotch contraction mapping is illustrated in Figure 1.

REMARK 3.4. When, for every $m \in \{1, 2, \ldots, M\}$ and $n \in \{1, 2, \ldots, N\}$, a mapping $F_{m,n}(\cdot, \cdot, z)$ satisfies Banach (or Rakotch) contraction condition, and the values of vertical scaling functions in all the boundaries are equal to zero, then the construction methods of all the FISs are equal to [**5**] and [**12**].

## Acknowledgments

The authors want to thank the reviewers for their helpful insights and assistance with literature review and manuscript preparation. Their expertise added credibility to the findings of our study.

## References

[1] M. F. Barnsley, *Fractal functions and interpolation*, Constr. Approx. **2** (1986), no. 4, 303–329, DOI 10.1007/BF01893434. MR892158

[2] ———, *Fractals everywhere*, 3rd ed., Dover Publications, Inc., New York, 2012.

[3] L. Dalla, *Bivariate fractal interpolation functions on grids*, Fractals **10** (2002), no. 1, 53–58, DOI 10.1142/S0218348X02000951. MR1894902

[4] V. Drakopoulos and S.-I. Ri, *Hölder continuous fractal interpolation functions*, New perspectives on nonlinear dynamics and complexity, Nonlinear Syst. Complex., vol. 35, Springer, Cham, [2023] ©2023, pp. 141–152, DOI 10.1007/978-3-030-97328-5_9. MR4487140

[5] Z. Feng, Y. Feng, and Z. Yuan, *Fractal interpolation surfaces with function vertical scaling factors*, Appl. Math. Lett. **25** (2012), no. 11, 1896–1900, DOI 10.1016/j.aml.2012.02.059. MR2957775

[6] P. R. Massopust, *Fractal surfaces*, J. Math. Anal. Appl. **151** (1990), no. 1, 275–290, DOI 10.1016/0022-247X(90)90257-G. MR1069462

[7] ———, *Fractal functions, fractal surfaces and wavelets*, Academic Press, San Diego, 1994.

[8] J. Matkowski, *Integrable solutions of functional equations*, Dissertationes Math. (Rozprawy Mat.) **127** (1975), 68. MR412650

[9] E. Rakotch, *A note on contractive mappings*, Proc. Amer. Math. Soc. **13** (1962), 459–465, DOI 10.2307/2034961. MR148046

[10] B. E. Rhoades, *A comparison of various definitions of contractive mappings*, Trans. Amer. Math. Soc. **226** (1977), 257–290, DOI 10.2307/1997954. MR433430

[11] S.-I. Ri, *A new nonlinear bivariate fractal interpolation function*, Fractals **26** (2018), no. 4, 1850054, 14, DOI 10.1142/S0218348X18500548. MR3858494

[12] S. Ri, *A new nonlinear fractal interpolation function*, Fractals **25** (2017), no. 6, 1750063, 12, DOI 10.1142/S0218348X17500633. MR3725302

[13] S. Ri, *New types of fractal interpolation surfaces*, Chaos Solitons Fractals **119** (2019), 291–297, DOI 10.1016/j.chaos.2019.01.010. MR3900757

[14] N.-A. Secelean, *Generalized iterated function systems on the space $l^\infty(X)$*, J. Math. Anal. Appl. **410** (2014), no. 2, 847–858, DOI 10.1016/j.jmaa.2013.09.007. MR3111871

[15] F. Strobin, *Attractors of generalized IFSs that are not attractors of IFSs*, J. Math. Anal. Appl. **422** (2015), no. 1, 99–108, DOI 10.1016/j.jmaa.2014.08.029. MR3263451

[16] F. Strobin and J. Swaczyna, *On a certain generalisation of the iterated function system*, Bull. Aust. Math. Soc. **87** (2013), no. 1, 37–54, DOI 10.1017/S0004972712000500. MR3011940

[17] N. Van Dung and A. Petruşel, *On iterated function systems consisting of Kannan maps, Reich maps, Chatterjea type maps, and related results*, J. Fixed Point Theory Appl. **19** (2017), no. 4, 2271–2285, DOI 10.1007/s11784-017-0419-z. MR3720450

[18] Manuj Verma and Amit Priyadarshi, *Fractal surfaces involving rakotch contraction for countable data sets*, Fractals **32** (2024), no. 02, 2440002.

[19] H. Xie and H. Sun, *The study on bivariate fractal interpolation functions and creation of fractal interpolated surfaces*, Fractals **5** (1997), no. 4, 625–634, DOI 10.1142/S0218348X97000504. MR1611075

Department of Computer Science and Biomedical Informatics, University of Thessaly, Lamia, Greece
*Email address*: vdrakop@uth.gr

Department of Mathematics, University of Science, Pyongyang, DPR Korea
*Email address*: si1974730@star-co.net.kp

Contemporary Mathematics
Volume **825**, 2025
https://doi.org/10.1090/conm/825/16509

# Inhomogeneous attractors and box dimension

Jonathan M. Fraser

ABSTRACT. Iterated function systems (IFSs) are one of the most important tools for building examples of fractal sets exhibiting some kind of 'approximate self-similarity'. Examples include self-similar sets, self-affine sets etc. A beautiful variant on the standard IFS model was introduced by Barnsley and Demko in 1985 where one builds an *inhomogeneous* attractor by taking the closure of the orbit of a fixed compact condensation set under a given standard IFS. In this expository article I will discuss the dimension theory of inhomogeneous attractors, giving several examples and some open questions. I will focus on the upper box dimension with emphasis on how to derive good estimates, and when these estimates fail to be sharp.

## 1. Introduction

**1.1. Inhomogeneous attractors.** Let $X$ denote a compact subset of Euclidean space (often the closed unit cube or unit ball). An iterated function system (IFS) is a finite collection $\mathbb{I} = \{S_i\}_{i=1}^N$ of contraction mappings which map $X$ into itself. It is a fundamental result in fractal geometry due to Hutchinson (see [**F03**, **H81**]) that for every IFS there exists a unique non-empty compact set, $F$, called the attractor, which satisfies

$$F = \bigcup_{i=1}^{N} S_i(F). \tag{1.1}$$

We call such attractors *homogeneous* attractors.

2020 *Mathematics Subject Classification.* Primary 28A80, 26A18, 37F32.
*Key words and phrases.* Inhomogeneous attractor, iterated function system, orbital set, box dimension, self-similarity.
The author was financially supported by a *Leverhulme Trust Research Project Grant* (RPG-2019-034) and an *EPSRC Standard Grant* (EP/Y029550/1).

FIGURE 1. Two homogeneous attractors of classical IFSs. On the left is the Sierpiński triangle, which is a self-similar set generated by three contractions all contracting uniformly by 1/2. On the right is a self-affine set generated by two affine contractions.

Inhomogeneous attractors are built from standard IFSs in the following way. The ingredients are a standard IFS and a fixed compact set $C \subseteq X$, which we call the *condensation set*. Analogous to the classical setting above, there is a unique non-empty compact set, $F_C$, satisfying

$$F_C = \bigcup_{i=1}^{N} S_i(F_C) \cup C, \tag{1.2}$$

which we refer to as the *inhomogeneous* attractor (with condensation $C$). Setting $C = \varnothing$, we recover the associated homogeneous attractor as $F_\varnothing$. When the maps in the IFS are similarities, $F_C$ is called an *inhomogeneous self-similar set* and $F_\varnothing$ is called a *self-similar set* with similar naming conventions for conformal maps, affine maps, etc. Inhomogeneous attractors were introduced and studied in [**BD85**] (see also [**H85**]) and are also discussed in detail in [**B06**] where, among other things, Barnsley gives applications of these schemes to image compression. Roughly speaking, the idea here is that one only needs to store the data $\{\mathbb{I}, C\}$ to recover the set $F_C$ which is much more complicated, e.g. containing infinitely many distorted and scaled copies of $C$.

Define the *orbital set*, $\mathcal{O}$, by

$$\mathcal{O} = C \cup \bigcup_{k \in \mathbb{N}} \bigcup_{i_1, \dots, i_k \in \{1, \dots, N\}} S_{i_1} \circ \cdots \circ S_{i_k}(C),$$

that is, $\mathcal{O}$ is the union of the condensation set, $C$, together with all images of $C$ under compositions of maps from the IFS. The term orbital set was introduced in [**B06**] and this set plays an important role in the structure of inhomogeneous attractors. Indeed,

$$F_C = F_\varnothing \cup \mathcal{O} = \overline{\mathcal{O}}, \tag{1.3}$$

where $F_\varnothing$ is the *homogeneous* attractor of the IFS, $\mathbb{I}$. The formulations (1.3) are straightforward to prove, see for example [**S08**, Lemma 3.9] in the case where the defining maps are similarities, but we encourage the reader to try to prove them for themselves.

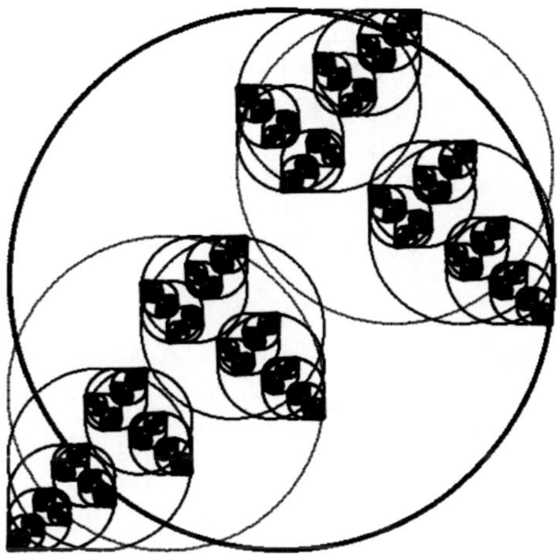

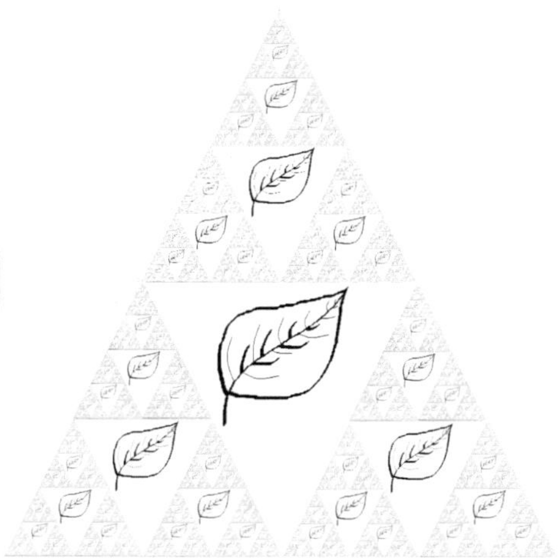

FIGURE 2. Two inhomogeneous attractors. On the left, the IFS consists of two similarity contractions (one of which has a non-trivial rotational component). The condensation set $C$ is the largest circle in the picture. You can see the homogeneous attractor emerging as the circles cluster towards it. On the right is an inhomogeneous attractor based on the Sierpiński triangle. The condensation set is the large central leaf and one can imagine this attractor depicts leaves falling to the forest floor, seen from above.

**1.2. Dimension theory of inhomogeneous attractors.** Inhomogeneous attractors are often fractal and (unlike standard IFS attractors) both the fractal behaviour of the condensation set $C$ and the underlying IFS come in to play. One of the central concepts in fractal geometry is that of dimension. There are many different ways to define the dimension of a fractal and part of the beauty of the subject is to understand how these different notions relate to each other. For now, we work with a general dimension dim. If dim is monotone, countably stable, and stable under Lipschitz maps, then it follows immediately from (1.3) that

$$\dim F_C = \max\{\dim F_\varnothing, \dim C\}. \tag{1.4}$$

See [**F03**] for more information about general dimensions and the properties mentioned above. In particular, (1.4) holds in complete generality for the Hausdorff and packing dimensions. Therefore, it is perhaps more interesting to consider notions of dimension which are not countably stable, such as the upper and lower box dimension or the Assouad dimension. Even though the formula (1.4) is not immediate for these dimensions, one might still expect it to hold; at least in certain situations. However, we will see that (1.4) can fail in several different ways, and exploring such examples—and identifying situations where (1.4) does hold—is our main goal.

We first recall the definition of the upper and lower box dimensions, discuss potential differences between them in our context, and then briefly describe some situations where other dimensions have been considered.

Given a bounded set $E \subseteq \mathbb{R}^d$ and a scale $\delta > 0$, write $N_\delta(E)$ to denote the smallest number of sets of diameter $\delta$ which are needed to cover $E$. Thus, by fixing a scale, we effectively discretize the problem of quantifying the 'size' of $E$ and we can then extract a meaningful notion of dimension as the scale tends to 0. More precisely, the *upper* and *lower box dimensions* of $E$ are defined by

$$\overline{\dim}_{\mathrm{B}} E = \limsup_{\delta \to 0} \frac{\log N_\delta(E)}{-\log \delta}$$

and

$$\underline{\dim}_{\mathrm{B}} E = \liminf_{\delta \to 0} \frac{\log N_\delta(E)}{-\log \delta},$$

respectively. If the upper and lower box dimensions agree, then we refer to the common value as the *box dimension*, denoted by $\dim_{\mathrm{B}} E$. We refer the reader to [**F03**] for more details and background on the upper and lower box dimensions.

The lower box dimension of inhomogeneous attractors is rather more challenging to study than the upper box dimension but was considered in some detail in [**F12**]. In particular, it was shown that (1.4) can fail in even the simplest settings for lower box dimension. To see why it is difficult to study the lower box dimension, consider a simple situation where the IFS consists of just two similarities but that the condensation set $C$ has a complicated scaling behaviour leading to $\underline{\dim}_{\mathrm{B}} C < \overline{\dim}_{\mathrm{B}} C$. Then, immediately we get

$$\max\{\underline{\dim}_{\mathrm{B}} F_\varnothing,\ \underline{\dim}_{\mathrm{B}} C\} \leqslant \underline{\dim}_{\mathrm{B}} F_C \leqslant \overline{\dim}_{\mathrm{B}} F_C$$

and one might try to prove that the first inequality is in fact an equality (that is, try to verify (1.4)). But this is not true in general because the orbital set consists of many different copies of $C$ appearing at different scales which means that the 'liminf behaviour' leading to $\underline{\dim}_{\mathrm{B}} C$ may not be witnessed by $F_C$ and is instead 'smoothened out'. Frustrated, one may then try to prove that the second inequality is an equality but, again, this is certainly not true in general. It is true that many copies of $C$ appear at different scales and that *some* of these scales may witness the 'limsup behaviour' leading to $\overline{\dim}_{\mathrm{B}} C$, but there is no reason to expect the 'smoothening' to take the lower box dimension of $F_C$ all the way up to the right hand side.

An interesting connection between the study of the lower box dimension of inhomogeneous self-similar sets and the lower box dimension of infinitely generated (homogeneous) self-similar or self-conformal sets has been identified in [**BR24+**]. For infinitely generated attractors, the upper and lower box dimensions of the countable set of fixed points of the defining maps play a role in the dimension theory and this set of fixed points is reproduced at many scales. This creates a similar phenomenon as we described above for inhomogeneous attractors and leads to similar subtle features of the lower box dimension in comparison with the upper box dimension.

We can be more optimistic concerning the *upper* box dimension of inhomogeneous attractors, as we shall see later. For definiteness, our main focus is now on verifying (or refuting)

$$\overline{\dim}_{\mathrm{B}} F_C = \max\{\overline{\dim}_{\mathrm{B}} F_\varnothing, \overline{\dim}_{\mathrm{B}} C\} \tag{1.5}$$

although we will also make some reference to (1.4) for the lower box dimension.

As mentioned above, there are many other interesting dimension theoretic questions one can ask about inhomogeneous attractors. For brevity, we only briefly summarise some of them here, encouraging the reader to refer to the growing body of literature. The Assouad dimension of inhomogeneous attractors was studied by Käenmäki and Lehrbäck [**KL17**], where (1.4) was verified for inhomogeneous self-similar sets satisfying some natural conditions. Although (1.4) is immediate for Hausdorff dimension, one might still consider the behaviour of the Hausdorff measure of $F_C$ in the Hausdorff dimension. This problem was comprehensively studied by Burrell [**B19**] for inhomogeneous self-similar sets. A graph-directed model for inhomogeneous attractors was considered by Dubey and Verma [**DV23+**] which built on work in the inhomogeneous self-similar case from [**F12**]. This was further studied by Bárány–Käenmäki–Nissinen [**BKN23+**] where finer properties of the covering function were described. There is also a connection with the complex dimensions studied by Lapidus–Radunović–Žubrinić [**LRZ17, LRZ18**], where inhomogeneous attractors emerge as notable examples in the development of the general theory.

There is also a wealth of literature studying inhomogeneous self-similar *measures*. These are defined similarly, but with $C$ replaced by a condensation measure $\nu$ and the IFS further equipped with a (sub-)probability vector. One may then ask dimension theoretic questions about the associated inhomogeneous measures $\mu_\nu$. Such questions include the study of the $L^q$ dimensions, considered by [**L13, OS07, ZGX22**], and the multifractal formalism, studied by Olsen and Snigireva [**OS08b**]. In another direction, one may study the Fourier analytic behaviour of the inhomogeneous self-similar measures $\mu_\nu$. This was considered in [**B11, OS08a, ZGX23**].

We ask for forgiveness from the reader for any glaring omissions in the brief history given above, but also for omitting further discussion of the many interesting results and questions which surround inhomogeneous constructions. From now on we focus only on the upper box dimensions of $F_C$ and consider several different settings.

**1.3. Notation.** For real-valued functions $A$ and $B$, we will write $A(x) \lesssim B(x)$ if there exists a constant $c > 0$ independent of the variable $x$ such that $A(x) \leqslant cB(x)$, $A(x) \gtrsim B(x)$ if $B(x) \lesssim A(x)$ and $A(x) \approx B(x)$ if $A(x) \lesssim B(x)$ and $A(x) \gtrsim B(x)$. In our setting, $x$ is normally $\delta > 0$ from the definition of box dimension or a parameter $k \in \mathbb{N}$ related to $\delta$ by, for example, $\delta = 2^{-k}$. The implicit constant $c$ above can depend on fixed quantities only, such as the condensation set $C$ or the IFS.

## 2. Inhomogeneous self-similar sets

In the case where the defining contractions in the IFS $\mathbb{I}$ are non-degenerate similarities, recall that the homogeneous attractor $F_\varnothing$ is called a *self-similar set* and the inhomogeneous attractor $F_C$ is called an *inhomogeneous self-similar set.* In this case, for each $i = 1, \ldots, N$, there exists a well-defined contraction ratio $\mathrm{Lip}(S_i) \in (0,1)$ such that for all $x, y \in X$

$$|S_i(x) - S_i(y)| = \mathrm{Lip}(S_i)|x - y|,$$

that is, $S_i$ contracts all distances uniformly by $\text{Lip}(S_i)$. The *similarity dimension* of the self-similar set (or perhaps, more accurately, of the IFS) is defined to be the unique real solution $s$ to the Hutchinson–Moran formula

$$\sum_{i=1}^{N} \text{Lip}(S_i)^s = 1. \tag{2.1}$$

The similarity dimension is in some sense the 'best guess' for the dimension of the self-similar set $F_\varnothing$. In fact if the 'pieces' $S_i(F_\varnothing)$ do not overlap too much (for example, they are pairwise disjoint), then

$$\dim_{\text{H}} F_\varnothing = \underline{\dim}_{\text{B}} F = \overline{\dim}_{\text{B}} F_\varnothing = s.$$

It is important to note that this holds much more generally, even when the pieces overlap in non-trivial ways. On the other hand, it does not always hold, for example if there are repeated maps in the defining IFS or 'exact overlaps' which emerge later in the construction.

The following theorem was obtained in [**F12**]. We give the full proof, which will hopefully serve as an instructive introduction to these type of arguments.

THEOREM 2.1. *Let $F_C$ be an inhomogeneous self-similar set with compact condensation set $C$ and similarity dimension $s$. Then*

$$\max\{\overline{\dim}_{\text{B}} F_\varnothing, \overline{\dim}_{\text{B}} C\} \leqslant \overline{\dim}_{\text{B}} F_C \leqslant \max\{s, \overline{\dim}_{\text{B}} C\}.$$

PROOF. The idea is to decompose the orbital set into 'small pieces' and 'big pieces' (relative to a fixed covering scale $\delta > 0$. The big pieces must be covered individually and the key observation is that covering a scaled down copy of $C$ by a factor $r$ at scale $\delta$ is the same as covering $C$ at the inflated scale $\delta/r$. On the other hand, the (infinitely many) small pieces can be grouped together with one group for every piece which is roughly of size $\delta$. Both of these covering strategies connect the covering number of the orbital set to sums of contraction ratios and the covering number of $C$ itself.

Fix an IFS $\mathbb{I} = \{S_1, \dots, S_N\}$, where each $S_i$ is a similarity map, and a compact condensation set $C$. Write $\mathcal{I} = \{1, \dots, N\}$ and $L_{\min} = \min_{i \in \mathcal{I}} \text{Lip}(S_i)$. Let

$$\mathcal{I}^* = \bigcup_{k \in \mathbb{N}} \mathcal{I}^k$$

denote the set of all finite words over $\mathcal{I}$. For $\mathbf{i} = (i_1, \dots, i_k) \in \mathcal{I}^*$, write $S_{\mathbf{i}} = S_{i_1} \circ \cdots \circ S_{i_k}$ for the composition of maps associated to the finite word $\mathbf{i}$. Thus the 'pieces' of the orbital set are the sets $S_{\mathbf{i}}(C)$. Further, write $\mathbf{i}_- = (i_1, \dots, i_{k-1})$ and $|\mathbf{i}| = k$ to denote the length of the word $\mathbf{i}$. For $\delta \in (0, 1]$, define a $\delta$-stopping, $\mathcal{I}(\delta)$, by

$$\mathcal{I}(\delta) = \big\{\mathbf{i} \in \mathcal{I}^* : \text{Lip}(S_{\mathbf{i}}) < \delta \leqslant \text{Lip}(S_{\mathbf{i}_-})\big\},$$

where we assume for convenience that $\text{Lip}(S_\omega) = 1$, where $\omega$ is the empty word. It is easy to see that, for all $\delta \in (0, 1]$,

$$\delta^{-s} \leqslant |\mathcal{I}(\delta)| \leqslant L_{\min}^{-s}\, \delta^{-s}. \tag{2.2}$$

Indeed, repeated application of Hutchinson's formula (2.1) gives

$$\sum_{\mathbf{i} \in \mathcal{I}(\delta)} \text{Lip}(S_{\mathbf{i}})^s = 1$$

from which we deduce

$$1 = \sum_{\mathbf{i} \in \mathcal{I}(\delta)} \mathrm{Lip}(S_{\mathbf{i}})^s \geqslant \sum_{\mathbf{i} \in \mathcal{I}(\delta)} (\delta\, L_{\min})^s = |\mathcal{I}(\delta)|\, (\delta\, L_{\min})^s$$

and

$$1 = \sum_{\mathbf{i} \in \mathcal{I}(\delta)} \mathrm{Lip}(S_{\mathbf{i}})^s \leqslant \sum_{\mathbf{i} \in \mathcal{I}(\delta)} \delta^s = |\mathcal{I}(\delta)|\, \delta^s.$$

Moreover, for all $t > s$,

$$\sum_{\mathbf{i} \in \mathcal{I}^*} \mathrm{Lip}(S_{\mathbf{i}})^t = \sum_{k=1}^{\infty} \sum_{\mathbf{i} \in \mathcal{I}^k} \mathrm{Lip}(S_{\mathbf{i}})^t = \sum_{k=1}^{\infty} \left( \sum_{i \in \mathcal{I}} \mathrm{Lip}(S_i)^t \right)^k < \infty, \tag{2.3}$$

by (2.1).

By monotonicity of upper box dimension, $\max\{\overline{\dim}_{\mathrm{B}} F_{\varnothing}, \overline{\dim}_{\mathrm{B}} C\} \leqslant \overline{\dim}_{\mathrm{B}} F_C$ and so we prove the other inequality. Since upper box dimension is finitely stable, it suffices to show that

$$\overline{\dim}_{\mathrm{B}} \mathcal{O} \leqslant \max\{s, \overline{\dim}_{\mathrm{B}} C\}.$$

Let $t > \max\{s, \overline{\dim}_{\mathrm{B}} C\}$. It follows from the definition of upper box dimension that

$$N_\delta(C) \lesssim \delta^{-t} \tag{2.4}$$

for all $\delta \in (0, 1]$, (with the implicit constant depending on $t$ but not $\delta$). Also note that since $X$ is compact, the smallest number of balls of radius 1 required to cover $X$ is a finite constant $N_1(X)$.

We are now ready to complete the proof. For $\delta \in (0, 1]$,

$$\begin{aligned}
N_\delta(\mathcal{O}) &= N_\delta\left( C \cup \bigcup_{\mathbf{i} \in \mathcal{I}^*} S_{\mathbf{i}}(C) \right) \\
&\leqslant \sum_{\substack{\mathbf{i} \in \mathcal{I}^*: \\ \delta \leqslant \mathrm{Lip}(S_{\mathbf{i}})}} N_\delta(S_{\mathbf{i}}(C)) \;+\; N_\delta\left( \bigcup_{\substack{\mathbf{i} \in \mathcal{I}^*: \\ \delta > \mathrm{Lip}(S_{\mathbf{i}})}} S_{\mathbf{i}}(C) \right) \;+\; N_\delta(C) \\
&\leqslant \sum_{\substack{\mathbf{i} \in \mathcal{I}^*: \\ \delta \leqslant \mathrm{Lip}(S_{\mathbf{i}})}} N_{\delta/\mathrm{Lip}(S_{\mathbf{i}})}(C) \;+\; N_\delta\left( \bigcup_{\mathbf{i} \in \mathcal{I}(\delta)} S_{\mathbf{i}}(X) \right) \;+\; N_\delta(C) \\
&\lesssim \sum_{\substack{\mathbf{i} \in \mathcal{I}^*: \\ \delta \leqslant \mathrm{Lip}(S_{\mathbf{i}})}} \big(\delta/\mathrm{Lip}(S_{\mathbf{i}})\big)^{-t} \;+\; \sum_{\mathbf{i} \in \mathcal{I}(\delta)} N_{\delta/\mathrm{Lip}(S_{\mathbf{i}})}(X) \;+\; \delta^{-t} \qquad \text{by (2.4)} \\
&\leqslant \delta^{-t} \sum_{\substack{\mathbf{i} \in \mathcal{I}^*: \\ \delta \leqslant \mathrm{Lip}(S_{\mathbf{i}})}} \mathrm{Lip}(S_{\mathbf{i}})^t \;+\; N_1(X)\, |\mathcal{I}(\delta)| \;+\; \delta^{-t} \\
&\lesssim \delta^{-t} \sum_{\mathbf{i} \in \mathcal{I}^*} \mathrm{Lip}(S_{\mathbf{i}})^t \;+\; \delta^{-s} \;+\; \delta^{-t} \qquad \text{by (2.2)} \\
&\lesssim \delta^{-t}
\end{aligned}$$

by (2.3). This proves the upper bound and the theorem. □

In [**OS07**, Corollary 2.6] and [**S08**, Theorem 3.10 (2)] it was proved that if each of the $S_i$ are similarities, and the sets $S_1(F_C), \ldots, S_N(F_C)$, $C$ are pairwise disjoint, then

$$\overline{\dim}_B F_C = \max\{\overline{\dim}_B F_\varnothing, \overline{\dim}_B C\}.$$

This follows from Theorem 2.1 above and the fact that, under this strong separation condition,

$$\overline{\dim}_B F_\varnothing = s. \tag{2.5}$$

However, (2.5) holds far more generally than this. For example, the rather weaker *open set condition* will suffice (see [**F03**]) but, moreover, the celebrated 'exact overlaps conjecture' asserts that (2.5) holds for self-similar sets in the line provided $s \leqslant 1$ and there are no exact overlaps in the system; that is, the semigroup generated by the defining IFS is free, see [**PS00**]. Whilst still open, this conjecture is known to hold provided the contractions are algebraic and it also holds outside of a very small exceptional set for suitably parametrised IFSs. We omit further discussion of this, but refer the reader to the extensive and impressive recent literature on this problem, including but not limited to [**H14**, **R22**]. We state one simple corollary of Theorem 2.1 in this direction, which combines Theorem 2.1 with the main result in [**R22**].

COROLLARY 2.2. *Let $F_C$ be an inhomogeneous self-similar set in $\mathbb{R}$. Suppose the linear parts of the maps in the IFS are defined with algebraic parameters and that the semigroup generated by the IFS is free. Then* (1.5) *holds, that is,*

$$\overline{\dim}_B F_C = \max\{\overline{\dim}_B F_\varnothing, \overline{\dim}_B C\}.$$

FIGURE 3. Inhomogeneous self-similar mushroom patch.

## 3. More general attractors

Theorem 2.1 provides a roadmap to study more general inhomogeneous attractors where there is a 'best guess' for the dimension of the *homogeneous* attractor. For example, a similar result was proved in [**BF20**] for general self-affine sets with the $s$ given by the 'affinity dimension' and a similar result was proved in [**B19**] for general self-conformal sets with the $s$ given by Bowen's formula; the unique zero of the associated topological pressure. Similar to Theorem 2.1 and Corollary 2.2, these results allow precise formulae to be deduced provided the 'best guess' actually gives the upper box dimension of the homogeneous attractor. This is known to hold in many situations and we briefly recall some results of this type below.

For self-affine sets there is again a wealth of recent literature, building on earlier work of Falconer [**F88**], which says that the dimension of a self-affine set is given by the affinity dimension under only very mild assumptions; see [**F92**, **S98**, **BHR19**].

However, different to the self-similar case, there are important classes of exceptional set where the dimension is smaller than the affinity dimension due to, for example, excessive alignment of cylinders; see Bedford–McMullen carpets [**B84, M84**]. A detailed study of inhomogeneous Bedford–McMullen carpets was provided by [**F16**] and general families of sets where (1.5) fails were described. We provide an explicit example later in Section 5.

For self-conformal sets, the appropriately formulated open set condition will suffice for the dimension of the homogeneous attractor to be given by the zero of the topological pressure; thus verifying (1.5). Such examples include inhomogeneous 'cookie-cutters'.

Burrell [**B19**] also gave bounds of the form provided by Theorem 2.1 for general IFSs. In this case, the $s$ was given by the 'upper Lipschitz dimension'. In general this is a poor upper bound for the upper box dimension of the homogeneous attractor, but more can be said if the system satisfies some bounded distortion type estimates.

## 4. Inhomogeneous self-similar sets with overlaps: number theoretic counterexamples

Given the optimism provided by Theorem 2.1 and Corollary 2.2, one might start to believe that (1.5) *always* holds for inhomogeneous self-similar sets. In fact, it does not. This was first proved in [**BFM19**] and two distinct types of counterexample were given, both of which utilised the ambient space in a crucial way. The first, which we exhibit below, uses an IFS which is 'trapped' in a line but a condensation set which 'releases' it into the plane. The second is an example in $\mathbb{R}^3$ where one utilises the algebraic properties of $SO(3)$, specifically the existence of subgroups with the spectral gap property. For this family of (counter)examples, both the condensation set $C$ and the homogeneous attractor were singletons, but the (lower) box dimension of the inhomogeneous self-similar set could be made arbitrarily close to 2. This latter construction is not possible in $\mathbb{R}^2$. Given that neither of these counterexamples work in the line, this leaves what is our favourite open problem on inhomogeneous attractors; first stated as [**BFM19**, Conjecture 5.3].

QUESTION 4.1. Consider an inhomogeneous self-similar set $F_C \subseteq \mathbb{R}$. Is it necessarily true that (1.5) holds, that is,

$$\overline{\dim}_B F_C = \max\{\overline{\dim}_B F_\varnothing, \overline{\dim}_B C\}?$$

We now describe the first family of counterexamples provided by [**BFM19**]. The construction is based on Bernoulli convolutions. Fix $\lambda \in (0,1)$, let $X = [0,1]^2$ and let $S_1, S_2 : X \to X$ be defined by

$$S_1(x) = \lambda x \qquad \text{and} \qquad S_2(x) = \lambda x + (1-\lambda, 0).$$

To the homogeneous IFS $\{S_1, S_2\}$, associate the condensation set

$$C = \{0\} \times [0,1]$$

and observe that $F_\varnothing = [0,1] \times \{0\}$ and so $\dim_B F_\varnothing = \dim_B C = 1$. We will denote the inhomogeneous attractor of this system by $F_C^\lambda$ to emphasise the dependence on $\lambda$.

In order to effectively release the dimension of the IFS into the plane, we need to use number theoretic properties of $\lambda$. In particular, it must be chosen so that the orbital set spreads out the images of $C$ as quickly and uniformly as possible. This is most dramatically achieved by a well known class of algebraic integers known as *Garsia numbers*. Recall that a Garsia number is a positive real algebraic integer with norm $\pm 2$, whose conjugates are all of modulus strictly greater than 1. Examples of Garsia numbers include $\sqrt[n]{2}$ for (integers $n \geqslant 2$) and $1.76929\ldots$; the appropriate root of $x^3 - 2x - 2 = 0$. In [**G62**] Garsia proved that if $\lambda$ is the reciprocal of a Garsia number, then the associated Bernoulli convolution is absolutely continuous with a bounded density. This 'smoothness' property of the Bernoulli convolution is another manifestation of the the 'spreading out power' of Garsia numbers. In the other direction, *Pisot numbers* are the polar opposite of Garsia numbers. They lead to excessive piling up of copies of $C$ and singular Bernoulli convolutions.

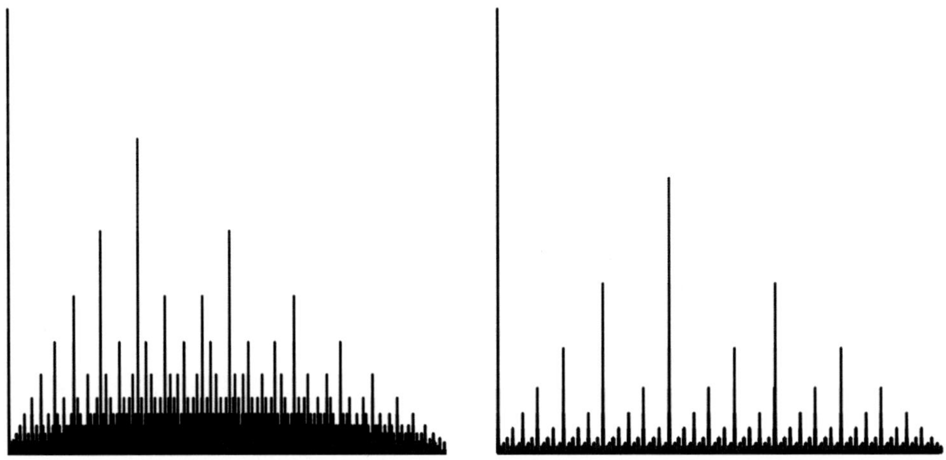

FIGURE 4. Two plots of $F_C^\lambda$ where $\lambda$ is chosen to be the reciprocal of $\sqrt{2}$, which is a Garsia number (left); and the reciprocal of the golden mean, which is Pisot (right). You can see that the images of $C$ are more uniformly spread out in the example on the left.

THEOREM 4.2. *If $\lambda \in (1/2, 1)$ is the reciprocal of a Garsia number, then*

$$\dim_{\mathrm{B}} F_C^\lambda = \frac{\log(4\lambda)}{\log 2} > 1.$$

For every $\lambda \in (1/2, 1)$ which is the reciprocal of a Garsia number, the set $F_C^\lambda$ thus provides a counterexample to (1.5) for the upper (and lower) box dimension. Furthermore, this example shows that $\overline{\dim}_{\mathrm{B}} F_C^\lambda$ does not just depend on the sets $F_\varnothing$ and $C$, but also depends on the IFS itself. To see this, simply observe that $F_\varnothing$ and $C$ do not depend on $\lambda$, but $\overline{\dim}_{\mathrm{B}} F_C^\lambda$ does.

Before we get to the proof we state a useful separation property that holds for the reciprocals of Garsia numbers and demonstrate the relevance to our situation via (4.1) below.

LEMMA 4.3 (Garsia [**G62**]). *Let $\lambda \in (1/2, 1)$ be the reciprocal of a Garsia number and $(i_k)_{k=1}^n, (i'_k)_{k=1}^n \in \{1,2\}^n$ be distinct words of length $n$. Then*

$$\left|(1-\lambda)\sum_{k=1}^{n} i_k \lambda^{k-1} - (1-\lambda)\sum_{k=1}^{n} i'_k \lambda^{k-1}\right| > \frac{K}{2^n}$$

*for some strictly positive constant $K$ that only depends on $\lambda$.*

Observe that, for any $\mathbf{i} = (i_1, \ldots, i_n) \in \{1,2\}^n$,

$$S_{\mathbf{i}}(C) = \{S_{\mathbf{i}}(0,0)\} \times [0, \lambda^n] = \left\{(1-\lambda)\sum_{k=1}^{n} i_k \lambda^{k-1}\right\} \times [0, \lambda^n]. \tag{4.1}$$

Combining Lemma 4.3 with (4.1), we see that whenever $\lambda$ is the reciprocal of a Garsia number the images of $C$ will be separated by a factor $K \cdot 2^{-n}$. This property is the main tool we use in the proof of Theorem 4.2.

PROOF OF THEOREM 4.2. Fix $\delta > 0$ and partition the unit square into horizontal strips of the form $[0,1] \times (\lambda^{k+1}, \lambda^k]$ for $k$ ranging from 0 to $k(\lambda, \delta)$, defined to be the largest integer satisfying $\lambda^{k(\lambda,\delta)+1} > \delta$. Observe that the only part of $F_C^\lambda$ which intersect the interior of the $k$th horizontal strip is

$$\bigcup_{l=0}^{k} \bigcup_{\mathbf{i} \in \{1,2\}^l} S_{\mathbf{i}}(C)$$

which is a union of vertical lines. Within the $k$th horizontal strip, each vertical line $S_{\mathbf{i}}(C)$ appearing in the above expression intersects roughly $\lambda^k/\delta$ many squares from the $\delta$-mesh. However, if two of these lines are too close to each another, they may not both contribute to the total intersections for $F_C^\lambda$. In fact, the number of lines which make a contribution to the total intersections is roughly $N_\delta(\Lambda(k))$, where

$$\Lambda(k) \;=\; \bigcup_{l=0}^{k} \bigcup_{\mathbf{i} \in \{1,2\}^l} S_{\mathbf{i}}(0,0),$$

that is, the number of base points of the lines intersecting the $k$th horizontal strip which lie in different $\delta$-intervals. This yields

$$N_\delta(F_C^\lambda) \approx \delta^{-1} \;+\; \sum_{k=0}^{k(\lambda,\delta)} (\lambda^k/\delta)\, N_\delta(\Lambda(k))$$

where the $\delta^{-1}$ comes from the intersections below the $k(\lambda, \delta)$th horizontal strip. It follows from Lemma 4.3 and subsequent discussion that

$$N_\delta(\Lambda(k)) \;\approx\; \min\{2^k, \delta^{-1}\}$$

which is the maximum value possible and where the 'comparison constants' are independent of $\delta$ and $k$, but do depend on $\lambda$, which is fixed. Let $k_0(\delta)$ be the largest integer satisfying $2^{k_0(\delta)} < \delta^{-1}$. It follows that

$$N_\delta(F_C^\lambda) \approx \delta^{-1} + \sum_{k=0}^{k_0(\delta)} (\lambda^k/\delta)\, 2^k + \sum_{k=k_0(\delta)+1}^{k(\lambda,\delta)} (\lambda^k/\delta)\, \delta^{-1}$$

$$= \delta^{-1} + \delta^{-1} \sum_{k=0}^{k_0(\delta)} (2\lambda)^k + \delta^{-2} \sum_{k=k_0(\delta)+1}^{k(\lambda,\delta)} \lambda^k$$

$$\approx \delta^{-1} + \delta^{-1} (2\lambda)^{k_0(\delta)} + \delta^{-2}\big(\lambda^{k(\lambda,\delta)} - \lambda^{k_0(\delta)}\big)$$

$$\approx \delta^{-1} + \delta^{-1}\delta^{-\log(2\lambda)/\log 2} + \delta^{-2}\big(\delta - \delta^{-\log\lambda/\log 2}\big)$$

$$\approx \delta^{-1-\log(2\lambda)/\log 2}$$

which yields

$$\overline{\dim}_{\mathrm{B}} F_C^\lambda = \underline{\dim}_{\mathrm{B}} F_C^\lambda = 1 + \log(2\lambda)/\log 2 = \frac{\log(4\lambda)}{\log 2}$$

as required. □

The key reason that the sets $F_C^\lambda$ provide counterexamples to (1.5) is that the set $F_\varnothing$ is trapped in a proper subspace of the plane. The underlying IFS has potential to give rise to an attractor with dimension bigger than 1, but cannot because the attractor is forced to lie in a 1-dimensional line. However, since the condensation set does not lie in this subspace, it 'releases' some of this potential dimension.

## 5. Inhomogeneous self-affine sets: fractal combs and counterexamples

We give a construction of an inhomogeneous Bedford–McMullen carpet, which we refer to as an *inhomogeneous fractal comb* which exhibits some interesting properties; in particular, failure of (1.5). These examples were first constructed in [**F16**]. The underlying homogeneous IFS is a Bedford–McMullen construction where the unit square has been divided into 2 columns of width $1/2$, and $n > 2$ rows of height $1/n$. The IFS is then made up of all the maps which correspond to the left hand column. The condensation set for this construction is taken as $C = [0,1] \times \{0\}$, i.e. the base of the unit square and the *homogeneous* attractor is $\{0\} \times [0,1]$, that is, the left hand side of the unit square. In particular, $\max\{\overline{\dim}_{\mathrm{B}} F_\varnothing, \overline{\dim}_{\mathrm{B}} C\} = 1$. The inhomogeneous attractor is termed the inhomogeneous fractal comb and is denoted by $F_C^n$, to emphasise the dependence on $n$.

THEOREM 5.1. *For $n \geqslant 3$,*

$$\underline{\dim}_{\mathrm{B}} F_C^n = \overline{\dim}_{\mathrm{B}} F_C^n = 2 - \log 2/\log n > 1.$$

For every $n \geqslant 3$, the set $F_C^n$ thus provides a counterexample to (1.5) for the upper (and lower) box dimension. Notable here is that the underlying IFS satisfies the open set condition; this is not possible for inhomogeneous self-similar sets by Theorem 2.1. A minor modification of the construction of $F_C^n$ also permits failure of (1.5) under the strong separation condition. We leave the details to the reader.

Furthermore, this example again shows that $\overline{\dim}_{\mathrm{B}} F_C^n$ does not just depend on the sets $F_\varnothing$ and $C$, but also depends on the IFS itself. To see this, observe that $F_\varnothing$ and $C$ do not depend on $n$, but $\overline{\dim}_{\mathrm{B}} F_C^n$ does. Finally, observe that, although the inhomogeneous fractal combs are subsets of $\mathbb{R}^2$ and the expected box dimension is 1, we can find examples where the achieved box dimension is arbitrarily close to 2 by letting $n \to \infty$. This demonstrates that, even in this simple case, there is no limit to how 'badly' the relationship (1.5) can fail.

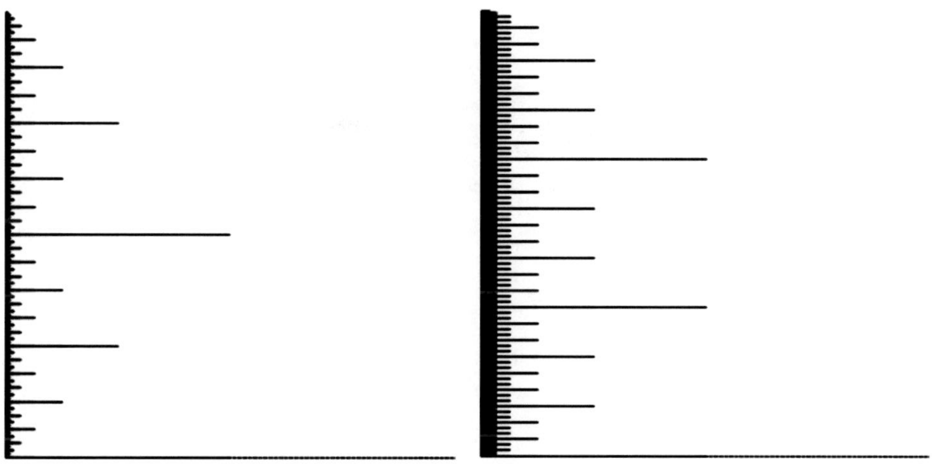

FIGURE 5. Two fractal combs: the inhomogeneous fractal combs $F_C^2$, with box dimension 1 (left); and $F_C^3$, with box dimension $2 - \log 2/\log 3 > 1$ (right).

Theorem 5.1 follows from a special case of the main result in [**F16**] but we give a simple direct proof here.

PROOF OF THEOREM 5.1. We again write $\mathcal{I} = \{1, \ldots, N\}$ for the set of indices describing the maps in the defining IFS and $\mathcal{I}^k$ for the set of words of length $k$ over $\mathcal{I}$. In this particular case $N = n$.

Let $\delta > 0$ be a small scale and choose $m(\delta)$ to be the largest integer value of $m$ such that

$$\delta < n^{-m}.$$

We choose $m(\delta)$ in this way so that the heights of the cylinders $S_{\mathbf{i}}([0,1]^2)$ are roughly $\delta$ for $\mathbf{i} \in \mathcal{I}^{m(\delta)}$. The choice of $m(\delta)$ also means that the vertical separation of the $m(\delta)$th level images of $C$ is also roughly $\delta$. This means that we may build an efficient cover of $F_C^n$ by treating images of $C$ up to the $m(\delta)$th level independently. Covering such an image of $C$ at scale $\delta$ is simply the length of the line divided by

$\delta$ and so we get

$$\begin{aligned}
N_\delta(F_C^n) &\approx \sum_{k=1}^{m(\delta)} \#\mathcal{I}^k\, 2^{-k}/\delta \\
&= \delta^{-1} \sum_{k=1}^{m(\delta)} (n/2)^k \\
&\approx \delta^{-1}(n/2)^{m(\delta)} \\
&\approx \delta^{-1}\delta^{-1+\log 2/\log n}
\end{aligned}$$

which proves that $\dim_B F_C^n = 2 - \log 2/\log n$ as required. □

## 6. Kleinian orbital sets: the original inhomogeneous attractors?

Let $n \geqslant 2$ be an integer and consider the Poincaré ball

$$\mathbb{D}^n = \{z \in \mathbb{R}^n : |z| < 1\}$$

equipped with the hyperbolic metric $d$ given by

$$|ds| = \frac{2|dz|}{1-|z|^2}.$$

This provides a model of $n$-dimensional hyperbolic space. The group of orientation preserving isometries of $(\mathbb{D}^n, d)$ is the group of conformal automorphisms of $\mathbb{D}^n$, which we denote by $\mathrm{Con}^+(n)$. A group $\Gamma \leqslant \mathrm{Con}^+(n)$ is called *Kleinian* if it is a discrete subset of $\mathrm{Con}^+(n)$. Kleinian groups generate fractal limit sets living on the boundary $S^{n-1} = \partial\mathbb{D}^n$ as well as beautiful tessellations of hyperbolic space. Both of these objects are defined via *orbits* under the action of the Kleinian group. The *limit set* is defined by

$$L(\Gamma) = \overline{\Gamma(0)} \setminus \Gamma(0)$$

where $\Gamma(0) = \{g(0) : g \in \Gamma\}$ is the orbit of 0 under $\Gamma$ and $\overline{\Gamma(0)}$ is the Euclidean closure of $\Gamma(0)$. On the other hand, hyperbolic tessellations arise by taking the orbit of a fundamental domain for the group action.

The *Poincaré exponent* is a coarse measure of the rate of accumulation to the boundary. It is defined as the exponent of convergence of the *Poincaré series*

$$P_\Gamma(s) = \sum_{g\in\Gamma} \exp(-sd(0,g(0))) = \sum_{g\in\Gamma} \left(\frac{1-|g(0)|}{1+|g(0)|}\right)^s$$

for $s \geqslant 0$. That is, the *Poincaré exponent* is

$$\delta(\Gamma) = \inf\{s \geqslant 0 : P_\Gamma(s) < \infty\}.$$

A Kleinian group is called *non-elementary* if its limit set contains at least 3 points, in which case it is necessarily an uncountable perfect set. In the case $n = 2$, Kleinian groups are more commonly referred to as Fuchsian groups. For more background on hyperbolic geometry and Kleinian groups see [**B83, B93, S04**].

At this point, you may recognise a few familiar sounding concepts. First, the orbit $\Gamma(0)$ is an orbital set with condensation $C = \{0\}$, albeit the orbit is under a group action rather than a semigroup action as in the IFS case, but many similarities remain. Not least that the orbit accumulates on the fractal limit set, in this case $L(\Gamma)$. Furthermore, the Poincaré exponent shares many features with the similarity dimension, recall (2.1), or perhaps more accurately with the zero of the topological pressure associated to a conformal system. In particular, it provides a 'best guess' for the dimensions of the limit set.

In [**BF23**], Kleinian orbital sets were formally introduced and studied in the context of inhomogeneous attractors. Fix a non-empty set $C \subseteq \mathbb{D}^n$ and a Kleinian group $\Gamma$. The *orbital set* is defined to be

$$\Gamma(C) = \bigcup_{g \in \Gamma} g(C).$$

It is easy to see that the limit set is contained in the Euclidean closure of any orbital set.

There is a celebrated connection between hyperbolic geometry (especially Fuchsian groups) and the artwork of M. C. Escher. Orbital sets fall very naturally into this discussion since many of the memorable images from Escher's work are orbital sets (rather than tessellations). Here $C$ could be a large central bat or fish, which is then repeated many times on smaller and smaller scales towards the boundary of $\mathbb{D}^n$. In some sense this makes Kleinian orbital sets the original inhomogeneous attractors since they already appeared in the work of Escher.

FIGURE 6. Inhomogeneous tiling: a tessellation of the hyperbolic plane by a triangular fundamental domain. This is also an orbital set with $C$ being any one of the triangles. The triangles in the tessellation are coloured such that no two triangles sharing an edge are coloured the same.
Figure credit citation: By Tom Ruen and Roice Nelson - KaleidoTile (original), new 4k version [1], Public Domain, `https://commons.wikimedia.org/w/index.php?curid=23887885`

The main result of [**BF23**] was a complete characterisation of the upper box dimension of Kleinian orbital sets with $C$ bounded in the hyperbolic metric. Boundedness in the hyperbolic metric simply means that $C$ is uniformly bounded away from the boundary $S^{n-1}$. For clarity, we emphasise that we compute the dimension of the orbital set with respect to the *Euclidean* metric on $\mathbb{R}^n$.

THEOREM 6.1. *Let $\Gamma$ be a Kleinian group acting on $\mathbb{D}^n$ and $C$ be a non-empty subset of $\mathbb{D}^n$ which is bounded in the hyperbolic metric. Then*

$$\overline{\dim}_{\mathrm{B}}\Gamma(C) = \max\left\{\overline{\dim}_{\mathrm{B}}L(\Gamma), \overline{\dim}_{\mathrm{B}}C, \delta(\Gamma)\right\}.$$

This should be compared with Theorem 2.1. It is perhaps noteworthy that no assumptions are imposed on the Kleinian group and that each of the three terms appearing on the right hand side is necessary; see [**BF23**] for more information. If we impose the additional assumption that $\Gamma$ is non-elementary and geometrically finite, then $\overline{\dim}_{\mathrm{B}}L(\Gamma) = \delta(\Gamma)$ (see [**BJ97, SU96**]) and the result reduces to something more reminiscent of (1.5).

In the proof of the above theorem, essential use was made of the assumption that $C$ is bounded (in the hyperbolic metric). It turns out that this is necessary and we exhibit a family of counterexamples from [**BF23**] below.

THEOREM 6.2. *There exists a Kleinian (in fact, Fuchsian) group $\Gamma$ acting on $\mathbb{D}^2$ and a condensation set $C \subseteq \mathbb{D}^2$ such that*

$$\overline{\dim}_{\mathrm{B}}L(\Gamma) = \overline{\dim}_{\mathrm{B}}C = \delta(\Gamma) = 0$$

*but*

$$\underline{\dim}_{\mathrm{B}}\Gamma(C) = \overline{\dim}_{\mathrm{B}}\Gamma(C) = 1.$$

PROOF. The set $C$ and group $\Gamma$ are very simple. The work is in proving that the orbital set has large dimension, and this relies on some number theory. Let $\alpha > 1$ and $\beta \in (0,1)$ be such that $\log \alpha$ and $\log \beta$ are rationally independent, that is, $\log \alpha / \log \beta \notin \mathbb{Q}$. Here and throughout log is the natural logarithm. For example, $\alpha = 2$, and $\beta = 1/3$ suffices. Let

$$C = \{1 - \beta^n : n \in \mathbb{N}\} \subseteq \mathbb{D}^2$$

noting that $C$ is unbounded in $(\mathbb{D}^2, d)$. Let $h \in \mathrm{con}^+(\mathbb{D}^2)$ be the hyperbolic element with repelling fixed point $-1$ and attracting fixed point 1 given by

$$h(z) = \frac{(\alpha+1)z + (\alpha-1)}{(\alpha-1)z + (\alpha+1)}.$$

Let $\Gamma = \langle h \rangle$ be the elementary Fuchsian group generated by $h$. It is an elementary but instructive exercise to show that

$$\overline{\dim}_{\mathrm{B}}L(\Gamma) = \overline{\dim}_{\mathrm{B}}C = \delta(\Gamma) = 0.$$

We encourage the reader to try this for themselves. In order to prove that $\underline{\dim}_{\mathrm{B}}\Gamma(C) = \overline{\dim}_{\mathrm{B}}\Gamma(C) = 1$ we show that $\Gamma(C)$ is dense in $(-1,1) \subseteq \mathbb{D}^2$, recalling that the box dimensions are stable under taking closure. This shows that the box dimensions are at least 1, but the orbital set is contained in $(-1,1)$ and so they are also at most 1. The orbital set has a straightforward description due to the simplicity of $\Gamma$ and $C$. Indeed,

$$\begin{aligned}\Gamma(C) &= \{h^m(1-\beta^n) : m \in \mathbb{Z},\, n \in \mathbb{N}\} \\ &= \left\{\frac{(\alpha^m+1)(1-\beta^n) + (\alpha^m-1)}{(\alpha^m-1)(1-\beta^n) + (\alpha^m+1)} : m \in \mathbb{Z},\, n \in \mathbb{N}\right\} \\ &= \left\{\frac{2 - \alpha^m\beta^n - \beta^n}{2 + \alpha^m\beta^n - \beta^n} : m \in \mathbb{Z},\, n \in \mathbb{N}\right\}\end{aligned}$$

noting that we switch the role of $m$ and $-m$ in the final expression, which is fine since $m \in \mathbb{Z}$. Let $y \in (0, \infty)$ be such that $\log y \in \mathbb{Q}$. Since $\log \alpha / \log \beta \notin \mathbb{Q}$ we can find sequences $m_k \in \mathbb{Z}$, $n_k \in \mathbb{N}$ such that

$$\alpha^{m_k} \beta^{n_k} \to y$$

as $k \to \infty$. This is a standard application of Dirichlet's approximation theorem. Moreover, again since $\log \alpha$ and $\log \beta$ are rationally independent, we necessarily have $n_k \to \infty$ as $k \to \infty$. Therefore

$$\frac{2 - \alpha^{m_k} \beta^{n_k} - \beta^{n_k}}{2 + \alpha^{m_k} \beta^{n_k} - \beta^{n_k}} \to \frac{2 - y}{2 + y}$$

as $k \to \infty$. The set

$$\left\{ \frac{2 - y}{2 + y} : y \in (0, \infty) \text{ and } \log y \in \mathbb{Q} \right\}$$

is dense in $(-1, 1)$ and the density of $\Gamma(C)$ in $(-1, 1)$ follows. □

## Acknowledgments

I am grateful to Lars Olsen for first introducing me to inhomogeneous attractors. I was quite inspired by Nina Snigireva's PhD thesis on the topic, back when I first read it in 2010 or so. I am also grateful to Simon Baker, Amlan Banaji, Tom Bartlett, Stuart Burrell, Antti Käenmäki, Andras Mathé, and Alex Rutar for various helpful discussions over the years.

I wrote this expository article in May and June 2024 during pleasant trips to Madison, Banff and Warwick. I blame jet-lag or dodgy airport Wi-Fi for any typos or inconsistencies!

## References

[BFM19] S. Baker, J. M. Fraser, and A. Máthé, *Inhomogeneous self-similar sets with overlaps*, Ergodic Theory Dynam. Systems **39** (2019), no. 1, 1–18, DOI 10.1017/etds.2017.13. MR3881123

[BR24+] A. Banaji and A. Rutar, *Lower box dimension of infinitely generated self-conformal sets*, preprint, available at: `arXiv:2406.12821`.

[BHR19] B. Bárány, M. Hochman, and A. Rapaport, *Hausdorff dimension of planar self-affine sets and measures*, Invent. Math. **216** (2019), no. 3, 601–659, DOI 10.1007/s00222-018-00849-y. MR3955707

[BKN23+] B. Bárány, A. Käenmäki and P. Nissinen, *Covering number on inhomogeneous graph-directed self-similar sets*, preprint available at: `arXiv:2307.16263`

[B88] M. Barnsley, *Fractals everywhere*, Academic Press, Inc., Boston, MA, 1988. MR977274

[B06] M. F. Barnsley, *Superfractals*, Cambridge University Press, Cambridge, 2006, DOI 10.1017/CBO9781107590168. MR2254477

[BD85] M. F. Barnsley and S. Demko, *Iterated function systems and the global construction of fractals*, Proc. Roy. Soc. London Ser. A **399** (1985), no. 1817, 243–275. MR799111

[BF23] T. Bartlett and J. M. Fraser, *Dimensions of Kleinian orbital sets*, J. Fractal Geom. **10** (2023), no. 3-4, 267–278, DOI 10.4171/jfg/139. MR4651740

[B83] A. F. Beardon, *The geometry of discrete groups*, Graduate Texts in Mathematics, vol. 91, Springer-Verlag, New York, 1983, DOI 10.1007/978-1-4612-1146-4. MR698777

[B84] T. Bedford, *Crinkly curves, Markov partitions and box dimensions in self-similar sets, Ph.D dissertation, University of Warwick*, (1984).

[B11] A. Bisbas, *On the Fourier transforms of inhomogeneous self-similar measures*, Illinois J. Math. **55** (2011), no. 2, 675–684 (2012). MR3020701

[BJ97] C. J. Bishop and P. W. Jones, *Hausdorff dimension and Kleinian groups*, Acta Math. **179** (1997), no. 1, 1–39, DOI 10.1007/BF02392718. MR1484767

[B93] B. H. Bowditch, *Geometrical finiteness for hyperbolic groups*, J. Funct. Anal. **113** (1993), no. 2, 245–317, DOI 10.1006/jfan.1993.1052. MR1218098

[B19] S. A. Burrell, *On the dimension and measure of inhomogeneous attractors*, Real Anal. Exchange **44** (2019), no. 1, 199–215, DOI 10.14321/realanalexch.44.1.0199. MR3951342

[BF20] S. A. Burrell and J. M. Fraser, *The dimensions of inhomogeneous self-affine sets*, Ann. Acad. Sci. Fenn. Math. **45** (2020), no. 1, 313–324, DOI 10.5186/aasfm.2020.4516. MR4056541

[DV23+] S. Dubey and S. Verma, *Inhomogeneous graph-directed attractors and fractal measures*, J. Anal. **32** (2024), no. 1, 157–170, DOI 10.1007/s41478-023-00614-2. MR4706946

[F88] K. J. Falconer, *The Hausdorff dimension of self-affine fractals*, Math. Proc. Cambridge Philos. Soc. **103** (1988), no. 2, 339–350, DOI 10.1017/S0305004100064926. MR923687

[F92] K. J. Falconer, *The dimension of self-affine fractals. II*, Math. Proc. Cambridge Philos. Soc. **111** (1992), no. 1, 169–179, DOI 10.1017/S0305004100075253. MR1131488

[F03] K. Falconer, *Fractal geometry*, 2nd ed., John Wiley & Sons, Inc., Hoboken, NJ, 2003. Mathematical foundations and applications, DOI 10.1002/0470013850. MR2118797

[F12] J. M. Fraser, *Inhomogeneous self-similar sets and box dimensions*, Studia Math. **213** (2012), no. 2, 133–156, DOI 10.4064/sm213-2-2. MR3024316

[F16] J. M. Fraser, *Inhomogeneous self-affine carpets*, Indiana Univ. Math. J. **65** (2016), no. 5, 1547–1566, DOI 10.1512/iumj.2016.65.5903. MR3571439

[G62] A. M. Garsia, *Arithmetic properties of Bernoulli convolutions*, Trans. Amer. Math. Soc. **102** (1962), 409–432, DOI 10.2307/1993615. MR137961

[H85] M. Hata, *On some properties of set-dynamical systems*, Proc. Japan Acad. Ser. A Math. Sci. **61** (1985), no. 4, 99–102. MR796477

[H14] M. Hochman, *On self-similar sets with overlaps and inverse theorems for entropy*, Ann. of Math. (2) **180** (2014), no. 2, 773–822, DOI 10.4007/annals.2014.180.2.7. MR3224722

[H81] J. E. Hutchinson, *Fractals and self-similarity*, Indiana Univ. Math. J. **30** (1981), no. 5, 713–747, DOI 10.1512/iumj.1981.30.30055. MR625600

[KL17] A. Käenmäki and J. Lehrbäck, *Measures with predetermined regularity and inhomogeneous self-similar sets*, Ark. Mat. **55** (2017), no. 1, 165–184, DOI 10.4310/ARKIV.2017.v55.n1.a8. MR3711147

[LRZ17] M. L. Lapidus, G. Radunović, and D. Žubrinić, *Zeta functions and complex dimensions of relative fractal drums: theory, examples and applications*, Dissertationes Math. **526** (2017), 105, DOI 10.4064/dm757-4-2017. MR3764036

[LRZ18] M. L. Lapidus, G. Radunović, and D. Žubrinić, *Fractal tube formulas for compact sets and relative fractal drums: oscillations, complex dimensions and fractality*, J. Fractal Geom. **5** (2018), no. 1, 1–119, DOI 10.4171/JFG/57. MR3760302

[L13] P. Liszka, *On inhomogeneous self-similar measures and their $L^q$ spectra*, Ann. Polon. Math. **109** (2013), no. 1, 75–92, DOI 10.4064/ap109-1-6. MR3071730

[M84] C. McMullen, *The Hausdorff dimension of general Sierpiński carpets*, Nagoya Math. J. **96** (1984), 1–9, DOI 10.1017/S0027763000021085. MR771063

[OS07] L. Olsen and N. Snigireva, *$L^q$ spectra and Rényi dimensions of in-homogeneous self-similar measures*, Nonlinearity **20** (2007), no. 1, 151–175, DOI 10.1088/0951-7715/20/1/010. MR2285110

[OS08a] L. Olsen and N. Snigireva, *In-homogenous self-similar measures and their Fourier transforms*, Math. Proc. Cambridge Philos. Soc. **144** (2008), no. 2, 465–493, DOI 10.1017/S0305004107000771. MR2405903

[OS08b] L. Olsen and N. Snigireva, *Multifractal spectra of in-homogenous self-similar measures*, Indiana Univ. Math. J. **57** (2008), no. 4, 1789–1843, DOI 10.1512/iumj.2008.57.3622. MR2440882

[PS00] Y. Peres and B. Solomyak, *Problems on self-similar sets and self-affine sets: an update*, Fractal geometry and stochastics, II (Greifswald/Koserow, 1998), Progr. Probab., vol. 46, Birkhäuser, Basel, 2000, pp. 95–106. MR1785622

[R22] A. Rapaport, *Proof of the exact overlaps conjecture for systems with algebraic contractions* (English, with English and French summaries), Ann. Sci. Éc. Norm. Supér. (4) **55** (2022), no. 5, 1357–1377. MR4517689

[S08] N. Snigireva, *Inhomogeneous self-similar sets and measures*, *Ph.D Dissertation*, University of St Andrews, (2008).

[S98] B. Solomyak, *Measure and dimension for some fractal families*, Math. Proc. Cambridge Philos. Soc. **124** (1998), no. 3, 531–546, DOI 10.1017/S0305004198002680. MR1636589

[S04] B. O. Stratmann, *The exponent of convergence of Kleinian groups; on a theorem of Bishop and Jones*, Fractal geometry and stochastics III, Progr. Probab., vol. 57, Birkhäuser, Basel, 2004, pp. 93–107. MR2087134

[SU96] B. Stratmann and M. Urbański, *The box-counting dimension for geometrically finite Kleinian groups*, Fund. Math. **149** (1996), no. 1, 83–93, DOI 10.4064/fm-149-1-83-93. MR1372359

[WH19] C. Wang and P. Huang, *The box dimensions of inhomogeneous self-conformal sets with overlaps*, J. Math. Anal. Appl. **470** (2019), no. 1, 434–446, DOI 10.1016/j.jmaa.2018.10.013. MR3865148

[ZGX22] S. Zhang, B. Gao, and Y. Xiao, *On the $L^q$ spectra of in-homogeneous self-similar measures*, Forum Math. **34** (2022), no. 5, 1383–1410, DOI 10.1515/forum-2022-0142. MR4472890

[ZGX23] S. Zhang, B. Gao, and Y. Xiao, *The lower Fourier dimensions of in-homogeneous self-similar measures*, J. Fourier Anal. Appl. **29** (2023), no. 5, Paper No. 55, 29, DOI 10.1007/s00041-023-10037-z. MR4630521

Jonathan M. Fraser, University of St Andrews, Scotland
*Email address*: jmf32@st-andrews.ac.uk

Contemporary Mathematics
Volume **825**, 2025
https://doi.org/10.1090/conm/825/16510

# Fractal Measures and their induced Gaussian Processes

Palle E.T. Jorgensen and James Tian

*Dedicated to the memories of Professors Ka-Sing Lau, and Bob Strichartz.*

ABSTRACT. Our general theme is classes of measures and their induced Gaussian Processes. Our motivation is the class of IFS fractal measures. We outline our construction, and we prove new duality results for Gaussian processes. Two features are stressed, first the Gaussian processes presented here are indexed by general measures; and secondly, the link in the induction is a framework of reproducing kernel Hilbert spaces for this measurable category. We obtain a precise necessary and sufficient condition for equivalence of pairs of probability measures which determine the corresponding two Gaussian processes.

## 1. Introduction

We begin with an outline of the connection between the classes of measures we study, and their induced Gaussian Processes. This correspondence will be via new families of corresponding reproducing kernel Hilbert spaces (RKHSs).

While RKHSs play an important role in the analysis of Gaussian processes, there is a mismatch between the measurable setting of stochastic analysis on the one hand, and the more traditional setting of RKHS on the other. To see this, recall that a RKHS is a Hilbert space $\mathscr{H}$ of functions on some prescribed set, say $X$, such that the values $f(x)$ of functions $f$ from $\mathscr{H}$ are reproduced via the inner product in $\mathscr{H}$. By contrast, in stochastic analysis one studies system of random variables, which is of course formulated in a measurable category, details below. Hence the better framework for these applications is that of measure spaces $(X, \mathscr{B})$ and suitable Hilbert spaces of measurable functions defined on it. But since measurable functions are not defined in the pointwise sense, the more traditional and geometric approach of Aronszajn et al to RKHS theory is not amenable to the measurable category. Going back to the pioneering work of Aronszajn [**Aro43**, **Aro50**], the motivation for the study of RKHSs derived primarily from Greens function analysis in the study of partial differential equations (PDE), and from the study of Hilbert spaces of analytic functions (S. Bergman et al), and for this, the pointwise setting appears natural and fruitful. We note that, historically, the applications of RKHS theory to probability theory came later, and so it is not surprising that, by necessity, the use of Hilbert space geometry is different there.

---

2020 *Mathematics Subject Classification.* Primary 47B32, 46E22, 28A80, 60G15, 60H30.
*Key words and phrases.* Reproducing kernel Hilbert space (RKHS), measurable category, Gaussian processes, IFS fractal measures, generalized Ito-integration, random signed measures, random fields.

The purpose of the present paper is to offer a duality theory for Gaussian processes, on the one hand, and to develop a remedy to the above mentioned problem with the reproducing property for Hilbert spaces of measurable functions on prescribed measure spaces $(X, \mathscr{B})$, on the other.

**Discussion of the literature.** The theory of RKHS and their applications is vast, and below we only make a selection. Readers will be able to find more cited there. As for the general theory of RKHS in the pointwise category, we find useful [**LMP09**, **PR16**]. The applications include fractals (see e.g., [**Aro43**, **BH14**]); probability theory [**Sai16**, **MSF**$^+$**16**, **HE15**, **Jr68**, **EMESO17**],[**PVPK17**], [**CMPS17**]; and application to learning theory [**SZ04**, **SZ09a**, **SZ09b**, **JT15**].

## 2. RKHS in the Measurable Category

The focus of this section will be details of our correspondence principle for measure classes and associated reproducing kernel Hilbert spaces (RKHSs).

Motivated by applications to stochastic processes, we propose here the following generalization of RKHS in the usual sense of Aronszajn (see, e.g., [**Aro43**, **Aro50**, **SZ04**, **SZ09b**, **SZ09a**]), but now to a measurable framework. In the traditional approach, the starting point is a positive definite function $K : X \times X \to \mathbb{C}$, which is defined pointwise, and $X$ is a given set. The "positive definite" (p.d.) requirement is that, for all $p \in \mathbb{N}$, and all $x_i \in X$, $1 \leq i \leq p$ the $p \times p$ matrix $\left(K\left(x_i, x_j\right)\right)_{i,j=1}^{p}$ is positive semidefinite. The RKHS $\mathscr{H}$ is then defined as the Hilbert-completion of the space of functions (on $X$),

$$\sum\nolimits_i \xi_i K\left(\cdot, x_i\right), \; \xi_i \in \mathbb{C} \tag{2.1}$$

with respect to $\left\|\cdot\right\|_{\mathscr{H}}$ defined as follows:

$$\left\|\sum\nolimits_i \xi_i K\left(\cdot, x_i\right)\right\|_{\mathscr{H}}^2 = \sum\nolimits_i \sum\nolimits_j \overline{\xi}_i \xi_j K\left(x_i, x_j\right). \tag{2.2}$$

For recent applications, we refer to [**JT15**, **JPT15**, **JPT16**].

**The Measurable Setting.** Since stochastic processes are formulated in the language of probability, it is important that the elements in $\mathscr{H}$ be suitably determined measurable functions; and that the new inner product then be redefined in this measurable setting.

Definition 2.1. Let $X$ be a locally compact Hausdorff space, and let $\mathscr{B}$ be the corresponding $\sigma$-algebra of Borel measurable subsets of $X$. Let $\lambda$ be a positive measure on $(X, \mathscr{B})$, i.e., $\lambda$ is defined to be $\sigma$-additive on $\mathscr{B}$.

We shall consider Hilbert spaces $\mathscr{H} \subset L^2_{loc}(\lambda)$. (The subscript loc for "local" is important, i.e., the requirement is that

$$\int_B |f|^2 \, d\lambda < \infty \tag{2.3}$$

for all $B \in \mathscr{B}$ s.t. $\lambda(B) < \infty$, and all $f \in \mathscr{H}$.)

We shall denote by

$$\mathscr{B}_{fin} := \left\{B \in \mathscr{B} \mid \lambda(B) < \infty\right\}. \tag{2.4}$$

DEFINITION 2.2. A pair $(\mathscr{H}, \lambda)$ as above is said to be *positive definite* (p.d.) if and only if, for all $B \in \mathscr{B}_{fin}$,

$$\sup_{f \in \mathscr{H},\, \|f\|_{\mathscr{H}} \leq 1} \left| \int_B f d\lambda \right| < \infty. \tag{2.5}$$

REMARK 2.3. At the outset our setting is that of general $\sigma$-finite measure spaces $(X, \mathscr{B})$; but for some considerations it will be useful to assume that $X$ be a locally compact Hausdorff space, and in that case the prescribed $\sigma$-algebra $\mathscr{B}$ will be the Borel subsets of $X$. We shall then be working with $C_c(X) =$ the continuous functions of compact support.

REMARK 2.4 (**The reproducing property**). The RKHSs are of great use in part, because of what is called the *reproducing property.*

In the case when $K$ is given as in (2.1)–(2.2) above, then the corresponding RKHS $\mathscr{H}(K)$ is a Hilbert space of functions $f$ on $X$, and we have

$$f(x) = \langle f, K(\cdot, x) \rangle_{\mathscr{H}(X)}, \ \forall x \in X, \ \forall f \in \mathscr{H}(K). \tag{2.6}$$

Now in the measurable category, (2.6) does not make sense since functions are not defined pointwise.

REMARK. In the proofs below we shall mostly restrict attention to real valued functions, but all arguments easily extend to the complex case.

LEMMA 2.5. *Suppose $(\mathscr{H}, \lambda)$ is positive definite.*

(i) *Then, for $\forall B \in \mathscr{B}_{fin}$, $\exists! K^B \in \mathscr{H}$ such that*

$$\int_B f d\lambda = \left\langle f, K^B \right\rangle_{\mathscr{H}}, \ \forall f \in \mathscr{H}. \tag{2.7}$$

(ii) *Setting $K(\cdot, B) = K^B$, $B \in \mathscr{B}_{fin}$, note that $x \longmapsto K(x, \cdot)$ is then a measurable field of random signed measures on $(X, \mathscr{B})$, and that*

$$\begin{aligned} \rho_{\mathscr{H}}(A, B) &= \left\langle K^A, K^B \right\rangle_{\mathscr{H}} \\ &= \int_A K(x, B)\, d\lambda(x) = \int_B K(y, A)\, d\lambda(y). \end{aligned} \tag{2.8}$$

*Then $\rho_{\mathscr{H}}$ is p.d. on $\mathscr{B}_{fin} \times \mathscr{B}_{fin}$.*

PROOF. (i) By the assumption (2.5), the mapping $\mathscr{H} \ni f \longrightarrow \int_B f d\lambda$ is a bounded linear functional on $\mathscr{H}$, so by Riesz theorem, $\exists! K(\cdot, B) \in \mathscr{H}$ s.t. (2.7) holds.

For part (ii), note that (2.8) follows from (2.7) with $f = K^A$; and it is immediate that $\rho_{\mathscr{H}}$ is positive definite on $\mathscr{B}_{fin} \times \mathscr{B}_{fin}$. Indeed, for $\forall p \in \mathbb{N}$, $\forall (A_i)_{i=1}^p$ in $\mathscr{B}_{fin}$, and $\forall (\xi_i)_{i=1}^p$ in $\mathbb{R}$, we have

$$\begin{aligned} \sum_i \sum_j \xi_i \xi_j \rho_{\mathscr{H}}(A_i, A_j) &= \sum_i \sum_j \xi_i \xi_j \left\langle K^{A_i}, K^{A_j} \right\rangle_{\mathscr{H}} \\ &= \left\| \sum_i \xi_i K^{A_i} \right\|_{\mathscr{H}}^2 \geq 0. \end{aligned}$$

□

REMARK 2.6. Let $(\mathscr{H}, \lambda)$ be as specified in Definitions 2.1–2.2. In this case, the corresponding reproducing property takes any one of the following equivalent forms (see the setting $(X, \mathscr{B}, \mathscr{H}, \lambda, K)$ as in Definitions 2.1–2.2):

(1) For all $B \in \mathscr{B}_{fin}$, we have $\mathscr{H} \subset L^2_{loc}(\lambda)$, and

$$\int_B f d\lambda = \langle f(\cdot), K(\cdot, B)\rangle_{\mathscr{H}}, \ \forall B \in \mathscr{B}_{fin}, \forall f \in \mathscr{H}. \tag{2.9}$$

(2) Setting $K^{\varphi}(\cdot) = \int \varphi(y) K(\cdot, dy) (\in \mathscr{H})$, then we have

$$\int f\varphi d\lambda = \langle f, K^{\varphi}\rangle_{\mathscr{H}}, \ \forall \varphi \in C_c(X), \forall f \in \mathscr{H}. \tag{2.10}$$

When $(\mathscr{H}, \lambda)$ is given as above, and $f \in \mathscr{H}$, we set $\mu_f$ to be the signed measure

$$\mu_f(B) := \langle f(\cdot), K(\cdot, B)\rangle_{\mathscr{H}}, \ B \in \mathscr{B}_{fin}. \tag{2.11}$$

We shall show below that $d\mu_f = f d\lambda$ locally, see Lemma 2.8.

DEFINITION 2.7. Let $(X, \lambda)$ be as above. Let $\rho_{\mathscr{H}}$ be the positive definite function on $\mathscr{B}_{fin} \times \mathscr{B}_{fin}$ as in (2.8). We denote $\mathscr{H}^{(Aronsz)}(\rho_{\mathscr{H}})$ the corresponding RKHS in the sense of Aronszajn.

LEMMA 2.8. *Let $(\mathscr{H}, \lambda)$ be as above, i.e., positive definite (p.d.). For $f \in \mathscr{H}$, we set $\mu_f$ to be the signed measure* (2.11) *on $\mathscr{B}_{fin}$, i.e., $\mu_f(B) := \langle f(\cdot), K(\cdot, B)\rangle_{\mathscr{H}}$, $B \in \mathscr{B}_{fin}$; then $\mu_f \ll \lambda$, and $d\mu_f/d\lambda = f$ in $L^1_{loc}(\lambda)$.*

PROOF. We now show that if $\lambda(A) = 0$ then $\mu_f(A) = 0$. But for $\forall B \in \mathscr{B}_{fin}$ we have

$$0 = \int_A K(x, B) d\lambda(x) = \int_B K(y, A) d\lambda(y); \tag{2.12}$$

and so by approximation

$$\int \varphi(y) K(y, A) d\lambda(y) = 0, \tag{2.13}$$

for $\forall \varphi \in C_c(X)$, and

$$\langle K^{\varphi}, K^{A}\rangle_{\mathscr{H}} = 0. \tag{2.14}$$

Since $\{K^{\varphi} \mid \varphi \in C_c(X)\}$ is dense in $\mathscr{H}$, we get $\langle f(\cdot), K(\cdot, A)\rangle_{\mathscr{H}} = \mu_f(A) = 0$ which is the desired conclusion.

But, by (2.13)–(2.14), we have

$$\begin{aligned}\int_X \varphi f d\lambda = \langle f, K^{\varphi}\rangle_{\mathscr{H}} &= \int_X \varphi(y) \langle f, K(\cdot, dy)\rangle_{\mathscr{H}} \\ &= \int_X \varphi(y) d\mu_f(y), \ \forall \varphi \in C_c(X),\end{aligned}$$

and so the conclusion $d\mu_f/d\lambda = f$ in $L^1_{loc}(\lambda)$ follows. □

REMARK 2.9 (The Radon-Nikodym-derivative). Fix $(X, \mathscr{B})$, assume $(\mathscr{H}, \lambda)$ is positive definite. Thus, $\lambda$ is a positive measure on $(X, \mathscr{B})$ s.t. (2.5) holds; set $\mu_f(A) = \langle f(\cdot), K(\cdot, A)\rangle_{\mathscr{H}}$, $\forall A \in \mathscr{B}_{fin}$, $\forall f \in \mathscr{H}$.

Then, for a.e. $x$ w.r.t $\lambda$, there exists $A_i \in \mathscr{B}_{fin}$, $i = 1, 2, \cdots$, s.t. $\lambda(A_i) \to 0$, $x \in \cap A_i$, so that (see [**Rud87**], and Figure 2.1)

$$\lim_{i\to\infty} \frac{\mu_f(A_i)}{\lambda(A_i)} = f(x).$$

This is a way to make precise the local Radon-Nikodym derivative, $d\mu_f = f d\lambda$.

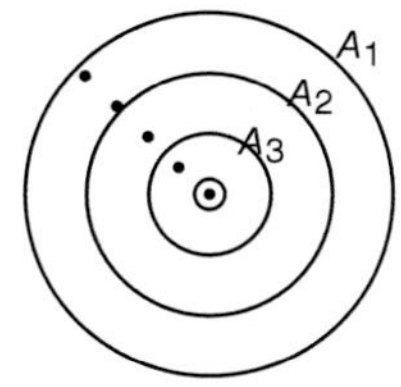


FIGURE 2.1. The sequence $(A_i)_{i=1}^{\infty}$ in $\mathscr{B}_{fin}$ s.t. $\lambda(A_i) \to 0$, as $i \to \infty$.

COROLLARY 2.10. *Let $\lambda$ be a positive regular measure on a locally compact Hausdorff space $X$ with Borel $\sigma$-algebra $\mathscr{B}$; and let $\mathscr{H}$ be a Hilbert space s.t.* (2.5) *holds; i.e. $(\mathscr{H}, \lambda)$ is positive definite. Then the functions $f \in \mathscr{H}$ are in $L^2_{loc}(\lambda)$.*

PROOF. Given $f \in \mathscr{H}$, we proved that the signed measures
$$\mu_f(B) = \langle f, K(\cdot, B)\rangle_{\mathscr{H}},\ B \in \mathscr{B}_{fin}$$
are well defined, and locally $\mu_f \ll \lambda$, with $d\mu_f/d\lambda = f$ (locally), i.e., for $\forall A \in \mathscr{B}_{fin}$, we have
$$\mu_f(A) = \int_A f d\lambda.$$
□

COROLLARY 2.11. *Let $(\mathscr{H}, \lambda)$ be as above; and let $\{f_i\}_{i\in\mathbb{N}}$ be a fixed ONB in $\mathscr{H}$, or a Parseval Frame, then, for all $B \in \mathscr{B}_{fin}$, we have:*

$$K^B(x) = \sum_{i=1}^{\infty}\left(\int_B f_i d\lambda\right) f_i(x); \tag{2.15}$$

*with convergence in the $\mathscr{H}$-norm, and*

$$\|K^B\|^2_{\mathscr{H}} = \sum_{i=1}^{\infty}\left|\int_B f_i d\lambda\right|^2. \tag{2.16}$$

PROOF. Note that
$$\begin{aligned}\|K^B\|^2_{\mathscr{H}} &= \sum_{i=1}^{\infty}\langle K^B, f_i\rangle_{\mathscr{H}}\langle f_i, K^B\rangle_{\mathscr{H}} \qquad \text{(Parseval)}\\ &= \sum_{i=1}^{\infty}|\mu_{f_i}(B)|^2 = \sum_{i=1}^{\infty}\left|\int_B f_i d\lambda\right|^2.\end{aligned}$$
□

DEFINITION 2.12. Let $(X, \mathscr{B})$ be a measure space as above, and let $\lambda$ be a positive measure on $(X, \mathscr{B})$. Let $X \ni x \longmapsto K(x, A)$, $A \in \mathscr{B}$, be a *random signed measure.* We say that $(\lambda, K)$ is *positive definite* (p.d.) iff, for $\forall k$, $\forall\{A_i\}_{i=1}^k$, $A_i \in \mathscr{B}_{fin}$, $\{\xi_i\}_{i=1}^k$, $\xi_i \in \mathbb{C}$, we have:

$$\sum_i\sum_j \overline{\xi}_i\xi_j\int_{A_i} K(x, A_j)\, d\lambda(x) \geq 0. \tag{2.17}$$

When (2.17) is satisfied for a fixed pair $(\lambda, K)$, let $\mathscr{H}$ be the corresponding Hilbert space completion. In this case, of course, $\lambda \in \mathscr{L}_+(\mathscr{H})$, see Definition 5.1.

THEOREM 2.13. *Let $(\mathscr{H}, \lambda)$ be as above, i.e., positive definite (p.d.).*

(i) *For all* $f \in \mathscr{H}$, *set* $\mu_f(B) := \langle f(\cdot), K(\cdot, B)\rangle_{\mathscr{H}}$, $\forall B \in \mathscr{B}_{fin}$, *i.e.,* $\mu_f$ *is a signed measure for* $\forall f \in \mathscr{H}$. *Then* $\mu_f \ll \lambda$, $f \in L^2_{loc}(\lambda)$, *and*

$$\frac{d\mu_f}{d\lambda} = f \tag{2.18}$$

*defined in* $L^1_{loc}(\lambda)$, *as a Radon-Nikodym-derivative.*

(ii) *Let* $\rho_{\mathscr{H}}$ *be as in* (2.8), *i.e., p.d. on* $\mathscr{B}_{fin} \times \mathscr{B}_{fin}$, *and let* $\mathscr{H}^{(Aronsz)}(\rho_{\mathscr{H}})$ *be the corresponding RKHS. Then, for all* $f \in \mathscr{H}$, *we have* $\mu_f \in \mathscr{H}^{(Aronsz)}(\rho_{\mathscr{H}})$, *with*

$$\|\mu_f\|_{\mathscr{H}^{(Aronsz)}(\rho_{\mathscr{H}})} = \|f\|_{\mathscr{H}}, \; \forall f \in \mathscr{H}. \tag{2.19}$$

*More precisely, the functions in* $\mathscr{H}^{(Aronsz)}(\rho_{\mathscr{H}})$ *are signed measures on* $(X, \mathscr{B})$; *and the map*

$$\mathscr{H} \ni f \xrightarrow[\simeq]{T} \mu_f = f d\lambda \in \mathscr{H}^{(Aronsz)}(\rho_{\mathscr{H}}), \tag{2.20}$$

*is an isometric isomorphism from* $\mathscr{H}$ *onto* $\mathscr{H}^{(Aronsz)}(\rho_{\mathscr{H}})$.

PROOF. For part (i), see the proof of Lemma 2.8.
To prove part (ii), we will need the following: □

LEMMA 2.14. *Set* $Tf = \mu_f$, *for all* $f \in \mathscr{H}$, *where* $\mu_f$ *is the signed measure on* $(X, \mathscr{B})$, *given by* $\mu_f(A) = \langle f, K(\cdot, A)\rangle_{\mathscr{H}}$ $\forall A \in \mathscr{B}_{fin}$; *then*

$$\mu_f \in \mathscr{H}^{(Aronsz)}(\rho_{\mathscr{H}}), \; \forall f \in \mathscr{H}. \tag{2.21}$$

PROOF. Given $f \in \mathscr{H}$, we must prove that: $\forall p \in \mathbb{N}$, $\forall \xi_i \in \mathbb{R}$, and $\forall A_i \in \mathscr{B}_{fin}$, $1 \le i \le p$, there is a constant $C_f < \infty$ s.t.

$$\left|\sum\nolimits_i \xi_i \mu_f(A_i)\right|^2 \le C_f \sum\nolimits_i \sum\nolimits_j \xi_i \xi_j \rho_{\mathscr{H}}(A_i, A_j), \tag{2.22}$$

where $\rho_{\mathscr{H}}(A_i, A_j) = \int_{A_i} K(x, A_j)\, d\lambda(x)$, $\forall i, j = 1, \cdots, p$.

Proof of (2.22). Set $\psi(x) = \sum_{i=1}^p \xi_i \chi_{A_i}(x)$, then

$$\begin{aligned}
\text{LHS}_{(2.22)} &= \left|\left\langle f(\cdot), \int \psi(y) K(\cdot, dy)\right\rangle_{\mathscr{H}}\right|^2 \\
&\le \|f\|^2_{\mathscr{H}} \left\|\int \psi(y) K(\cdot, dy)\right\|^2_{\mathscr{H}} \quad \text{(by Cauchy-Schwarz)} \\
&= \|f\|^2_{\mathscr{H}} \left|\sum\nolimits_i \sum\nolimits_j \xi_i \xi_j \int_{A_i} K(x, A_j)\, d\lambda(x)\right| \\
&= \|f\|^2_{\mathscr{H}} \sum\nolimits_i \sum\nolimits_j \xi_i \xi_j \rho_{\mathscr{H}}(A_i, A_j) \\
&= \|f\|^2_{\mathscr{H}} \|\psi\|^2_{\mathscr{H}^{(Aronsz)}} = \text{RHS}_{(2.22)}.
\end{aligned}$$

□

LEMMA 2.15. *Let* $T : \mathscr{H} \to \mathscr{H}^{(Aronsz)}(\rho_{\mathscr{H}})$ *be as in* (2.20). *Then*

$$T^*(\rho_{\mathscr{H}}(A, \cdot)) = K(\cdot, A), \; \forall A \in \mathscr{B}_{fin}.$$

PROOF. Let $f \in \mathscr{H}$, and $\rho_{\mathscr{H}}(A, \cdot) \in \mathscr{H}^{(Aronsz)}(\rho_{\mathscr{H}})$, $A \in \mathscr{B}_{fin}$; then

$$\langle Tf, \rho_{\mathscr{H}}(A, \cdot)\rangle_{\mathscr{H}^{(Aronsz)}} = \mu_f(A) = \langle f(\cdot), K(\cdot, A)\rangle_{\mathscr{H}}$$

so that the conclusion follows. □

LEMMA 2.16. *Let* $(\mathscr{H}, \lambda)$ *be as above, then span* $\{K(\cdot, A)\}_{A \in \mathscr{B}_{fin}}$ *is dense in* $\mathscr{H}$.

PROOF. To see this, let $g \in \mathscr{H}$ and assume that $\langle g(\cdot), K(\cdot, A)\rangle_{\mathscr{B}_{fin}} = 0$, $\forall A \in \mathscr{B}_{fin}$. Then $\langle g, K^{\varphi}\rangle_{\mathscr{H}} = 0$, $\forall \varphi \in C_c(X)$, where $K^{\varphi}(\cdot) = \int \varphi(y) K(\cdot, dy)$. It follows that

$$\int \varphi d\mu_g = \int \varphi g d\lambda = 0$$

which implies $gd\lambda = 0$, so that $g = 0$ a.e.-$\lambda$; thus $g = 0$. □

PROOF OF THEOREM 2.13, CONTINUED. It remains to prove that $T$ is an isometry, i.e., $\|\mu_f\|_{\mathscr{H}(Aronsz)} = \|f\|_{\mathscr{H}}$.

Note that every $f \in \mathscr{H}$ is $\|\cdot\|_{\mathscr{H}}$-limit of finite sums

$$\sum \xi_i K(\cdot, A_i), \; \xi_i \in \mathbb{R}, A_i \in \mathscr{B}_{fin}.$$

From the definition, $Tf = \mu_f$ is the $\|\cdot\|_{\mathscr{H}(Aronsz)}$-limit of finite sums

$$\sum \xi_i \rho_{\mathscr{H}}(\cdot, A_i), \; \xi_i \in \mathbb{R}, \; A_i \in \mathscr{B}_{fin}.$$

Now we have

$$\begin{aligned} \left\|\sum \xi_i K(\cdot, A_i)\right\|_{\mathscr{H}}^2 &= \sum_i \sum_j \xi_i \xi_j \langle K(\cdot, A_i), K(\cdot, A_j)\rangle \\ &= \sum_i \sum_j \xi_i \xi_j \rho_{\mathscr{H}}(A_i, A_j) \\ &= \left\|\sum \xi_i \rho_{\mathscr{H}}(\cdot, A_i)\right\|_{\mathscr{H}(Aronsz)}^2; \end{aligned}$$

which yields the desired conclusion. □

## 3. Applications to Gaussian Processes

The purpose of this section will be the link between the earlier results and the associated classes of Gaussian processes.

Given $(\mathscr{H}, \lambda)$, positive definite (see Definition 2.2), there is a centered Gaussian process $\{W_A\}_{A \in \mathscr{B}_{fin}}$, realized on a probability space $L^2(\Omega, \mathbb{P})$. The Gaussian process is determined by

$$\mathbb{E}[W_A] = 0, \; W_A \in L^2(\Omega, \mathbb{P}), \text{ and} \tag{3.1}$$

$$\mathbb{E}[W_A W_B] = \rho_{\mathscr{H}}(A, B) = \langle K^A, K^B\rangle_{\mathscr{H}}, \tag{3.2}$$

$\forall A, B \in \mathscr{B}_{fin}$. Note that

$$\mathbb{E}[\exp(iW_A)] = \exp\left(-\frac{1}{2}\rho_{\mathscr{H}}(A, A)\right), \; A \in \mathscr{B}_{fin}, \tag{3.3}$$

determines the process.

For all measurable functions $\varphi$ on $(X, \mathscr{B})$, we may define the corresponding Ito-integral $\int \varphi dW$; details below.

THEOREM 3.1. *If* $\{f_i\}_{i\in\mathbb{N}}$ *is an ONB in* $\mathscr{H}$, *and* $Z_i(\cdot) \sim N(0,1)$, *i.i.d., then*

$$W_A = \sum_{i=1}^{\infty} \left( \int_A f_i d\lambda \right) Z_i, \text{ and} \tag{3.4}$$

$$\int \varphi dW = \sum_{i=1}^{\infty} \left( \int \varphi f_i d\lambda \right) Z_i. \tag{3.5}$$

PROOF. This is an immediate application of the new RKHS $\mathscr{H} = \mathscr{H}^{(Aronsz)}$. Apply Kolmogorov's extension principle to $\rho_{\mathscr{H}}$, we get a Gaussian process $\{W_A\}_{A\in\mathscr{B}_{fin}}$ satisfying conditions (3.1)–(3.2), so that

$$\mathscr{H} \ni K^A = K(\cdot, A) \longmapsto W_A \in L^2(\Omega, \mathbb{P}) \tag{3.6}$$

is isometric from $\mathscr{H}$ into $L^2(\Omega,\mathbb{P})$, where $\|K^A\|_{\mathscr{H}}^2 = \mathbb{E}[|W_A|^2] = \rho_{\mathscr{H}}(A,A)$, $\forall A \in \mathscr{B}_{fin}$.

We introduce

$$K^{\varphi} = \int \varphi(y) K(\cdot, dy) \, (\in \mathscr{H})$$

as a dense space of vectors in $\mathscr{H}$; and then the new Ito isometry may be realized as

$$K^{\varphi} \longmapsto \int \varphi dW, \quad \|K^{\varphi}\|_{\mathscr{H}}^2 = \mathbb{E}\left[ \left| \int \varphi dW \right|^2 \right]. \tag{3.7}$$

Since $L^2(\Omega,\mathbb{P})$ is complete, we conclude that (3.7) extends to $\mathscr{H} \to L^2(\Omega,\mathbb{P})$ as an isometry. Hence $\mathscr{H}$ is realized as a closed subspace in $L^2(\Omega,\mathbb{P})$.

Now (3.4)–(3.5) follow from Corollary 2.11.

Details: Given $\epsilon > 0$, pick a *measurable partition* of $X$, i.e., $\{A_i\} \subset \mathscr{B}_{fin}$, $A_i \cap A_j = \emptyset$ if $i \neq j$, and $\cup A_i = X$, so that

$$\left\| K^{\varphi} - \sum\nolimits_i \varphi(y_i) K(\cdot, A_i) \right\|_{\mathscr{H}} < \epsilon, \; y_i \in A_i.$$

With a net $\alpha$ of partitions, $\max \lambda(A_i) \xrightarrow{\alpha} 0$, we get as an application of our $(\mathscr{H},\lambda)$ analysis:

$$\left\| \sum\nolimits_i \varphi(x_i) K(\cdot, A_i) - \sum\nolimits_j \varphi(y_j) K(\cdot, B_j) \right\|_{\mathscr{H}} < \epsilon,$$

for all $\alpha = (A_i)$, $\beta = (B_i)$. By (3.6), and Kolmogorov, we then get

$$\mathbb{E}\left[ \left| \sum\nolimits_i \varphi(x_i) W_{A_i} - \sum\nolimits_j \varphi(y_j) W_{B_j} \right|^2 \right] < \epsilon.$$

Hence the induced net $N_\alpha^\varphi = \sum_i \varphi(x_i) W_{A_i}$ is convergent in $L^2(\Omega,\mathbb{P})$, and the limit

$$\int \varphi dW = \lim_{\alpha} \sum_i \varphi(x_i) W_{A_i}$$

is the generalized Ito-integral; with this approach it is clear that (3.7) then yields

$$\begin{aligned} \mathbb{E}\left[ \left| \int \varphi dW \right|^2 \right] &= \|K^{\varphi}\|_{\mathscr{H}}^2 = \int \varphi(x) \int \varphi(y) K(x, dy) \, d\lambda(x) \\ &= \int \varphi(x) \int \varphi(y) \, \mathbb{E}[W_x dW_y]. \end{aligned}$$

∎

COROLLARY 3.2. *Let* $(X, \mathscr{B}, \mathscr{H}, \lambda)$ *be as above. Let* $\{W_A\}_{A\in\mathscr{B}_{fin}}$ *be the Gaussian process with* $\mathbb{E}[W_A] = 0$, $\mathbb{E}[W_A W_B] = \langle K^A, K^B \rangle_{\mathscr{H}} = \rho_{\mathscr{H}}(A, B)$. *Then we get an Ito-integral* $\int \varphi dW$ *s.t.*

$$\mathbb{E}\left[\left|\int \varphi dW\right|^2\right] = \|K^\varphi\|^2_{\mathscr{H}} = \varphi K_\lambda \varphi. \tag{3.8}$$

*More precisely,*

$$\begin{aligned} \varphi K_\lambda \varphi &= \int \varphi(x) \int \varphi(y) K(x, dy)\, d\lambda(x) \\ &= \int \varphi(x) \int \varphi(y)\, \mathbb{E}[W_x dW_y]. \end{aligned} \tag{3.9}$$

THEOREM 3.3. *Let* $(X, \mathscr{B}, \mathscr{H}, \lambda)$ *be as above, and let* $\{W_A\}_{A\in\mathscr{B}_{fin}}$ *be the corresponding Gaussian process. Denote by* $\int \varphi dW$ *the Ito-integral constructed from Theorem* 3.1. *Then*

$$\mathbb{E}\left[e^{i\int \varphi dW}\right] = e^{-\frac{1}{2}\varphi K_\lambda \varphi} \tag{3.10}$$

*where* $\varphi$ *is a measurable function on* $(X, \mathscr{B})$ *s.t.* $\varphi K_\lambda \varphi < \infty$, *see* (3.8).

PROOF. Let $A \in \mathscr{B}_{fin}$, then

$$\begin{aligned} \mathbb{E}\left[W_A^{2n+1}\right] &= 0, \\ \mathbb{E}\left[W_A^{2n}\right] &= (2n-1)!! \left(\int_A K(x, A)\, d\lambda(x)\right)^n, \end{aligned}$$

where

$$(2n-1)!! = \frac{(2n)!}{2^n n!} = (2n-1)(2n-3)\cdots 5 \cdot 3 \cdot 1.$$

It follows that

$$\mathbb{E}\left[\left|\int \varphi dW\right|^{2n}\right] = (2n-1)!! \left(\varphi K_\lambda \varphi\right)^n.$$

Now

$$\begin{aligned} \mathbb{E}\left[e^{i\int \varphi dW}\right] &= \sum_{n=0}^{\infty} \frac{(-1)^n}{(2n)!} \mathbb{E}\left[\left|\int \varphi dW\right|^{2n}\right] \\ &= \sum_{n=0}^{\infty} \frac{(-1)^n}{n! 2^n} \left(\varphi K_\lambda \varphi\right)^n = e^{-\frac{1}{2}\varphi K_\lambda \varphi} \end{aligned}$$

which is the desired conclusion. □

REMARK. We get the results, even when there is *not* an obvious Gelfand triple in the framework of (3.10).

THEOREM 3.4. *Let* $(X, \mathscr{B}, \mathscr{H}, \lambda)$ *be as usual and let* $\{W_A\}_{A\in\mathscr{B}_{fin}}$ *be the corresponding Gaussian process realized on* $L^2(\Omega, \mathbb{P})$. *Set* $K^\varphi(\cdot) = \int \varphi(y) K(\cdot, dy)$. *Let* $\{f_n\}_{n\in\mathbb{N}}$ *be an ONB in* $\mathscr{H}$. *Then TFAE:*

(i) $\int \varphi dW \in L^2(\Omega, \mathbb{P})$;
(ii) $K^\varphi \in \mathscr{H}$;

(iii) $\sum_{n=1}^{\infty} \left|\int \varphi f_n d\lambda\right|^2 < \infty$, *and then*

$$\mathbb{E}\left[e^{i\int \varphi dW}\right] = e^{-\frac{1}{2}\|K^{\varphi}\|^2_{\mathscr{H}}}. \tag{3.11}$$

*If any one of the conditions (i)-(iii) holds, then*

$$\mathbb{E}\left[\left|\int \varphi dW\right|^2\right] = \|K^{\varphi}\|^2_{\mathscr{H}} = \varphi K_{\lambda}\varphi;$$

*and if* $(Z_n(\cdot))$ *is a system of i.i.d* $N(0,1)$, *we then have*

$$\int \varphi dW = \sum_{n=1}^{\infty} \left(\int \varphi f_n d\lambda\right) Z_n. \tag{3.12}$$

PROOF. See Theorems 3.1, 3.3, and Corollary 3.2 for details. □

EXAMPLE 3.5. (i) Let $\lambda$ be a positive measure on $(X, \mathscr{B})$, and consider the function $\rho_\lambda$ on $\mathscr{B}_{fin}$ (see (2.4) and Section 2) defined by

$$\rho_\lambda(A,B) = \lambda(A\cap B), \ \forall A, B \in \mathscr{B}_{fin}. \tag{3.13}$$

It is immediate that $\rho$ is positive definite.

We denote by $\mathscr{H}^{(Aronsz)}(\rho_\lambda)$ the associated RKHS in the sense of Aronszajn. The corresponding pair $(\mathscr{H}, \lambda)$ in the measurable category is $\mathscr{H} = L^2(X,\lambda)$, and the isometric isomorphism $\mathscr{H} \to \mathscr{H}^{(Aronsz)}(\rho_\lambda)$, $f \to \mu_f$, with $\|f\|_{L^2(\lambda)} = \|\mu_f\|_{\mathscr{H}^{(Aronsz)}(\rho_\lambda)}$ is simply $f \to f d\lambda$.

(ii) In the special case where $X = [0,1]$ = the unit interval, and $\lambda = dx$ = the restriction of Lebesgue measure to $[0,1]$, then the RKHS $\mathscr{H}^{(Aronsz)}(\rho_\lambda)$ is the usual Cameron-Martin Hilbert space (see [**Hid80, AJV14, JT15**])

$$\left\{f \text{ measurable} \mid f' \in L^2(0,1),\ f(0)=0\right\},$$

and

$$\|f\|^2_{\mathscr{H}^{(Aronsz)}(\rho_\lambda)} = \int_0^1 |f'(x)|^2 dx.$$

Note that this $\mathscr{H}^{(Aronsz)}(\rho_\lambda)$ is the RKHS of Brownian motion.

COROLLARY 3.6. *Let* $(\Omega, \mathscr{F}, \mathbb{P})$ *be a probability space, and let* $(X, \mathscr{B})$ *be a measure space. Let* $\{W_A\}_{A\in\mathscr{B}_{fin}}$ *be a Gaussian process such that* $W_A \in L^2(\Omega, \mathscr{F}, \mathbb{P})$ *for* $\forall A \in \mathscr{B}_{fin}$, *and* $\mathbb{E}[W_A] = 0$. *For measurable functions* $\varphi$ *on* $(X,\mathscr{B})$, *let* $\int_X \varphi dW$ *be the corresponding Ito-integral, defined for measurable functions* $\varphi$ *such that* $\mathbb{E}\left[\left|\int \varphi dW\right|^2\right] < \infty$.

*Let* $\lambda$ *be a positive measure on* $(X,\mathscr{B})$. *Then there is a Hilbert space* $\mathscr{H} \subset L^2_{loc}(\lambda)$ *such that* $(\mathscr{H},\lambda)$ *is p.d. (see Definition 2.2), and*

$$\mathbb{E}[W_A W_B] = \rho_{\mathscr{H}}(A,B) = \left\langle K^A, K^B\right\rangle_{\mathscr{H}}, \ \forall A, B \in \mathscr{B}_{fin}, \tag{3.14}$$

*if and only if*

$$\mathbb{E}\left[W_A \int \varphi dW\right] = \int_A \varphi d\lambda, \ \forall A \in \mathscr{B}_{fin}, \ \forall \varphi \in C_c(X). \tag{3.15}$$

COROLLARY 3.7. *Let $(\mathscr{H},\lambda)$, $W$, and $(\Omega,\mathscr{F},\mathbb{P})$ be as above, i.e., $\{W_A\}_{A\in\mathscr{B}_{fin}}$ is a Gaussian process on $L^2(\Omega,\mathbb{P})$, indexed by $\mathscr{B}_{fin}$, such that $\mathbb{E}\left[W_A \int\varphi dW\right] = \int_A \varphi d\lambda$ holds for all $\varphi\in C_c(X)$, where $\int\varphi dW$ denotes the Ito-integral.*

*A function $F$ in $L^2(\Omega,\mathbb{P})$ has the form $F=\int\varphi dW$, for some $\varphi\in L^2_{loc}(\lambda)$ if and only if there is a finite constant $C=C_F$ such that, for $\forall p$, $\forall\{\xi_i\}_1^p$, $\xi_i\in\mathbb{R}$, $\forall\{A_i\}_1^p$, $A_i\in\mathscr{B}_{fin}$,*

$$\left|\sum_{i=1}^{p}\xi_i\mathbb{E}\left[W_{A_i}F\right]\right|^2 \le C\sum_i\sum_j \xi_i\xi_j\rho_{\mathscr{H}}(A_i,A_j). \tag{3.16}$$

## 4. Choice of Probability Space

In this section we return to the measures, including fractals, and make precise the way they link to associated Gaussian processes.

In Section 3, we make reference to an as of yet unspecified probability space $(\Omega,\mathscr{F},\mathbb{P})$, and a Gaussian process $\{W_B\}_{B\in\mathscr{B}_{fin}}$, indexed by a fixed measure space $(X,\mathscr{B})$. In the present section, we shall keep the above assumptions placed on $(X,\mathscr{B})$. The notation $\mathscr{B}_{fin}$ refers to a fixed positive measure $\lambda$ on $(X,\mathscr{B})$, so $\mathscr{B}_{fin} := \{B\in\mathscr{B}\mid\lambda(B)<\infty\}$.

We showed in Corollary 3.6, that when a Hilbert space $\mathscr{H}$ is given such that for all $B\in\mathscr{B}_{fin}$,

$$\sup_{f\in\mathscr{H},\,\|f\|_{\mathscr{H}}\le 1}\left|\int_B f d\lambda\right| < \infty, \tag{4.1}$$

i.e., with a finite constant in (4.1) depending on $\mathscr{B}$, we then get a RKHS in the measurable category, and $\mathscr{H}\subset L^2_{loc}(\lambda)$.

As an application of Kolmogorov's extension principle, we showed that a Gaussian process $W=W^{\mathscr{H}}$, indexed by $\mathscr{B}_{fin}$, exists, subject to the conditions from (3.1)–(3.2). However, in general, there are many choices of probability spaces $(\Omega,\mathscr{F},\mathbb{P})$ which realize this Gaussian process $W^{\mathscr{H}}$. To be more precise, the Gaussian process $W^{\mathscr{H}}$ satisfies the following conditions: Each random variable $W_A^{\mathscr{H}}$, $A\in\mathscr{B}_{fin}$ is Gaussian, with distribution $N\left(0,\|K^A\|^2_{\mathscr{H}}\right)$ where $K^A\in\mathscr{H}$ is as specified in Lemma 2.5, so $W_A^{\mathscr{H}}$ is Gaussian with mean 0, and variance $\|K^A\|^2_{\mathscr{H}}$. But it also follows that, for $\forall k\in\mathbb{N}$, $\forall\{A_i\}_{i=1}^k$ from $\mathscr{B}_{fin}$, the system

$$\left(W^{\mathscr{H}}_{A_1},W^{\mathscr{H}}_{A_2},\cdots,W^{\mathscr{H}}_{A_k}\right) \tag{4.2}$$

has a *joint* Gaussian distribution on $\mathbb{R}^k$ with mean 0, and with $k\times k$ covariance matrix, given by the entries

$$\mathbb{E}\left[W^{\mathscr{H}}_{A_i}W^{\mathscr{H}}_{A_j}\right] = \left\langle K_{A_i},K_{A_j}\right\rangle_{\mathscr{H}} = \rho_{\mathscr{H}}(A_i,A_j); \tag{4.3}$$

see Theorem 3.1 and Corollary 3.6.

Below we identify two choices of probability space $(\Omega,\mathscr{F},\mathbb{P})$ which serve to realize the process outlined in (4.2), and in Section 3 above.

CHOICE 1.

$$\Omega = \prod\nolimits_{\mathbb{N}} \mathbb{R} = \text{the infinite Cartesian product;} \tag{4.4}$$

$$\mathscr{F} = \text{the cylinder } \sigma\text{-algebra of subsets of } \Omega; \tag{4.5}$$

$$\mathbb{P} = \times_{\mathbb{N}} N(0,1), \tag{4.6}$$

i.e., $\mathbb{P}$ is the infinite product measure on $\Omega$ constructed from the fixed standard Gaussian

$$g(x) = \frac{1}{\sqrt{2\pi}} e^{-\frac{1}{2}x^2} \text{ on } \mathbb{R}, \tag{4.7}$$

and each coordinate a copy of $\mathbb{R}$.

In this realization, we must use the representation $\{W_B^{\mathscr{H}}\}_{B \in \mathscr{B}_{fin}}$ from Theorem 3.1. Indeed, it follows from (3.4)–(3.5), and the i.i.d. condition for $Z_1, Z_2, \cdots$, that $\{W_B^{\mathscr{H}}\}_{B \in \mathscr{B}_{fin}}$ will then satisfy the conditions outlined in (4.2)–(4.3). Note that we may use the same $\mathbb{P}$ (see (4.6)) for all the Gaussian processes $W^{\mathscr{H}}$, so for all possible Hilbert spaces $\mathscr{H}$ subject to (4.1).

CHOICE 2. This is the opposite extreme (of choice for the sample space $\Omega$); hence the probability measure $\mathbb{P}$ will depend on the choice of $\mathscr{H}$.

Details:

$$\Omega_{\mathfrak{M}} = \mathfrak{M}(X, \mathscr{B}) = \text{all signed measures on } (X, \mathscr{B}); \tag{4.8}$$

$$\mathscr{F}_{\mathfrak{M}} = \text{the cylinder } \sigma\text{-algebra of subsets of } \mathfrak{M}(X, \mathscr{B}); \tag{4.9}$$

$$\mathbb{P}^{(\mathscr{H})} = \text{a probability measure in } (\Omega_{\mathfrak{M}}, \mathscr{F}_{\mathfrak{M}}). \text{ (See (4.15) below.)} \tag{4.10}$$

We now proceed to describe $\mathbb{P}^{(\mathscr{H})}$. The construction of $\mathbb{P}^{(\mathscr{H})}$ will have the property that the process $\{W_A^{\mathscr{H}}\}_{A \in \mathscr{B}_{fin}}$ takes the form

$$W_A^{\mathscr{H}}(\mu) = \mu(A), \tag{4.11}$$

for $\forall A \in \mathscr{B}_{fin}$, and $\forall \mu \in \mathfrak{M}(X, \mathscr{B})$.

Now fix $(\mathscr{H}, \lambda)$, subject to condition (4.1). Then for each finite system

$$\alpha := (A_1, \cdots, A_k), \; A_i \in \mathscr{B}_{fin}, \tag{4.12}$$

we consider $\pi_\alpha : \mathfrak{M}(X, \mathscr{B}) \longrightarrow \mathbb{R}^k$,

$$\mathfrak{M}(X, \mathscr{B}) \ni \mu \xmapsto{\pi_\alpha} (\mu(A_1), \mu(A_1), \cdots, \mu(A_k)) \in \mathbb{R}^k. \tag{4.13}$$

Equip $\mathbb{R}^k$ with the joint Gaussian $g_\alpha$ with mean 0, and covariance matrix

$$[\rho_{\mathscr{H}}(A_i, A_j)]_{i,j=1}^k,$$

and we set $\mathbb{P}_\alpha$ the unique probability measure on $(\Omega_{\mathfrak{M}}, \mathscr{F}_{\mathfrak{M}})$ determined by

$$\mathbb{P}_\alpha \circ \pi_\alpha^{-1} = g_\alpha. \tag{4.14}$$

Using standard Kolmogorov extension theory, we note that the system of measures $\{P_\alpha\}$ is *consistent*, and that therefore, there is a unique probability measure $\mathbb{P}^{(\mathscr{H})}$ on $(\Omega_{\mathfrak{M}}, \mathscr{F}_{\mathfrak{M}})$ such that

$$\mathbb{P}^{(\mathscr{H})} \circ \pi_\alpha^{-1} = g_\alpha \tag{4.15}$$

for all $\alpha$ as in (4.12) above.

With this, it is then clear that (4.11) defines the desired Gaussian process realized in $(\Omega_{\mathfrak{M}}, \mathscr{F}_{\mathfrak{M}}, \mathbb{P}^{(\mathscr{H})})$ as per construction; see also (4.8)–(4.10).

**Equivalence of $\mathbb{P}^{(\mathscr{H}_1)}$ and $\mathbb{P}^{(\mathscr{H}_2)}$.** Let $(X, \mathscr{B})$ be a fixed measure space specified as above; and let $\lambda$ be a positive measure on $(X, \mathscr{B})$. Let $\mathscr{H}_i$, $i = 1, 2$, be two Hilbert spaces, $\mathscr{H}_i \subset L^2_{loc}(\lambda)$ such that both pairs $(\mathscr{H}_i, \lambda)$, $i = 1, 2$, are p.d.; in particular, $\mathscr{H}_i \in \mathscr{K}(\lambda)$, $i = 1, 2$.

Let $\mathbb{P}^{(\mathscr{H}_i)}$, $i = 1, 2$, be the corresponding probability measures on $(\Omega, \mathscr{F}) = (\Omega_{\mathfrak{M}}, \mathscr{F}_{\mathfrak{M}})$, as specified in (4.10). Recall, setting

$$W_B(\mu) := \mu(B), \; B \in \mathscr{B}_{fin}, \; \mu \in \mathfrak{M}(X, \mathscr{B}), \tag{4.16}$$

we get that $\{W_B\}_{B \in \mathscr{B}_{fin}}$ is a Gaussian process; but it has different distributions relative to the two probability spaces $\left(\Omega_{\mathfrak{M}}, \mathscr{F}_{\mathfrak{M}}, \mathbb{P}^{(\mathscr{H}_i)}\right)$, $i = 1, 2$; in particular,

$$\mathbb{E}_{\mathbb{P}^{(\mathscr{H}_i)}}[W_A W_B] = \rho_{\mathscr{H}_i}(A, B), \; i = 1, 2; \tag{4.17}$$

see (3.14) in Corollary 3.6.

The following result follows from our present Section 3, and a result of [**Jr68**].

COROLLARY 4.1. *Let $X, \mathscr{B}, \lambda, \mathscr{H}_i, \mathbb{P}^{(\mathscr{H}_i)}$, $i = 1, 2$, be as specified above. Then the two probability measures $\mathbb{P}^{(\mathscr{H}_i)}$ are equivalent if and only if there is an ONB $\{f_n\}_{n \in \mathbb{N}}$ for $\mathscr{H}_2$; see Corollary 2.11, and scalars $c_n > 0$ with $\sum_{n=1}^{\infty}(1 - c_n)^2 < \infty$, and*

$$K^{(\mathscr{H}_1)}(x, A) = \sum_{n=1}^{\infty} c_n \left(\int_A f_n d\lambda\right) f_n(x), \tag{4.18}$$

*for a.a. $x$ (w.r.t $\lambda$), $A \in \mathscr{B}_{fin}$.*

## 5. A Duality

In this final section we outline how the correspondences studied above take the form of a new duality principle.

Fix a measurable space $(X, \mathscr{B})$ as in Definition 2.1, where $X$ is locally compact and Hausdorff, and $\mathscr{B}$ is a Borel $\sigma$-algebra of subsets of $X$. We shall consider the following duality:

DEFINITION 5.1. Fix a positive measure $\lambda$ on $(X, \mathscr{B})$, and set

$$\begin{aligned} \mathscr{K}(\lambda) :&= \Bigg\{ \text{Hilbert space } \mathscr{H} \text{ on } (X, \mathscr{B}) \mid \text{For } \forall \varphi \in C_c(X) \\ & \sup\left\{ \left| \int \varphi f d\lambda \right|, \; \|f\|_{\mathscr{H}} \leq 1 \right\} < \infty \Bigg\}. \end{aligned} \tag{5.1}$$

If $\mathscr{H}$ is a Hilbert space of measurable functions on $(X, \mathscr{B})$, set

$$\mathscr{L}(\mathscr{H}) := \left\{ \text{signed measures } \lambda \mid \sup\left\{ \left| \int \varphi f d\lambda \right|, \; \|f\|_{\mathscr{H}} \leq 1 \right\} < \infty \right\}, \tag{5.2}$$

and

$$\mathscr{L}_+(\mathscr{H}) = \{\lambda \in \mathscr{L}(\mathscr{H}) \mid \lambda \text{ is positive}\}. \tag{5.3}$$

LEMMA 5.2. *The following are equivalent:*

$$\sup\left\{\left|\int_A f d\lambda\right|,\ f\in\mathscr{H},\ \|f\|_{\mathscr{H}}\leq 1\right\}<\infty,\ \forall A\in\mathscr{B}_{fin}, \tag{5.4}$$

$$\Updownarrow$$

$$\sup\left\{\left|\int \varphi f d\lambda\right|,\ f\in\mathscr{H},\ \|f\|_{\mathscr{H}}\leq 1\right\}<\infty,\ \forall \varphi\in C_c\left(X\right). \tag{5.5}$$

PROOF. By Riesz' theorem, (5.4) $\Longleftrightarrow$ $\forall A\in\mathscr{B}_{fin}$ $\exists!$ $K^A\in\mathscr{H}$ s.t.

$$\int_A f d\lambda = \left\langle f, K^A\right\rangle_{\mathscr{H}}; \tag{5.6}$$

and (5.5) $\Longleftrightarrow$ $\forall\varphi\in C_c\left(X\right)$ $\exists! K^{\varphi}$ s.t.

$$\int \varphi f d\lambda = \left\langle f, K^{\varphi}\right\rangle_{\mathscr{H}}. \tag{5.7}$$

To show that (5.4) $\Longrightarrow$ (5.5): The RHS of (5.6) is $\sigma$-additive, so it is a signed measure on $(X,\mathscr{B})$, $\mu_f\left(A\right) := \left\langle f, K^A\right\rangle_{\mathscr{H}}$. We see that $\mu_f \ll \lambda$, $f\in L^1_{loc}\left(\lambda\right)$, $d\mu_f = f d\lambda$, and from (5.6),

$$\begin{aligned}\int f\varphi d\lambda &= \int \varphi d\mu_f\\ &= \int \varphi\left(y\right)\left\langle f\left(\cdot\right), K\left(\cdot, dy\right)\right\rangle_{\mathscr{H}}\\ &= \left\langle f\left(\cdot\right), \int\varphi\left(y\right)K\left(\cdot,dy\right)\right\rangle_{\mathscr{H}},\quad \forall\varphi\in C_c\left(X\right);\end{aligned}$$

so that $K^{\varphi}\left(\cdot\right) = \int\varphi\left(y\right)K\left(\cdot,dy\right)\in\mathscr{H}$, we see that (5.7) holds. It is clear that $K^{\varphi}$ in (5.7) is unique.

The converse direction (5.5) $\Longrightarrow$ (5.4) is clear. $\square$

DEFINITION 5.3 (Multipliers). Let $(X,\mathscr{B})$ be as above. Fix $(\mathscr{H},\lambda)$ such that $\lambda\in\mathscr{L}\left(\mathscr{H}\right)$. We then have $\mathscr{H}\subset L^2_{loc}\left(\lambda\right)$. A measurable function $g$ on $(X,\mathscr{B})$ is said to be a *multiplier* if the operator $M_g f = gf$, $f\in\mathscr{H}$, is bounded $\mathscr{H}\to\mathscr{H}$. The corresponding adjoint operator will be denoted $M_g^*$. The set of all multipliers will be denoted $Multp\left(\mathscr{H}\right)$.

THEOREM 5.4. *Let $(X,\mathscr{B},\mathscr{H},\lambda)$ be as above, and let $g$ be a measurable function on $(X,\mathscr{B})$, then TFAE:*

(i) *$g\in Multp\left(\mathscr{H}\right)$, and*
(ii) *$g\,d\lambda\in\mathscr{L}\left(\mathscr{H}\right)$; see Definition 5.1.*

PROOF. If $g\in Multp\left(\mathscr{H}\right)$, then for all $A\in\mathscr{B}_{fin}$, and all $f\in\mathscr{H}$, we have

$$\begin{aligned}\int_A fg d\lambda &= \left\langle gf, K\left(\cdot, A\right)\right\rangle_{\mathscr{H}}\\ &= \left\langle f, M_g^*\left(K\left(\cdot,A\right)\right)\right\rangle_{\mathscr{H}}.\end{aligned} \tag{5.8}$$

But $M_g^*\left(K\left(\cdot,A\right)\right)\in\mathscr{H}$ by assumption, and so the condition for $gd\lambda\in\mathscr{L}\left(\mathscr{H}\right)$ is satisfied.

Using (5.8), the reader will note that the argument works in reverse, so we also have (ii) $\Longrightarrow$(i). $\square$

COROLLARY 5.5. *Let* $(X, \mathscr{B}, \mathscr{H}, \lambda)$ *be as above, and suppose* $g \in Multp(\mathscr{H})$, *then*

$$K_{gd\lambda}(\cdot, A) = M_g^*(K_\lambda(\cdot, A)), \ \forall A \in \mathscr{B}_{fin}.$$

Given $(X, \mathscr{B}, \lambda)$, let $\mathscr{H} \in \mathscr{K}(\lambda)$. Then we may define a random signed measure $K(\cdot, B)$, for all $B \in \mathscr{B}_{fin}$, see Lemma 2.5.

There are two associated RKHSs:

- $\mathscr{H}$: Hilbert space of measurable functions on $(X, \mathscr{B})$;
- $\mathscr{H}^{(Aronsz)}$: The RKHS determined by the positive definite kernel $\rho_{\mathscr{H}}$ on $\mathscr{B}_{fin}$ (see (2.8).) , where

$$\rho_{\mathscr{H}}(A, B) = \left\langle K^A, K^B \right\rangle_{\mathscr{H}}, \ \forall A, B \in \mathscr{B}_{fin}.$$

  Note that $\mathscr{H}^{(Aronsz)}$ consists of functions on $\mathscr{B}_{fin}$, and it depends on both $\mathscr{H}$ and $\lambda$.

There are many solutions to $\mathscr{H} \in \mathscr{K}(\lambda)$, i.e., to the Hilbert factorization problem (see (5.1)), but they are all isomorphic to an associated Aronszajn RKHS. Thus, the two RKHSs are *isometrically isomorphic* via

$$\mathscr{H} \xrightarrow{\simeq} \mathscr{H}^{(Aronsz)}, \quad f \longmapsto \mu_f. \tag{5.9}$$

See Figure 5.1 below.

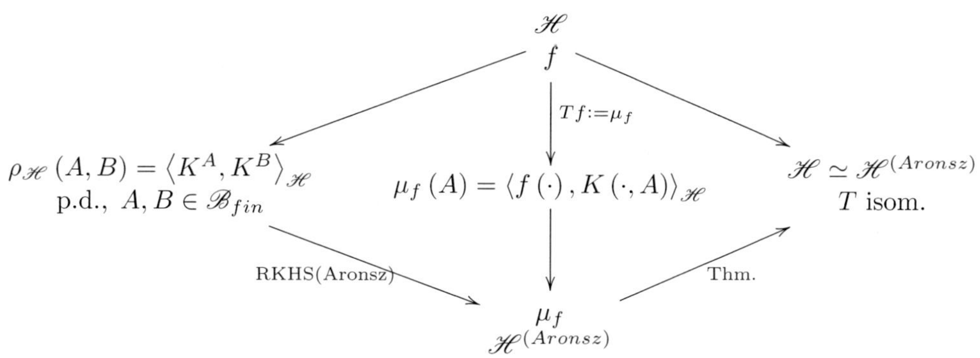


FIGURE 5.1. The RKHSs $\mathscr{H} \simeq \mathscr{H}^{(Aronsz)}$, $\mathscr{H} \ni f \longmapsto \mu_f \in \mathscr{H}^{(Aronsz)}$.

The converse problem:

Given $(X, \mathscr{B})$ and a positive definite (p.d.) function $\rho : \mathscr{B} \times \mathscr{B} \longrightarrow \mathbb{R}$ (or $\mathbb{C}$), i.e.,

$$\sum_i \sum_j \xi_i \xi_j \rho(A_i, A_j) \geq 0 \tag{5.10}$$

holds for $\forall p \in \mathbb{N}$, $\forall \xi_i \in \mathbb{R}$, $\forall A_i \in \mathscr{B}$, $1 \leq i \leq p$, then we get an Aronszajn RKHS $\mathscr{H}(\rho)$, consisting of functions on $\mathscr{B}$.

*Not* every p.d. function on $\mathscr{B}$ corresponds to a pair $(\mathscr{H}, \lambda)$. Given $\rho : \mathscr{B} \times \mathscr{B} \longrightarrow \mathbb{R}$ p.d., we place two additional conditions on it:

(i) $\forall A \in \mathscr{B}$, $\rho(\cdot, A)$ is a signed measure on $(X, \mathscr{B})$.

(ii) There exists a positive measure $\lambda$ on $(X, \mathscr{B})$ s.t. $\rho(\cdot, A) \ll \lambda$ for all $A \in \mathscr{B}$, and denote the Radon-Nikodym derivative

$$K(\cdot, A) = \frac{d\rho(\cdot, A)}{d\lambda} \tag{5.11}$$

or equivalently,

$$d\rho(x, A) = K(x, A)\, d\lambda(x). \tag{5.12}$$

Then for $A, B \in \mathscr{B}_{fin}$ we have

$$\rho_{\mathscr{H}}(A, B) = \int_A K(y, B)\, d\lambda(y) = \int_B K(x, A)\, d\lambda(x).$$

This is a p.d. function on $\mathscr{B}_{fin} \times \mathscr{B}_{fin}$.

We get $\mathscr{H}$ from $(K, \lambda)$ as defined in (5.1); and the isomorphism $\mathscr{H} \longrightarrow \mathscr{H}^{(Aronsz)}(\rho_{\mathscr{H}})$, $f \longmapsto \mu_f$.

THEOREM 5.6. *$\lambda \in \mathscr{L}(\mathscr{K}(\lambda))$, and $\mathscr{H} \in \mathscr{K}(\mathscr{L}(\mathscr{H}))$.*

PROOF. See Section 2, especially Lemmas 2.5, 2.8, and Theorem 2.13. □

## References


[AJV14] D. Alpay, P. Jorgensen, and D. Volok, *Relative reproducing kernel Hilbert spaces*, Proc. Amer. Math. Soc. **142** (2014), no. 11, 3889–3895, DOI 10.1090/S0002-9939-2014-12121-6. MR3251728

[Aro43] N. Aronszajn, *La théorie des noyaux reproduisants et ses applications. I* (French), Proc. Cambridge Philos. Soc. **39** (1943), 133–153. MR8639

[Aro50] N. Aronszajn, *Theory of reproducing kernels*, Trans. Amer. Math. Soc. **68** (1950), 337–404, DOI 10.2307/1990404. MR51437

[BH14] S. Bezuglyi and D. Handelman, *Measures on Cantor sets: the good, the ugly, the bad*, Trans. Amer. Math. Soc. **366** (2014), no. 12, 6247–6311, DOI 10.1090/S0002-9947-2014-06035-2. MR3267010

[CMPS17] R. X. Cortes, T. G. Martins, M. O. Prates, and B. A. Silva, *Inference on dynamic models for non-Gaussian random fields using INLA*, Braz. J. Probab. Stat. **31** (2017), no. 1, 1–23, DOI 10.1214/15-BJPS300. MR3601658

[EMESO17] M. El Machkouri, K. Es-Sebaiy, and I. Ouassou, *On local linear regression for strongly mixing random fields*, J. Multivariate Anal. **156** (2017), 103–115, DOI 10.1016/j.jmva.2017.02.002. MR3624688

[HE15] T. Hsing and R. Eubank, *Theoretical foundations of functional data analysis, with an introduction to linear operators*, Wiley Series in Probability and Statistics, John Wiley & Sons, Ltd., Chichester, 2015, DOI 10.1002/9781118762547. MR3379106

[Hid80] T. Hida, *Brownian motion*, Applications of Mathematics, vol. 11, Springer-Verlag, New York-Berlin, 1980. Translated from the Japanese by the author and T. P. Speed. MR562914

[JPT15] P. E. T. Jorgensen, S. Pedersen, and F. Tian, *Harmonic analysis of a class of reproducing kernel Hilbert spaces arising from groups*, Trends in harmonic analysis and its applications, Contemp. Math., vol. 650, Amer. Math. Soc., Providence, RI, 2015, pp. 157–197, DOI 10.1090/conm/650/13009. MR3441738

[JPT16] P. Jorgensen, S. Pedersen, and F. Tian, *Extensions of positive definite functions*, Lecture Notes in Mathematics, vol. 2160, Springer, [Cham], 2016. Applications and their harmonic analysis, DOI 10.1007/978-3-319-39780-1. MR3559001

[Jr68] O. G. Jørsboe, *Equivalence or singularity of Gaussian measures on function spaces*, Various Publications Series, No. 4, Aarhus Universitet, Matematisk Institut, Aarhus, 1968. MR277027

[JT15] P. Jorgensen and F. Tian, *Discrete reproducing kernel Hilbert spaces: sampling and distribution of Dirac-masses*, J. Mach. Learn. Res. **16** (2015), 3079–3114. MR3450534

[LMP09] S. Lata, M. Mittal, and V. I. Paulsen, *An operator algebraic proof of Agler's factorization theorem*, Proc. Amer. Math. Soc. **137** (2009), no. 11, 3741–3748, DOI 10.1090/S0002-9939-09-09928-6. MR2529882

[MSF+16] K. Muandet, B. Sriperumbudur, K. Fukumizu, A. Gretton, and B. Schölkopf, *Kernel mean shrinkage estimators*, J. Mach. Learn. Res. **17** (2016), Paper No. 48, 41. MR3504608

[PR16] V. I. Paulsen and M. Raghupathi, *An introduction to the theory of reproducing kernel Hilbert spaces*, Cambridge Studies in Advanced Mathematics, vol. 152, Cambridge University Press, Cambridge, 2016, DOI 10.1017/CBO9781316219232. MR3526117

[PVPK17] L. Parussini, D. Venturi, P. Perdikaris, and G. E. Karniadakis, *Multi-fidelity Gaussian process regression for prediction of random fields*, J. Comput. Phys. **336** (2017), 36–50, DOI 10.1016/j.jcp.2017.01.047. MR3622604

[Rud87] W. Rudin, *Real and complex analysis*, 3rd ed., McGraw-Hill Book Co., New York, 1987. MR924157

[Sai16] S. Saitoh, *A reproducing kernel theory with some general applications*, Mathematical analysis, probability and applications—plenary lectures, Springer Proc. Math. Stat., vol. 177, Springer, [Cham], 2016, pp. 151–182, DOI 10.1007/978-3-319-41945-9_6. MR3571702

[SZ04] S. Smale and D.-X. Zhou, *Shannon sampling and function reconstruction from point values*, Bull. Amer. Math. Soc. (N.S.) **41** (2004), no. 3, 279–305, DOI 10.1090/S0273-0979-04-01025-0. MR2058288

[SZ09a] S. Smale and D.-X. Zhou, *Geometry on probability spaces*, Constr. Approx. **30** (2009), no. 3, 311–323, DOI 10.1007/s00365-009-9070-2. MR2558684

[SZ09b] S. Smale and D.-X. Zhou, *Online learning with Markov sampling*, Anal. Appl. (Singap.) **7** (2009), no. 1, 87–113, DOI 10.1142/S0219530509001293. MR2488871

Department of Mathematics, The University of Iowa, Iowa City, Iowa 52242-1419
*Email address*: palle-jorgensen@uiowa.edu

Mathematical Reviews, 416 4th Street Ann Arbor, Michigan 48103-4816
*Email address*: jft@ams.org

Contemporary Mathematics
Volume **825**, 2025
https://doi.org/10.1090/conm/825/16511

# Ćirić contractions and Banach-valued fractal interpolation functions

María A. Navascués, R. Miculescu, B. C. Anghelina, and Ram N. Mohapatra

ABSTRACT. In this paper we study two types of contractions proposed by Ćirić, extending them to the framework of b-metric spaces. We provide first some Collage Theorems for these mappings. Then we give sufficient conditions on the ratios and the space for the existence of fixed points, and common fixed points of two maps. We analyze two iterative algorithms to approximate the fixed points, and establish their convergence and stability. We define also Iterated Function Systems in abstract spaces, that enable the fractal interpolation of data belonging to Banach spaces and algebras. Thus, we define fractal continuous curves whose points may be measures, functions, matrices, etc. The existence and approximation of these interpolants are considered. In the last section of the paper we provide a theorem on the existence of fixed point for multivalued Ćirić mappings.

## 1. Introduction

L.B. Ćirić defined several contractive maps on metric spaces, and proved the existence of unique and non-unique fixed points for such maps (see for instance [**8**], [**9**], [**10**]). Based on his ideas, a great number of authors have generalized these results, considering more general structures or mappings. Detailed accounts can be found in the references [**1**], [**6**], [**7**], [**13**], [**17**], [**27**], [**28**], [**29**], [**30**], [**33**], that compose a very short and incomplete sample of works on the topic.

In particular Ćirić proposed two types of contractions concerning a self-map $T : X \to X$ of a metric space $(X, d)$. The contractivities that we call of Type I assume that there exist positive constants $a, b, c, e, f$ such that $a+b+c+e+f < 1$, and for all $x, y \in X$

$$d(Tx, Ty) \leq ad(x, y) + bd(x, Tx) + cd(y, Ty) + ed(x, Ty) + fd(y, Tx). \tag{1.1}$$

2020 *Mathematics Subject Classification.* Primary 26A18, 28A80; Secondary 47H10, 47J26, 54H25, 37C25.

*Key words and phrases.* Ćirić contractions, iterated function systems, fixed point theory, S-iteration, multivalued mappings, b-metric spaces.

The first author was supported by Mohapatra Foundation.

For the contractivities of Type II, assume that there exists a positive constant $\mu$ such that $\mu < 1$ and for all $x, y \in X$

$$d(Tx,Ty) \leq \mu \max\{d(x,y), d(x,Tx), d(y,Ty), d(x,Ty), d(y,Tx)\} \tag{1.2}$$

The latter maps were called by the author quasi contractions. Ćirić proved that this kind of contractivities on a complete metric space owns a fixed point (see [**10**]). The inequality (1.1) implies (1.2) when $\mu = a + b + c + e + f$.

We will consider here both types of contractions (I and II) in the context of b-metric spaces. Our purpose is to study the existence and properties of fixed points for this kind of contractivities in a larger scenario, and the convergence and stability of two different iterations for the approximation of the fixed point.

After this introductory part, section 2 is devoted to the study of some properties of the Ćirić maps of type I. We prove first a Collage Theorem for a contraction of this kind whose constants satisfy a specific condition. In particular, the result holds for Ćirić contractions defined in metric spaces. We find bounds for the orbit of a point as well. We propose also a new definition for two maps on a b-metric space to be mutually contractive of type I, and prove that they have a common fixed point if the space is complete and the ratios satisfy some conditions related to the index of the space. Section 3 deals with Ćirić contractivities of type II, and contains proofs of some theorems of fixed point on complete and orbitally complete b-metric spaces. The convergence of the ordered orbits to the fixed point is also established.

In Section 4, we study the convergence and stability of Picard and S-iterations (see [**3**]) with respect to fixed points of Ćirić contractions of type II defined on quasi-Banach spaces. We analyze also the asymptotically stability and convergence rates of the classical Picard (Banach) iterations for Ćirić maps defined on b-metric spaces.

Section 5 presents new types of Iterated Function Systems, whose attractors are fractal continuous curves interpolating data belonging to Banach spaces and algebras, thus extending the standard case of real data. We prove the existence of such interpolating maps, and study their approximation by means of the algorithms treated in Section 4, concerning their convergence and stability.

Section 6 provides a theorem of fixed point for multivalued Ćirić maps. The maps of type Kannan ($a = e = f = 0$, $b = c$), Chatterjea ($a = b = c = 0$, $e = f$), Reich ($e = f = 0$), and Banach ($b = c = e = f = 0$) are included in these models. Thus, all the results proved are applicable to them.

## 2. Ćirić contractions of type I

Definition 2.1. A $b$-metric space $X$ is a set endowed with a mapping $d : X \times X \to \mathbb{R}^+$ with the following properties

(1) $d(x,y) \geq 0$, $d(x,y) = 0$ if and only if $x = y$.
(2) $d(x,y) = d(y,x)$ for any $x, y \in X$.
(3) There exists $s \geq 1$ such that $d(x,y) \leq s(d(x,z) + d(z,y))$ for any $x, y, z \in X$.

The constant $s$ is the index of the $b$-metric space, and $d$ is called a $b$-metric.

Unlike the metric case a b-metric is not continuous in general (see for instance [**26**]). If $d$ is continuous then $(X, d)$ is called a continuous b-metric space.

A useful inequality is the following: Given any $x_0, x_1, \ldots x_n \in X$, where $X$ is a b-metric space with index $s$, one has the following:

$$
\begin{aligned}
d(x_0, x_n) &\leq \sum_{k=0}^{n-2} s^{k+1} d(x_k, x_{k+1}) + s^{n-1} d(x_{n-1}, x_n) \\
&\leq \sum_{k=0}^{n-1} s^{k+1} d(x_k, x_{k+1}).
\end{aligned}
\tag{2.1}
$$

This is a straightforward consequence of the third condition of a b-metric (see [**34**]).

Let us start proving a Collage Theorem for this kind of contractivities, fulfilling certain conditions on the constants. In general, the Collage theorems provide bounds for the distance between any element of the space and a fixed point of a self-map.

For the image of a point $x \in X$ by the map $T$ we will use the notations $Tx$ and $T(x)$ without making any distinction.

THEOREM 2.2. *Let $X$ be a b-metric space and $T : X \to X$ be a Ćirić contraction of type I with a fixed point $x^*$. If the constants of the contractivity are such that $(a + e + fs) < s^{-1}$ then, for any $x \in X$,*

$$\frac{1 - bs - fs^2}{s + as + es + fs^2} d(x, Tx) \leq d(x, x^*) \leq \frac{s + bs + fs^2}{1 - as - es - fs^2} d(x, Tx).$$

PROOF. We assume that $T$ has a fixed point $x^*$. Applying the third property of a b-metric and the contractivity condition, for any $x \in X$,

$$d(x, x^*) \leq sd(x, Tx) + sd(Tx, Tx^*) \leq sd(x, Tx) + sM(x, x^*),$$

where

$$M(x, x^*) = ad(x, x^*) + bd(x, Tx) + ed(x^*, x) + fd(x^*, Tx).$$

Applying the condition (3) of the b-metric in the last summand, with intermediate point $x$, we have

$$(1 - as - es)d(x, x^*) \leq (s + bs)d(x, Tx) + sf(sd(x, x^*) + sd(x, Tx))$$

obtaining that

$$(1 - as - es - fs^2)d(x, x^*) \leq (s + bs + fs^2)d(x, Tx).$$

For the lower bound, let us consider

$$d(x, Tx) \leq sd(x, x^*) + sd(x^*, Tx)$$

and thus

$$d(x, Tx) \leq sd(x, x^*) + sad(x, x^*) + sbd(x, Tx) + esd(x, x^*) + sfd(x^*, Tx).$$

Using the third condition of the b-metric in the last summand we obtain that

$$(1 - bs - fs^2)d(x, Tx) \leq (s + as + es + fs^2)d(x, x^*)$$

and the result is complete. ☐

REMARK 2.3. In particular, for a metric space $X$ we would obtain, for any $x \in X$,

$$\frac{1-b-f}{1+a+e+f} d(x,Tx) \le d(x,x^*) \le \frac{1+b+f}{1-a-e-f} d(x,Tx),$$

and no additional conditions on the constants are needed.

We give in the following the definition of two mutually Ćirić contractive self-maps $T_1, T_2 : X \to X$, where $X$ is a b-metric space.

DEFINITION 2.4. The operators $T_1, T_2 : X \to X$, where $X$ is a b-metric space, are mutually Ćirić contractive of type I, if there exist positive constants $a$, $b$, $c$, $e$, $f$ such that $a+b+c+e+f<1$ and for all $x, y \in X$

$$\begin{aligned} d(T_1x, T_2y) \le\ & ad(x,y) + bd(x,T_1x) + cd(y,T_2y) \\ & + ed(x,T_2y) + fd(y,T_1x). \end{aligned} \tag{2.2}$$

The next result proves that two mutually Ćirić contractive maps of type I on a complete b-metric space have a common fixed point if the constants satisfy some restrictions. We need first a previous result due to Miculescu et al. (see [**16**]).

PROPOSITION 2.1. *Let* $(X,d)$ *be a b-metric space. Any sequence* $(x_n) \subseteq X$ *such that there exists* $\alpha \in [0,1)$ *satisfying the inequality*

$$d(x_{n+1}, x_n) \le \alpha d(x_n, x_{n-1})$$

*for all* $n \ge 1$ *is a Cauchy sequence.*

THEOREM 2.5. *Let* $(X,d)$ *be a complete b-metric space and* $T_1, T_2$ *two mutually Ćirić contractions of type I, with constants* $a, b, c, e, f$ *such that* $a+b+c+e+f<1$, $c+es^2<1$ *and* $\Lambda := \frac{a+b+es}{1-(c+es)} < 1$, *where* $s$ *is the index of the b-metric. Then* $T_1$ *and* $T_2$ *have a unique common fixed point.*

PROOF. Let us consider any element $x \in X$ and construct the sequence: $x_0 := x$, $x_1 := T_1x_0$, $x_2 := T_2x_1$, $x_3 := T_1x_2$, and so on. Using inequality (2.2), we have

$$\begin{aligned} d(x_1,x_2) = d(T_1(x_0), T_2(x_1)) \le\ & ad(x_0,x_1) + bd(x_0,T_1(x_0)) + cd(x_1,T_2(x_1)) \\ & + ed(x_0,T_2(x_1)) + fd(x_1,T_1(x_0)). \end{aligned}$$

Thus, applying the third condition of a b-metric in the fourth summand with intermediate point $x_1$, we get

$$d(x_1,x_2) \le (a+b+es)d(x_0,x_1) + (c+es)d(x_1,x_2),$$

and hence

$$d(x_1,x_2) \le \frac{a+b+es}{1-(c+es)} d(x_0,x_1).$$

Using the previous argument repeatedly, we conclude that, in general

$$d(x_k, x_{k+1}) \le \Lambda d(x_{k-1}, x_k), \tag{2.3}$$

and consequently,

$$d(x_k, x_{k+1}) \le \Lambda^k d(x_0, x_1) \tag{2.4}$$

for all $k \ge 0$. Applying the Proposition 2.1, with the condition given on $\Lambda$, we obtain that the sequence $(x_n)$ is Cauchy and hence convergent. Let $x^*$ be the limit of the sequence $(x_n)$, $x^* = \lim x_n$. Let us see that $x^*$ is the common fixed point of $T_1, T_2$. If $n$ is an even number,

$$\begin{aligned} d(x^*, T_1(x^*)) &\leq sd(x^*, x_n) + sd(x_n, T_1(x^*)) \\ &= sd(x^*, x_n) + sd(T_2(x_{n-1}), T_1(x^*)) \end{aligned}$$

and applying the condition of contractivity in the last term, for some even $n \geq 1$,

$$d(x^*, T_1(x^*)) \leq sd(x^*, x_n) + sM(x_{n-1}, x^*),$$

where

$$\begin{aligned} M(x_{n-1}, x^*) = ad(x_{n-1}, x^*) &+ bd(x_{n-1}, T_2(x_{n-1})) + cd(x^*, T_1(x^*)) \\ &+ ed(x_{n-1}, T_1(x^*)) + fd(x^*, T_2(x_{n-1})). \end{aligned}$$

Then

$$\begin{aligned} (1 - c - es^2)d(x^*, T_1(x^*)) \leq (s + sf)d(x^*, x_n) &+ (sa + s^2 e)d(x_{n-1}, x^*) \\ &+ sbd(x_{n-1}, x_n). \end{aligned}$$

Since all the terms on the right hand side tend to zero and $1 - c - es^2 > 0$, $x^* = T_1(x^*)$. Let us see that $x^*$ is also a fixed point of $T_2$. Note that

$$d(x^*, T_2(x^*)) = d(T_1(x^*), T_2(x^*)) \leq (c + e)d(x^*, T_2(x^*)).$$

Since $c + e < 1$ we have that $x^* = T_2(x^*)$. If there is another common fixed point $\bar{x}$, using the contractivity condition, it is easy to show that they are equal. □

REMARK 2.6. This Theorem includes the usual Kannan, Banach, Chatterjea, Reich and Ćirić contractions of type I in the b-metric context, taking $T_1 = T_2$, and the mutual Kannan, Banach and Reich contractions (see [**18**], [**20**]). Thus the result obtained is very general.

Let us define the orbit of $x := x_0 \in X$ with respect to a Ćirić contractivity $T$ of type I:

$$\mathcal{O}^+(x) = \{T^n(x) : n \geq 0\}$$

Applying the previous results we can bound the distance between $x_n := T^n(x_0)$ and the fixed point $x^*$, if $\Lambda < s^{-1}$:

$$d(x_n, x^*) \leq sd(x_n, x_0) + sd(x_0, x^*).$$

Using inequality (2.1)

$$d(x_n, x^*) \leq s\left(\sum_{k=0}^{n-1} s^{k+1} d(x_k, x_{k+1})\right) + sd(x_0, x^*).$$

Applying the inequality (2.4) and the Collage Theorem, if $\max\{\Lambda, a+e+fs\} < s^{-1}$,

$$\begin{aligned} d(x_n, x^*) &\leq s^2 \left(\sum_{k=0}^{\infty} s^k \Lambda^k\right) d(x_0, x_1) + sd(x_0, x^*) \\ &\leq \left(\frac{s^2}{1 - \Lambda s} + s\frac{s + bs + fs^2}{1 - as - es - fs^2}\right) d(x_0, x_1), \end{aligned}$$

for all $n \in \mathbb{N}$.

## 3. Ćirić contractions of type II

In this Section we consider Ćirić contractions of type II, that is to say, satisfying the condition (1.2).

We will prove the existence and other properties of a fixed point for a Ćirić contraction of type II on a complete b-metric space.

Let us define, for a b-metric space $(X, d)$, a self-map (or operator) $T : X \to X$ and $x \in X$ the orbit of $x$, as before

$$\mathcal{O}^+(x) := \{T^j(x) : j = 0, 1, \dots\},$$

where $T^0 := Id_X$ and $T^j$ denotes the composition of $T$ with itself $j$ times for $j \geq 1$. Let us define the $n$-orbit of $x$ as

$$\mathcal{O}_n^+(x) := \{T^j(x) : j = 0, 1, \dots, n\}. \tag{3.1}$$

DEFINITION 3.1. If $X$ is a b-metric space and $T : X \to X$, then $X$ is $T$-orbitally complete if for any $x \in X$ and a Cauchy sequence $(x_n) \subseteq \mathcal{O}^+(x)$, this sequence is convergent.

L. B. Ćirić (see [**10**]) proved the next Theorem about the existence of fixed point for a self-map $T : M \to M$, where $(M, \rho)$ is a metric space, having the property that there exists $\mu \in (0, 1)$ such that for all $x, y \in X$

$$\rho(Tx, Ty) \leq \mu \max\{\rho(x, y), \rho(x, Tx), \rho(y, Ty), \rho(x, Ty), \rho(y, Tx)\}. \tag{3.2}$$

THEOREM 3.2. *Let $T$ be a self-map satisfying the condition* (3.2) *on a metric space $(M, \rho)$ and let $M$ be $T$-orbitally complete. Then*

(1) *$T$ has a unique fixed point $u \in M$.*
(2) $\lim_{n\to\infty} T^n(x) = u$ *and*
(3) $\rho(T^n(x), u) \leq (\mu^n/(1-\mu))\rho(x, T(x))$ *for every $x \in M$.*

Let us consider the following result due to Schroeder (see [**11**] and [**31**]).

THEOREM 3.3. *If $(X, d)$ is a b-metric space with index $s \leq 2$, then there exists a metric $\rho$ on $X$ such that*

$$\rho(x, y) \leq d(x, y) \leq 2s\rho(x, y), \tag{3.3}$$

*for any $x, y \in X$.*

Using inequalities (3.3) it is an easy exercise to prove that if $(X, d)$ is complete as b-metric space, then $(X, \rho)$ is a complete metric space.

THEOREM 3.4. *If $(X, d)$ is a complete b-metric space with index $s \leq 2$, and $T : X \to X$ is a Ćirić contraction of type II with $\mu < (2s)^{-1}$ then $T$ is a Picard operator, that is to say, there exists a fixed point $x^* \in X$ of $T$ and $\lim T^n(x) = x^*$ for any $x \in X$.*

PROOF. According to the previous Theorem there exists a metric $\rho$ satisfying the inequalities (3.3). These imply that $T$ is also a Ćirić contraction with constant $2s\mu$ with respect to $\rho$ since

$$\rho(Tx, Ty) \leq d(Tx, Ty) \leq \mu M(x, y) \leq 2s\mu M^+(x, y),$$

where

$$M(x, y)) = \max\{d(x, y), d(x, Tx), d(y, Ty), d(x, Ty), d(y, Tx)\}$$

and

$$M^+(x,y)) = \max\{\rho(x,y), \rho(x,Tx), \rho(y,Ty), \rho(x,Ty), \rho(y,Tx)\}.$$

Since $(X,\rho)$ is a complete space, according to Ćirić's Theorem, $T$ has a fixed point $x^*$ and is a Picard operator with respect to $\rho$. Since, due to (3.3), for any $x \in X$

$$d(T^n(x), x^*) \leq 2s\rho(T^n(x), x^*) \to 0.$$

then $(d(T^n(x), x^*))_{n\in\mathbb{N}}$ tends to zero, and $T$ is Picard with respect to $d$ as well. □

We will use the following notations: $diam(Y)$ will denote the diameter of $Y$ defined as $diam(Y) := \sup\{d(y,y') : y, y' \in Y\}$ and the set of natural numbers with zero will be denoted as $\mathbb{N}_0$. Let us start with the following result.

PROPOSITION 3.1. *Let $(X,d)$ be a b-metric space and let $T : X \to X$ be a Ćirić contraction of type II. Then, for all $x \in X$ and $n \in \mathbb{N}_0$ there exists $k \in \{0,1,\ldots,n\}$ such that*

$$d(x, T^k(x)) = diam(\mathcal{O}_n^+(x)). \tag{3.4}$$

*Additionally if $\mu < s^{-1}$, for any $x \in X$ the orbit of $x$ is bounded as*

$$diam(\mathcal{O}^+(x)) \leq \frac{s}{1-s\mu} d(x,Tx). \tag{3.5}$$

PROOF. If $n = 0$ then $\mathcal{O}_n^+(x) = \{x\}$ and $k = 0$ satisfies the equation (3.4). Otherwise, for $k, k'$ such that $1 \leq k, k' \leq n$, the contractivity condition of $T$ implies that

$$\begin{aligned} d(T^k(x), T^{k'}(x)) \leq \mu \max\{&d(T^{k-1}(x), T^{k'-1}(x)), d(T^{k-1}(x), T^k(x)), \\ &d(T^{k'-1}(x), T^{k'}(x)), d(T^{k-1}(x), T^{k'}(x)), \\ &d(T^k(x), T^{k'-1}(x))\}, \end{aligned}$$

and consequently

$$d(T^k(x), T^{k'}(x)) \leq \mu \ diam(\mathcal{O}_n^+(x)). \tag{3.6}$$

On the other hand, since $\mathcal{O}_n^+(x)$ is a finite set, there exist $i, j \in \{0,1,\ldots,n\}$ such that

$$d(T^i(x), T^j(x)) = diam(\mathcal{O}_n^+(x)).$$

If $i, j \geq 1$ the inequality (3.6) contradicts the last expression, thus $i$ or $j$ must be null, and we obtain the first result.

For the second part of the Proposition, for $x \in X$ and $n \in \mathbb{N}_0$, let $k$ be the natural number satisfying (3.4). Applying the third property of the b-metric and using (3.6),

$$d(x, T^k(x)) \leq sd(x,Tx) + sd(Tx, T^k(x)) \leq sd(x,Tx) + s\mu \ diam(\mathcal{O}_n^+(x)).$$

Then, for all $n$,

$$diam(\mathcal{O}_n^+(x)) = d(x, T^k(x)) \leq \frac{s}{1-s\mu} d(x,Tx),$$

and consequently

$$diam(\mathcal{O}^+(x)) = \sup_n diam(\mathcal{O}_n^+(x)),$$

has the same bound. □

The purpose of the following Theorems is to provide sufficient conditions for the existence of fixed point of a self-map $T$ of Ćirić type on a b-metric space.

THEOREM 3.5. *If $X$ is a b-metric space whose distance $d$ is continuous and $T : X \to X$ is a Ciric contraction of type II with $\mu < s^{-1}$ such that $X$ is $T$-orbitally complete, then $T$ is Picard.*

PROOF. Let us consider any element $x \in X$ and $1 \leq m < n$. Let us prove that the ordered orbit of $x$ is a Cauchy sequence. Considering the $(n-m+1)$-orbit of $T^{m-1}(x)$, we have

$$\begin{aligned} d(T^m(x), T^n(x)) &= d(T(T^{m-1}(x)), T^{n-m+1}(T^{m-1}(x))) \\ &\leq \mu \ diam(\mathcal{O}^+_{n-m+1}(T^{m-1}(x))). \end{aligned}$$

Using the relation (3.4) for $T^{m-1}(x)$, there exists a natural number $k$ such that $0 \leq k \leq (n-m+1)$ and

$$\begin{aligned} d(T^m(x), T^n(x)) &\leq \mu \ diam(\mathcal{O}^+_{n-m+1}(T^{m-1}(x))) \\ &= \mu \ d(T^{m-1}(x), T^k(T^{m-1}(x))). \end{aligned}$$

Repeating the process with $T^{m-2}(x)$, $T^{m-3}(x) \ldots$ and using Proposition 3.1,

$$\begin{aligned} d(T^m(x), T^n(x)) &\leq \mu^m \ diam(\mathcal{O}^+_n(x)) \leq \mu^m \ diam(\mathcal{O}^+(x)) \\ &\leq \frac{\mu^m s}{1-\mu s} d(x, Tx). \end{aligned} \tag{3.7}$$

Consequently the sequence $(T^n(x))$ is Cauchy and thus convergent. Let us check that $x^* = \lim T^n(x)$ is a fixed point of $T$. Using the third condition of a b-metric and the inequality associated with the contractivity of $T$, we get

$$\begin{aligned} d(x^*, T(x^*)) &\leq sd(x^*, T^{n+1}(x)) + sd(T^{n+1}(x), T(x^*)) \\ &\leq sd(x^*, T^{n+1}(x)) + s\mu \ M_n(x, x^*), \end{aligned}$$

where

$$\begin{aligned} M_n(x, x^*) = \max\{&d(T^n(x), x^*), d(T^n(x), T^{n+1}(x)), \\ &d(x^*, T(x^*)), d(T^n(x), T(x^*)), d(T^{n+1}(x), x^*)\}. \end{aligned}$$

When $n$ tends to infinity $M_n(x, x^*) \to d(x^*, Tx^*)$. Thus

$$(1 - s\mu) d(x^*, T(x^*)) \leq 0$$

and $x^*$ is a fixed point of $T$ since $1 - s\mu > 0$. If there is another fixed point $x^+$ then

$$d(x^*, x^+) = d(Tx^*, Tx^+) \leq \mu d(x^*, x^+),$$

and both points must be equal. Thus we obtain that the fixed point is unique and the Banach iterations starting on any element converge to the fixed point. □

REMARK 3.6. Let us note that we do not need the continuity of the operator $T$ to obtain the result.

REMARK 3.7. Theorem 3.5 is true for complete b-metric spaces in particular.

The next result does not require the continuity of the b-metric $d$.

THEOREM 3.8. *If $X$ is a $T$-orbitally complete b-metric space for a Ćirić contraction of type II, $T : X \to X$, with $\mu < s^{-1}(1+s)^{-1}$, then $T$ is a Picard operator.*

Proof. Let us consider any element $x \in X$. The proof for the convergence of the sequence $(T^n x)$ is similar to the one outlined in the previous Theorem.

Let us check that $x^* = \lim T^n(x)$ is a fixed point of $T$. Using the third condition of a b-metric and the inequality associated with the contractivity of the operator $T$,

$$\begin{aligned} d(x^*, T(x^*)) &\leq sd(x^*, T^{n+1}(x)) + sd(T^{n+1}(x), T(x^*)) \\ &\leq sd(x^*, T^{n+1}(x)) + s\mu\, M_n(x, x^*), \end{aligned}$$

where

$$\begin{aligned} M_n(x, x^*) = \max\{&d(T^n(x), x^*), d(T^n(x), T^{n+1}(x)), \\ &d(x^*, T(x^*)), d(T^n(x), T(x^*)), d(T^{n+1}(x), x^*)\}. \end{aligned}$$

Hence

$$\begin{aligned} M_n(x, x^*) \leq\ & d(T^n(x), x^*) + d(T^n(x), T^{n+1}(x)) \\ &+ d(x^*, T(x^*)) + d(T^n(x), T(x^*)) + d(T^{n+1}(x), x^*). \end{aligned}$$

In the fourth term of the last expression we consider that

$$d(T^n(x), T(x^*)) \leq sd(T^n(x), x^*) + sd(x^*, T(x^*)).$$

Finally, one obtains that

$$\begin{aligned} (1 - s\mu - s^2\mu) d(x^*, T(x^*)) \leq\ & s(1+\mu) d(x^*, T^{n+1}(x)) \\ &+ \mu s(s+1) d(x^*, T^n(x)) \\ &+ s\mu\, d(T^n(x), T^{n+1}(x)). \end{aligned}$$

Since all the terms in the right hand side tend to zero we obtain $x^* = T(x^*)$. If there is another fixed point $x^+$ then

$$d(x^*, x^+) = d(Tx^*, Tx^+) \leq \mu d(x^*, x^+),$$

and both fixed points must be equal. Thus we obtain that the fixed point is unique and the Picard iterations starting on any element converge to the fixed point. □

Remark 3.9. Theorem 3.8 is true for complete b-metric spaces in particular.

## 4. Picard and S-iterations of Ćirić contractions

In this section we analyze the convergence and stability of two different iterative methods for the approximation of a fixed point. The first one requires the structure of a quasi-normed space.

Definition 4.1. If $X$ is a real linear space, the mapping $|\cdot|_s : X \times X \to \mathbb{R}^+$ is a quasi-norm of index $s$ if

(1) $|x|_s \geq 0$; $x = 0$ if and only if $|x|_s = 0$.
(2) $|\lambda x|_s = |\lambda||x|_s$.
(3) There exists $s \geq 1$ such that $|x + y|_s \leq s(|x|_s + |y|_s)$ for any $x, y \in X$.

The space $(X, |\cdot|_s)$ is a quasi-normed space.

A quasi-normed space is a particular case of a b-metric space, with respect to the distance induced by the quasi-norm. If $X$ is complete with respect to this b-metric, then $X$ is a quasi-Banach space.

In the case of a quasi-normed space, there are alternatives to the Picard iterations. We will prove here that for some values of the parameter $\mu$ of the contraction defined by the inequality (1.2), the S-iterative scheme (see [**3**]) does converge in the framework of a quasi-normed space. This is given by the following two-step procedure:

$$x_{n+1} = (1 - a_n)Tx_n + a_n Ty_n, \tag{4.1}$$

$$y_n = (1 - b_n)x_n + b_n Tx_n, \tag{4.2}$$

for $n \in \mathbb{N}_0$, $0 \le a_n, b_n \le 1$.

In the following we prove that for some values of $a_n, b_n$ and the ratio $\mu$ of the contraction defined in (1.2), the S-iteration associated with a Ćirić contraction $T$ of type II converges to the fixed point. We need a previous proposition. The next definition was introduced in the reference [**21**].

DEFINITION 4.2. Given a b-metric space $(X, d)$, $T : X \to X$ is a partial contractivity if there exist constants $\lambda$, $0 < \lambda < 1$, and $L \ge 0$ such that

$$d(Tx, Ty) \le \lambda d(x, y) + L d(x, Tx), \tag{4.3}$$

for any $x, y \in X$.

In the same reference, it has been proved that if a partial contractivity has a fixed point, this is unique ([**21**]).

PROPOSITION 4.1. *If $(X, d)$ is a b-metric space, and $T : X \to X$ is a Ćirić contraction of type II with $\mu < s^{-1}(1 + s)^{-1}$ then $T$ is a partial contractivity with constants*

$$\lambda = \frac{\mu s}{1 - \mu s^2}$$

*and*

$$L = \frac{\mu s^2}{1 - \mu s^2}.$$

PROOF. It is an easy exercise considering the different cases of the maximum of the set

$$M = \{d(x, y), d(x, Tx), d(y, Ty), d(x, Ty), d(y, Tx)\}.$$

□

THEOREM 4.3. *Let $X$ be a quasi-normed space with index $s$ and $T : X \to X$ a partial contractivity such that $\lambda s^2 < 1$ with a fixed point $x^*$. Let us choose the sequences $(a_n)$, $(b_n)$ such that*

$$\frac{1 - s^{-2}}{1 - \lambda} < B := \inf b_n \le 1 \tag{4.4}$$

*and*

$$\frac{1 - s^{-1}}{1 - sk_1} < A := \inf a_n \le 1, \tag{4.5}$$

*where $k_1 = 1 - B(1 - \lambda)$. Then the S-iteration given by (4.1) and (4.2) converges to the fixed point starting at any point $x \in X$ and it is asymptotically stable.*

PROOF. Applying (4.2) we have

$$|y_n - x^*|_s = |(1-b_n)(x_n - x^*) + b_n(T(x_n) - x^*)|_s.$$

The contractivity condition (4.3) implies that

$$|y_n - x^*|_s \le s(1 - b_n(1-\lambda))|x_n - x^*|_s \le sk_1|x_n - x^*|_s. \tag{4.6}$$

The condition (4.4) implies that $sk_1 < s^{-1} \le 1$. Using (4.1), for $n \ge 0$,

$$|x_{n+1} - x^*|_s \le s(1-a_n)|Tx_n - x^*|_s + sa_n|Ty_n - x^*|_s. \tag{4.7}$$

The inequalities (4.3) and (4.6) yield

$$|x_{n+1} - x^*|_s \le s\lambda(1-a_n)|x_n - x^*|_s + s^2 a_n \lambda k_1 |x_n - x^*|_s, \tag{4.8}$$

$$|x_{n+1} - x^*|_s \le s\lambda(1 - a_n(1-sk_1))|x_n - x^*|_s \le s\lambda(1 - A(1-sk_1))|x_n - x^*|_s. \tag{4.9}$$

and

$$|x_{n+1} - x^*|_s \le \lambda k|x_n - x^*|_s, \tag{4.10}$$

where $k = s(1 - A(1-sk_1))$. The condition on $(a_n)$ implies that $k < 1$. Consequently

$$|x_n - x^*|_s \le (\lambda k)^n |x_0 - x^*|_s. \tag{4.11}$$

Since $\lambda k < 1$, the last inequality assures the convergence of the algorithm and its asymptotical stability. □

REMARK 4.4. For a normed space the conditions on the constants reduce to $A, B > 0$.

REMARK 4.5. The previous theorem is proved for partial contractivities. According to Proposition 4.1, for a contraction of Ćirić type the former Theorem is true if $\mu < s^{-2}(s+1)^{-1}$. If $X$ is complete this condition ensures also the existence of fixed point according to Theorem 3.8.

Let us consider now the stability of the Picard iterations of fixed points of Ćirić maps defined on b-metric spaces.

THEOREM 4.6. *Let $X$ be a complete b-metric space, and $T : X \to X$ a Ćirić contraction of type II such that $\mu < s^{-1}(s+1)^{-1}$. Then any ball $B(x^*, r)$, where $r > 0$ and $x^*$ is the fixed point of $T$, is invariant ($T(B(x^*, r)) \subseteq B(x^*, r)$). Moreover, $x^*$ is globally asymptotically stable for the Picard iterations.*

PROOF. The hypotheses given for $X$ and $T$ assure the existence of a fixed point $x^*$ and the convergence of the Picard iteration, according to Theorem 3.8. Now, for any $x \in X$ let us define $x_0 := x$ and $x_n := T^n(x)$. According to the Proposition 4.1, for $x = x^*$ and $y = x_0$, we have

$$d(x_1, x^*) \le \lambda d(x_0, x^*),$$

where $\lambda = \mu s/(1 - \mu s^2) < 1$. Iteratively we find that for all $n \ge 1$,

$$d(x_n, x^*) \le \lambda^n d(x_0, x^*).$$

The statement on the stability and the invariance of the balls centered at $x^*$ are consequences of the last inequality (see for instance the reference [**18**]). □

THEOREM 4.7. *Let $X$ be a complete b-metric space with index $s \le 2$ and $T : X \to X$ a Ćirić contraction of type II such that $\mu < (2s)^{-1}$. Then the rate of convergence of the Picard iterations to the fixed point is $O(\gamma^n)$, where $\gamma = 2\mu s$.*

PROOF. For the case where $s \leq 2$, according to Schroeder's Theorem 3.3, there exists a metric $\rho$ satisfying the inequality (3.3). By Theorem 3.4, $\rho$ is a Ćirić contraction with constant $\overline{\mu} = 2\mu s$. Then, considering the third item of Ćirić's Theorem 3.2 and (3.3), we have

$$d(x_n, x^*) \leq 2s\overline{\mu}^n/(1-\overline{\mu})d(x_1, x_0),$$

for all $n \in \mathbb{N}$, where $(x_n)$ is the Picard orbit of $x \in X$. □

REMARK 4.8. The results obtained on the existence of fixed points, convergence and stability of the Picard and S-iterations are applicable in particular to Kannan, Reich, Chatterjea and Ćirić contractions of type I.

## 5. Banach-valued fractal interpolation functions

Currently there is a vast literature dealing with the theory of iterated function systems, that reflects an increasing interest for the topic. A sample of classical and new articles on this subject is given by the references [**2**], [**12**], [**25**], [**32**], [**36**]. An important particular case of iterated system was proposed initially by Barnsley ([**5**]), giving rise to single and multi-variable fractal interpolation functions (FIFs), that represent an essential progress in function theory (see for instance [**4**], [**15**], [**22**], [**23**], [**24**], [**35**]). We extend in this section the framework of these mappings, defining continuous Banach-valued fractal interpolation functions. Thus, we construct interpolatory maps of abstract data as, for instance, measures, operators, matrices, etc., whose graphs have a fractal character.

Let us consider a Banach algebra $\mathcal{A}$ and a collection of data $D = \{(t_i, A_i)\}_{i=0}^N$, $N > 1$, such that $t_0 < t_1 < \ldots < t_N$ are real numbers, and $A_i \in \mathcal{A}$ for $i = 1, \ldots, N$. Let $I = [t_0, t_N]$ and $I_i = [t_{i-1}, t_i]$ for $i = 1, \ldots N$. Let us define contractive homeomorphisms $L_i : I \to I_i$ satisfying the "join-up" conditions:

$$L_i(t_0) = t_{i-1}, \qquad L_i(t_N) = t_i. \tag{5.1}$$

for any $i = 1, 2, \ldots, N$.

Let us define $F_n : I \times \mathcal{A} \to \mathcal{A}$ as

$$F_n(t, A) = S_n(t, A) + Q_n(t),$$

where $S_n : I \times \mathcal{A} \to \mathcal{A}$ and $Q_n : I \to \mathcal{A}$ are continuous mappings for all $n = 1, \ldots, N$, and satisfy the conditions:

$$F_n(t_0, A_0) = A_{n-1}, \quad F_n(t_N, A_N) = A_n. \tag{5.2}$$

Let us assume that $S_n$ are uniform $\phi_n$-contractions in the second variable, that is to say,

$$\|S_n(t, A) - S_n(t, A')\| \leq \phi_n(\|A - A'\|), \tag{5.3}$$

for any $t \in I$ and $n = 1, 2, \ldots N$. Let $\phi_n : \mathbb{R}^+ \to \mathbb{R}^+$ be increasing and such that the mapping $\phi(\delta) := \sup_n\{\phi_n(\delta)\}$, for $\delta \in \mathbb{R}^+$, satisfies the property $\sum_{j=1}^{\infty} \phi^j(\delta) < \infty$ for all $\delta \geq 0$.

We will denote as $\mathcal{CB}(I)$ the space of continuous functions of type $w : I \to \mathcal{A}$ endowed with the supremum norm $\|\cdot\|_{sup}$. Let $\mathcal{CB}^*(I)$ be defined as

$$\mathcal{CB}^*(I) = \{w \in \mathcal{CB}(I) : w(t_0) = A_0, w(t_N) = A_N\}.$$

This set is convex, closed and complete with respect to the supremum norm.

Let us define an operator $T$ on $\mathcal{CB}^*(I)$ as

$$Tw(t) = F_n(L_n^{-1}(t), w \circ L_n^{-1}), \tag{5.4}$$

for $t \in I_n$ and any $w \in \mathcal{CB}^*(I)$. We have the following result:

THEOREM 5.1. *The operator $T$ has the following properties:*

(1) $T : \mathcal{CB}^*(I) \to \mathcal{CB}^*(I)$.
(2) *$T$ has a unique fixed point $w^* \in \mathcal{CB}^*(I)$.*
(3) *The Picard iterations $T^m(w)$ converge to $w^*$ as $m$ tends to infinity for all $w \in \mathcal{CB}^*(I)$.*
(4) *$w^*$ is an interpolant of the data $\{(t_n, A_n)\}_{n=1}^N$, that is to say, for all $n = 0, 1, \ldots, N$,*

$$w^*(t_n) = A_n.$$

(5) *The graph of $w^*$ has a fractal structure.*
(6) *The rate of convergence of the iterates of any map $w \in \mathcal{CB}^*(I)$ ($w_m := T^m w$) to $w^*$ is*

$$\|w_m - w^*\|_{sup} \leq \sum_{j=1}^{\infty} \phi^{m+j-1}(\|Tw - w\|_{sup}).$$

PROOF. The items 1 and 4 are based on the join-up conditions (5.1) and (5.2), and their proof is analogous to the real case (see for instance [**23**]). For the second item let us consider, for $t \in I_n$,

$$\|Tw(t) - Tw'(t)\| = \|S_n(L_n^{-1}(t), w \circ L_n^{-1}(t) - S_n(L_n^{-1}(t), w' \circ L_n^{-1}(t)\|.$$

Applying (5.3), and the properties of $\phi_n$ one has

$$\|Tw - Tw'\|_{sup} \leq \phi(\|w - w'\|_{sup}),$$

where $\phi(t) := \sup_n\{\phi_n(t)\}$, for $t \in \mathbb{R}^+$. The conditions required on the maps $\phi_n$ imply that $T$ is a Matkowski contraction on a complete space. As a consequence it has a fixed point $w^*$ and the Picard iterations of any map $w \in \mathcal{CB}^*(I)$ converge to $w^*$.

Regarding the rate of convergence, let us define for $w := w_0 \in \mathcal{CB}^*(I)$, $w_m = Tw_{n-1}$ for $n \geq 1$. Inductively one has:

$$\|w_{m+1} - w_m\|_{sup} \leq \phi^m(\|w_1 - w_0\|_{sup}).$$

Then

$$\|w_{m+p} - w_m\|_{sup} \leq \sum_{j=1}^{p} \|w_{m+j} - w_{m+j-1}\|_{sup} \leq \sum_{j=1}^{p} \phi^{m+j-1}(\|w_1 - w_0\|_{sup}).$$

Taking limits when $p$ tends to infinity,

$$\|w_m - w^*\|_{sup} \leq \sum_{j=1}^{\infty} \phi^{m+j-1}(\|w_1 - w_0\|_{sup}).$$

The proof of the item 5 follows from the definition of $T$ (see Theorem 5 of the reference [**19**]). □

**Particular cases:**

(1) For $S_n(t,A) = A_n(t) \cdot A$ and $Q_n(t) = tC_n + D_n$, where $A_n \in \mathcal{CB}(I)$ and $C_n, D_n \in \mathcal{A}$ are chosen to satisfy the conditions (5.2), the map $w^*$ is the affine fractal interpolation function in Banach algebras. Considering that $\mathcal{A}$ is a Banach space, and $S_n(t,A) = \alpha_n(t)A$ where $\alpha_n \in \mathcal{C}(I,\mathbb{R})$, $w^*$ is an affine fractal interpolation in a Banach space.

(2) For $S_n(t,A) = A_n(t) \cdot A$ and $Q_n(t) = u \circ L_n(t) - A_n(t) \cdot v_n(t)$, where $u, A_n, v_n \in \mathcal{CB}(I)$, such that $u(t_i) = v_n(t_i)$ for $i = 0, N$, $n = 1, \ldots, N$, we obtain the $\alpha$-fractal case in Banach algebras. Substituting $A_n(t)$ by $\alpha_n(t) \in \mathbb{R}$ we have $\alpha$-fractal continuous functions in Banach spaces, provided that $\mathcal{A}$ is only a Banach space.

According to (5.3), we can take in both cases the comparison function $\phi_n(\delta) = c_n\delta$, where $c_n := \sup\{\|A_n(t)\| : t \in I\}$ (or $c_n := \sup\{|\alpha_n(t)| : t \in I\}$), and $\phi(\delta) = c\delta$, where $c := \sup c_n < 1$. Thus $T$ is a partial contractivity (Definition 4.2) for $\lambda = c$ and $L = 0$.

The rate of convergence of the Picard iterations is, according to the last item of the previous Theorem,

$$\|w_m - w^*\|_{sup} \leq \frac{c^m}{1-c}(\|Tw_0 - w_0\|_{sup}).$$

The procedure is clearly convergent and stable.

Concerning the S-iteration, for $w_0 \in \mathcal{CB}^*(I)$ we have the iterative scheme:

$$\begin{aligned} w_{m+1} &= (1-a_m)Tw_m + a_m Ty_m, \\ y_m &= (1-b_m)w_m + b_m Tw_m. \end{aligned}$$

Though the set $\mathcal{CB}^*(I)$ is not a normed space, it is closed and convex, and consequently all the iterates $w_m$ will remain in $\mathcal{CB}^*(I)$ if $w_0$ does. The S-algorithm will converge with asymptotic stability to $w^*$ for values of the scalars $a_m, b_m$ such that $A = \inf a_m > 0$, $B = \inf b_m > 0$ (Theorem 4.3). The rate of convergence is, according to (4.11),

$$\|w_m - w^*\|_{sup} \leq (ck)^m \|w_0 - w^*\|_{sup}, \tag{5.5}$$

where $k = 1 - A(1-k_1)$, $k_1 = 1 - B(1-c)$, and $k, k_1 < 1$. s In the $\alpha$-fractal case (2), the fixed point $w^*$ is a perturbation of $u$. Consequently, starting at $w_0 = u$ seems a good choice to begin the algorithms.

## 6. A fixed point theorem for multivalued Ćirić maps

In the following we will prove a theorem of Ćirić's type for multivalued mappings. Let us start with some notations: $B(X)$ represents the family of non-empty bounded subsets of a b-metric space $X$. If $A \subseteq X$ and $x \in X$

$$d(x,A) = \inf\{d(x,a) : a \in A\}.$$

For $A, B \subseteq X$, let us define

$$\nu(A,B) = \sup\{d(a,b) : a \in A, b \in B\}.$$

$\mathcal{P}(X)$ denotes the set of all subsets of $X$, and for a multivalued mapping $\tau : X \to \mathcal{P}(X)$, a fixed point is an element $x \in X$ such that $x \in \tau(x)$.

THEOREM 6.1. *Let $X$ be a complete b-metric space, $\tau : X \to B(X)$ and let $\mu < s^{-1}(s+1)^{-1}$ such that for any $x, y \in X$*

$$\nu(\tau(x), \tau(y)) \leq \mu \max\{d(x,y), d(x,\tau(x)), d(y,\tau(y)), d(x,\tau(y)), d(y,\tau(x))\}. \tag{6.1}$$

*Then*

(1) *$\tau$ has a fixed point $x^*$ and $\tau(x^*) = \{x^*\}$.*
(2) *There exists $T : X \to X$ such that for any $x \in X$ the orbit $(x_n)$ of $x$ with respect to $T$ converges to the fixed point $x^*$ and $d(x_n, x^*) \leq \lambda^n d(x_0, x^*)$ for any $n \in \mathbb{N}_0$, where $\lambda = \mu s/(1-\mu s^2)$.*

PROOF. Let us define $T : X \to X$ such that for any $x \in X$, $T(x) \in \tau(x)$, where $\tau$ is the mapping considered in the statement. Let us see that $T$ is a Ćirić contraction of type II:

$$d(T(x), T(y)) \leq \nu(\tau(x), \tau(y)) \leq \mu \max M_\tau(x,y),$$

where $M_\tau(x,y) = \{d(x,y), d(x,\tau(x)), d(y,\tau(y)), d(x,\tau(y)), d(y,\tau(x))\}$. Then, by definition of $T$,

$$\begin{aligned} d(T(x), T(y)) &\leq \mu \max M_\tau(x,y) \\ &\leq \mu \max\{d(x,y), d(x,T(x)), d(y,T(y)), d(x,T(y)), d(y,T(x))\}. \end{aligned}$$

Thus we obtain that $T$ is a Ćirić contraction of type II with constant $\mu$. According to Theorem 3.8, $T$ has a fixed point $x^*$. By definition of $T$, $x^* \in \tau(x^*)$ and this element is a fixed point of $\tau$. Applying the contractivity condition (6.1) we obtain that $\nu(\tau(x^*), \tau(x^*)) = 0$ and consequently $\tau(x^*) = \{x^*\}$. The second item is a consequence of the Theorem 4.6. □

## Acknowledgments

The first author acknowledges the invitation of University of Central Florida for several research stays and the fourth author is grateful to the Mohapatra Family Foundation for supporting the visit of Professor Navascués which made this work possible.

## References

[1] M. Abbas, I. Altun, and S. Romaguera, *Common fixed points of Ćirić-type contractions on partial metric spaces*, Publ. Math. Debrecen **82** (2013), no. 2, 425–438, DOI 10.5486/PMD.2013.5342. MR3034356

[2] I. Abraham, R. Miculescu, *Generalized iterated function systems on b-metric spaces*, Mathematics, **11**(13), 2826 (2023) 1-16.

[3] R. P. Agarwal, D. O'Regan, and D. R. Sahu, *Iterative construction of fixed points of nearly asymptotically nonexpansive mappings*, J. Nonlinear Convex Anal. **8** (2007), no. 1, 61–79. MR2314666

[4] B.-C. Anghelina and R. Miculescu, *On the localization of Hutchinson-Barnsley fractals*, Chaos Solitons Fractals **173** (2023), Paper No. 113674, 7 pp., DOI 10.1016/j.chaos.2023.113674. MR4606054

[5] M. F. Barnsley, *Fractal functions and interpolation*, Constr. Approx. **2** (1986), no. 4, 303–329, DOI 10.1007/BF01893434. MR892158

[6] V. Berinde, *A common fixed point theorem for quasi contractive type mappings*, Ann. Univ. Sci. Budapest. Eötvös Sect. Math. **46** (2003), 101–110 (2004). MR2102690

[7] S.-H. Cho, *Fixed point theorems for Ćirić type $\mathcal{Z}$-contractions in generating spaces of quasi-metric family*, J. Funct. Spaces, posted on 2019, Art. ID 8290686, 8 pp., DOI 10.1155/2019/8290686. MR4050898

[8] L. B. Ćirić, *Generalized contractions and fixed-point theorems*, Publ. Inst. Math. (Beograd) (N.S.) **12(26)** (1971), 19–26. MR309092

[9] L. B. Ćirić, *On some maps with a nonunique fixed point*, Publ. Inst. Math. (Beograd) (N.S.) **17(31)** (1974), 52–58. MR356012

[10] Lj. B. Ćirić, *A generalization of Banach's contraction principle*, Proc. Amer. Math. Soc. **45** (1974), 267–273, DOI 10.2307/2040075. MR356011

[11] A. H. Frink, *Distance functions and the metrization problem*, Bull. Amer. Math. Soc. **43** (1937), no. 2, 133–142, DOI 10.1090/S0002-9904-1937-06509-8. MR1563501

[12] J. E. Hutchinson, *Fractals and self-similarity*, Indiana Univ. Math. J. **30** (1981), no. 5, 713–747, DOI 10.1512/iumj.1981.30.30055. MR625600

[13] E. Karapinar, F. Khojasteh, and W. Shatanawi, *Revisiting Ćirić-type contraction with Caristi's approach*, Symmetry, **11**, 726 (2019) 1–7.

[14] M. A. Krasnosel′skiĭ, *Two remarks on the method of successive approximations* (Russian), Uspehi Mat. Nauk (N.S.) **10** (1955), no. 1(63), 123–127. MR68119

[15] R. Lal, B. Selmi, and S. Verma, *On dimension of fractal functions on product of the Sierpiński gaskets and associated measures*, Results Math. **79** (2024), no. 2, Paper No. 73, 20 pp., DOI 10.1007/s00025-023-02108-w. MR4691939

[16] R. Miculescu and A. Mihail, *New fixed point theorems for set-valued contractions in b-metric spaces*, J. Fixed Point Theory Appl. **19** (2017), no. 3, 2153–2163, DOI 10.1007/s11784-016-0400-2. MR3692446

[17] G. Mınak, A. Helvacı, and I. Altun, *Ćirić type generalized $F$-contractions on complete metric spaces and fixed point results*, Filomat **28** (2014), no. 6, 1143–1151, DOI 10.2298/FIL1406143M. MR3360088

[18] R. N. Mohapatra, M. A. Navascués, S. Jha, and A. K. B. Chand, *Iteration of operators with mutual relations of Kannan type*, Mathematics, **10**(15), 2632 (2022) 1–14.

[19] M. A. Navascués, *Fractal curves in Banach algebras*, Fractal Fract., **6(12)**, 722 (2022), 1–17.

[20] M. A. Navascués, *Iterative schemes involving several mutual contractions*, Mathematics, **11:2019** (2023) 1–18.

[21] M. A. Navascués, *Approximation of fixed points and fractal functions by means of different iterative algorithms*, Chaos Solitons Fractals **180** (2024), Paper No. 114535, 7 pp., DOI 10.1016/j.chaos.2024.114535. MR4695574

[22] M. A. Navascués, R. N. Mohapatra, and M. N. Akhtar, *Construction of fractal surfaces*, Fractals, **28**(1) (2020) 2050033(1–13).

[23] M. A. Navascués, C. Pacurar, and V. Drakopoulos, *Scale-free fractal interpolation*, Fractal Fract., **6**, 602 (2022) 1–15.

[24] M. A. Navascués and S. Verma, *Non-stationary $\alpha$-fractal surfaces*, Mediterr. J. Math. **20** (2023), no. 1, Paper No. 48, 18 pp., DOI 10.1007/s00009-022-02242-9. MR4527888

[25] R. Pasupathi, A. K. B. Chand, and M. A. Navascués, *Cyclic iterated function systems*, J. Fixed Point Theory Appl. **22** (2020), no. 3, Paper No. 58, 17 pp., DOI 10.1007/s11784-020-00790-9. MR4113324

[26] G. Rano and T. Bag, *Quasi-metric space and fixed point theorems*, Int. J. Math. Sci. Comput. **3** (2013), no. 2, 27–31. MR3188702

[27] S. Reich, *Kannan's fixed point theorem*, Boll. Un. Mat. Ital. (4) **4** (1971), 1–11. MR305163

[28] S. Reich, *Some remarks concerning contraction mappings*, Canad. Math. Bull. **14** (1971), 121–124, DOI 10.4153/CMB-1971-024-9. MR292057

[29] S. Reich, *Fixed points of contractive functions* (English, with Italian summary), Boll. Un. Mat. Ital. (4) **5** (1972), 26–42. MR309095

[30] M. Sarwar and M. U. Rahman, *Fixed point theorems for Ćirić's and generalized contractions in b-metric spaces*, Int. J. An. Appl., **7**, 1 (2015) 70–78.

[31] V. Schroeder, *Quasi-metric and metric spaces*, Conform. Geom. Dyn. **10** (2006), 355–360, DOI 10.1090/S1088-4173-06-00155-X. MR2268484

[32] N.-A. Secelean, *Iterated function systems consisting of $F$-contractions*, Fixed Point Theory Appl., posted on 2013, 2013:277, 13, DOI 10.1186/1687-1812-2013-277. MR3213104

[33] W. Shatanawi, A. Pitea, and R. Lazović, *Contraction conditions using comparison functions on b-metric spaces*, Fixed Point Theory Appl., posted on 2014, 2014:135, 11, DOI 10.1186/1687-1812-2014-135. MR3259866

[34] T. Suzuki, *Basic inequality on a b-metric space and its applications*, J. Inequal. Appl., posted on 2017, Paper No. 256, 11 pp., DOI 10.1186/s13660-017-1528-3. MR3711630

[35] P. Viswanathan and M. A. Navascués, *A fractal operator on some standard spaces of functions*, Proc. Edinb. Math. Soc. (2) **60** (2017), no. 3, 771–786, DOI 10.1017/S0013091516000316. MR3674090

[36] E. R. Vrscay, *Iterated function systems: theory, applications and the inverse problem*, Fractal geometry and analysis (Montreal, PQ, 1989), NATO Adv. Sci. Inst. Ser. C: Math. Phys. Sci., vol. 346, Kluwer Acad. Publ., Dordrecht, 1991, pp. 405–468. MR1140728

Department of Mathematics, Universidad de Zaragoza, Zaragoza 50018, Spain
*Email address*: `manavas@unizar.es`

Faculty of Mathematics and Computer Science, Transilvania University of Braşov, Romania
*Email address*: `radu.miculescu@unitbv.ro`

Faculty of Mathematics and Computer Science, Transilvania University of Braşov, Romania
*Email address*: `bogdan.anghelina@unitbv.ro.`

Department of Mathematics, University of Central Florida, Florida 32816
*Email address*: `Ram.Mohaptra@ucf.edu`

Contemporary Mathematics
Volume **825**, 2025
https://doi.org/10.1090/conm/825/16512

# Comparison of Hausdorff dimension of $E(F_1)$ and $E(F_2)$ for $F_1, F_2 \subset \mathbb{N}$

R. D. Nussbaum

ABSTRACT. If $F$ is a set of two or more positive integers, $E(F)$ denotes the set of irrational real numbers $x \in (0,1)$ such that $x$ can be expressed as a continued fraction which only uses the integers in $F$. It is of interest to determine $\dim(E(F))$, the Hausdorff dimension of $F$. Methods of R. S. Falk and R. D. Nussbaum (see references in this paper) give rigorous high order upper and lower bounds for $\dim(E(F))$. Here, we give some elementary "comparison theorems," which sometimes allow one to prove that $\dim(E(F_1)) > \dim(E(F_2))$ without actually computing $\dim(E(F_1))$ or $\dim(E(F_2))$.

## 1. Background Information

We recall some notation, we shall write $B = \{m + in : m \in \mathbb{N}, n \in \mathbb{Z}\}$ and for $b \in B$ and complex numbers $z$ with $Re(z) > -1$, we shall write $\theta_b(z) = \frac{1}{z+b}$. If $F$ is a nonempty subset of $B$, there is a unique nonempty "invariant set" $E := E(F) \subset \{z \in \mathbb{C} : Re(z) \geq 0\}$ such that

$$E = \cup_{b \in F} \theta_b(E). \tag{1.1}$$

In general, if $Re(b) \geq r$ for all $b \in F$, we have that $E(F) \subset \{z \in \mathbb{C} : |z - \frac{1}{2r}| \leq \frac{1}{2r}\}$. If $F \subset \mathbb{N}$ and $m \geq r$ for all $m \in F$, $E(F) \subset \{x \in \mathbb{R} : |x - \frac{1}{2r}| \leq \frac{1}{2r}\}$. If $F \subset \mathbb{N}$, we say that "$E(F)$ is the set of continued fraction obtainable with alphabet $F$" and if $F$ is not contained in $\mathbb{N}$, "$E$(F) is the set of complex continued fractions obtainable with alphabet $F$." If $F$ is finite, $E(F)$ is known to be compact and contained in $\{z \in \mathbb{C} : |z - \frac{1}{2r}| < \frac{1}{2r}\}$ if $Re(b) \geq r$ for all $b \in F$.

We are interested in $\dim(E(F))$, the Hausdorff dimension of $E(F)$, and for notational simplicity, we shall write $\dim(F)$ instead of $\dim(E(F))$. Although the methods we shall use can also be applied to the case of complex continued fractions, we shall restrict our attention here to the case $F \subset \mathbb{N}$. It has been known for some time that rigorous computation of $\dim(F)$ is intimately connected with the facts about a class of bounded linear operators, which have sometimes been called "Perron-Frobenius operators" or "Transfer operators." Because our approach will use this operator theory connection, we need to recall some basic facts about Perron-Frobenius operator in our context. We refer to **[6]** for generalizations of the results

2020 *Mathematics Subject Classification.* Primary 28A80; Secondary 37C30, 11K55.

*Key words and phrases.* Hausdorff dimension, positive transfer operators, continued fractions.

below. Assume that $m \geq \gamma$ for all $m \in F$, where $F \subset \mathbb{N}$ and $F$ is finite, select any fixed $A \geq \frac{1}{\gamma}$ and for $k \in \mathbb{N} \cup \{0\}$ let $Y_{\mathbb{R}}^k := C_{\mathbb{R}}^k[0, A]$ (respectively, $Y_{\mathbb{C}}^k := C_{\mathbb{C}}^k[0, A]$) denote the real Banach space (respectively, complex Banach space) of $k$ times continuously differentiable maps $f : [0, A] \to \mathbb{R}$ (respectively, $f : [0, A] \to \mathbb{C}$). For $s \geq 0$ and $k \geq 0$, define a bounded linear map $L_{s,F} : Y_{\mathbb{R}}^k \to Y_{\mathbb{R}}^k$ by

$$(L_{s,F} f)(x) = \sum_{b \in F} \left(\frac{1}{x+b}\right)^{2s} f\left(\frac{1}{x+b}\right). \tag{1.2}$$

The same Equation (1.2) defines a bounded linear map of $Y_{\mathbb{C}}^k$ to $Y_{\mathbb{C}}^k$. From the functional analysis viewpoint, if $0 \leq j < k$, the map $L_{s,F}$ when considered as a map on $Y_{\mathbb{R}}^k$ or $Y_{\mathbb{C}}^k$ should be distinguished from $L_{s,F}$ when considered on $Y_{\mathbb{R}}^j$ or $Y_{\mathbb{C}}^j$. However, for all $s > 0$, there exists a $C^\infty$, strictly positive function $\nu_{s,F} : [0, \infty) \to [0, \infty)$ which is decreasing on $[0, \infty)$ and

$$(L_{s,F} \nu_{s,F})(x) = \lambda_{s,F} \nu_{s,F}(x), \text{ for } 0 \leq x \leq A, \tag{1.3}$$

so $\nu_{s,F}$ is an eigenvector with positive eigenvalue $\lambda_{s,F}$. The map $s \to \lambda_{s,F}$ is strictly decreasing and real analytic for $s \geq 0$, $\lambda_{0,F} = |F|$, and $\lim_{s\to\infty} \lambda_{s,F} = 0$. The functional analysis spectral radius of $L_{s,F}$, where $L_{s,F}$ is considered as a map on $Y_{\mathbb{R}}^k$ or $Y_{\mathbb{C}}^k$, equals $\lambda_{s,F}$. Recall that $L_{s,F}$, considered as a map from $Y_{\mathbb{C}}^k$ to $Y_{\mathbb{C}}^k$, has a spectrum $\sigma(L_{s,F})$ defined by

$$\sigma(L_{s,F}) = \{z \in \mathbb{C} \mid zI - L_{s,F} : Y_{\mathbb{C}}^k \to Y_{\mathbb{C}}^k \text{ is not one-one and onto}\}. \tag{1.4}$$

Here $I$ denotes the identity map on $Y_{\mathbb{C}}^k$. The spectrum $\sigma(L_{s,F})$ in Equation (1.4) may depend on $k$ but for $k \geq 1$

$$\sup\{|z| : z \in \sigma(L_{s,F}) \backslash \{\lambda_{s,F}\}\} < \lambda_{s,F}, \tag{1.5}$$

and $\lambda_{s,F}$ is an eigenvalue of algebraic multiplicity 1 in $Y_{\mathbb{C}}^k$, $k \geq 1$. Note that in general, Equation (1.5) may fail if $k = 0$. Furthermore, whether $L_{s,F}$ is considered as a map from $Y_{\mathbb{R}}^k$ to $Y_{\mathbb{R}}^k$ or $Y_{\mathbb{C}}^k$ to $Y_{\mathbb{C}}^k$ it is known that for $k \geq 0$

$$\lambda_{s,F} = \lim_{n\to\infty} \|L_{s,F}^n\|^{\frac{1}{n}} = \inf_{n \geq 1} \|L_{s,F}^n\|^{\frac{1}{n}},$$

so $\lambda_{s,F} = r(L_{s,F})$, where $r(L_{s,F})$ denotes the functional analytic "spectral radius of $L_{s,F}$" considered as a map from $Y_{\mathbb{R}}^k$ to $Y_{\mathbb{R}}^k$ and $k \in \mathbb{N} \cup \{0\}$. In general, if $s \geq 0$, $k \geq 0$ and $w \in Y_{\mathbb{R}}^k$ satisfies $w(x) > 0$ for $0 \leq x \leq A$ and $(L_{s,F} w) = \alpha w$ for some $\alpha > 0$, then $\alpha = \lambda_{s,F}$ and $w$ is a positive scalar multiple of $\nu_{s,F}$. In particular, if $w$ and $\nu_{s,F}$ are "normalized" so that $w(0) = 1$ and $\nu_{s,F}(0) = 1$, then $w = \nu_{s,F}$.

If $s = 0, v_{0,F} = 1$ for $0 \leq x \leq A$ and $\lambda_{0,F} = |F|$, so $\lambda_{0,F} > 1$ if $|F| > 1$. By our previous remarks, $s \to \lambda_{s,F}$ is strictly decreasing and $C^\infty$ and $\lim_{s\to\infty} \lambda_{s,F} = 0$, so if $|F| > 1$, there is a unique value of $s > 0$, call it $s_*$, such that $\lambda_{s,F} = 1$, In this case, it has been known for some time that $s_* = \dim(F)$. In fact, such a formula is known in a much more general context than continued fractions or complex continued fractions. See, for example, [**7**] and earlier papers. Note that any mathematically rigorous estimates for $\dim(F)$ must give rigorous upper and lower bounds for $s_*$, and if matrix "approximants" $B_s$ to $L_{s,F}$ are employed, the relationship of $r(B_s)$, the spectral radius of $B_s$ to $r(L_{s,F})$ must be rigorously justified. See, for example, [**1–3**].

If the set $F$ is not too large, it is possible to give a reasonably high order, rigorous approximation of $\dim(F)$, e.g., see [**1**–**3**]. Sometimes, one wants results which guarantee that $\dim(F_1) > \dim(F_2)$, even when $|F_1|$ and $|F_2|$ are large. The purpose of this note is to give some results which guarantee, without explicit computation, that $\dim(F_1) > \dim(F_2)$. We suspect that our method can be greatly generalized; see Section 3.

By our remarks above, we shall need some simple tool to estimate $r(L_{s,F})$. More generally, suppose that $M$ is a compact Hausdorff space and that $X = C_{\mathbb{R}}(M)$ is the real Banach space of continuous maps $f : M \to \mathbb{R}$ and $\widehat{X} = C_{\mathbb{C}}(M)$ is the corresponding complex Banach space of continuous maps $f : M \to \mathbb{C}$. If $B : X \to X$ is a bounded real linear map and $\widehat{B} : \widehat{X} \to \widehat{X}$ is the corresponding bounded complex linear map, it is well-known that $r(B) = r(\widehat{B})$ and

$$\lim_{n\to\infty} \|B^n\|^{\frac{1}{n}} = \lim_{n\to\infty} \|\widehat{B}^n\|^{\frac{1}{n}} = \inf_{n\geq 1} \|B^n\|^{\frac{1}{n}} = \inf_{n\geq 1} \|\widehat{B}^n\|^{\frac{1}{n}} = r(B) = r(\widehat{B}).$$

If $K(M) := \{f \in X | f(x) \geq 0 \,\forall x \in M\}$ and $B(K(M)) \subset K(M)$, $B$ is called positive (with respect to $K(M)$). If $B$ is a positive and $u(x) = 1$ for all $x \in M$ and $f \in X$ satisfies $f(x) > 0 \,\forall x \in M$, it is not hard to prove that

$$\lim_{n\to\infty} \|B^n(u)\|^{\frac{1}{n}} = \lim_{n\to\infty} \|B^n(f)\|^{\frac{1}{n}} = r(B) := \lim_{n\to\infty} \|B^n\|^{\frac{1}{n}}. \tag{1.6}$$

Note that $K(M)$ induces a partial ordering $f \leq g$ on $X$ by defining $f \leq g$ if and only if $g - f \in K(M)$, so if $f \leq g$ and $B$ is a positive then $Bf \leq Bg$. If $B$ is positive, $g \in K(M)\backslash\{0\}$ and $Bg \geq \mu g$ for some $\mu > 0$, one can easily prove that

$$r(B) \geq \mu. \tag{1.7}$$

If $h \in X$ and $h(x) > 0$ for all $x \in M$ and $Bh \leq \nu h$ for some $\nu > 0$, one can check that

$$r(B) \leq \nu. \tag{1.8}$$

If $F \subset \mathbb{N}$ and $F$ is finite, $s > 0$, $D = \frac{d}{dx}$ and $\gamma = \min\{m \mid m \in F\}$, we shall need an estimate for $\frac{D\nu_{s,F}(x)}{\nu_{s,F}(x)}$ on $[0, A]$, $A \geq \frac{1}{\gamma}$ :

$$\frac{-2s}{\gamma} \leq \frac{D\nu_{s,F}(x)}{\nu_{s,F}(x)} < 0, \text{ for } 0 \leq x \leq A. \tag{1.9}$$

Equation (1.9) is a special case of Corollary 6.4 in [**1**].

## 2. Comparing $\dim(F_1)$ and $\dim(F_2)$ for $F_1 \subset \mathbb{N}$ and $F_2 \subset \mathbb{N}$

We begin with a simple calculus lemma.

LEMMA 2.1. *Assume that $m_1$ and $m_2$ are positive integers with $m_1 < m_2$, $s$ is a positive real and $\gamma$ is a positive integer. Define $f(x)$ for $x \geq 0$ by*

$$f(x) = \Big(\frac{x + m_2}{x + m_1}\Big)^{2s} \exp\Big( - \frac{2s}{\gamma}\Big(\frac{1}{x + m_1} - \frac{1}{x + m_2}\Big)\Big). \tag{2.1}$$

*If $\gamma \geq 2$ or $m_1 \geq 2$, then $f'(x) < 0$ for all $x \geq 0$, so $\min\{f(x) : 0 \leq x \leq 1\} = f(1)$ if $\gamma \geq 2$ or $m_1 \geq 2$. If $\gamma = 1$ and $m_1 = 1$, then $\min\{f(x) : 0 \leq x \leq 1\} = f(1)$ if $m_2 > 3$ and $\min\{f(x) : 0 \leq x \leq 1\} = f(0)$ if $m_2 \in \{2, 3\}$.*

PROOF. A calculation gives

$$
\begin{aligned}
h(x) &:= \Big(\frac{1}{2s}\Big)\ln(f(x)) \\
&= \Big\{\ln(x+m_2)-\ln(x+m_1)-\frac{1}{\gamma}\Big(\frac{1}{x+m_1}-\frac{1}{x+m_2}\Big)\Big\}.
\end{aligned} \tag{2.2}
$$

So

$$
\begin{aligned}
h'(x) &= \Big[\frac{1}{x+m_2}-\frac{1}{x+m_1}\Big]\Big[1-\frac{1}{\gamma}\Big(\frac{1}{x+m_2}+\frac{1}{x+m_1}\Big)\Big] \\
&:= \Big[\frac{1}{x+m_2}-\frac{1}{x+m_1}\Big]g(x).
\end{aligned} \tag{2.3}
$$

Since $\left(\frac{1}{x+m_2}-\frac{1}{x+m_1}\right)<0$ for all $x>0$ and $\left[1-\frac{1}{\gamma}\left(\frac{1}{x+m_2}+\frac{1}{x+m_1}\right)\right]>0$ for all $x\geq 0$ if $\gamma\geq 2$ or $m_1\geq 2$, we see that $h'(x)<0$ for all $x\geq 0$ if $\gamma\geq 2$ or $m_1\geq 2$ and therefore

$$\min\{h(x):0\leq x\leq 1\}=h(1) \text{ if } \gamma\geq 2 \text{ or } m_1\geq 2.$$

If $\gamma=1$ and $m_1=1$, Equation (2.3) implies that $h'(x)<0$ for $x\geq 1$ and $h'(0)>0$ and $h'(x)=0$ for $x\geq 0$ if $0<x<1$ and

$$1-\frac{1}{x+m_2}-\frac{1}{x+1}:=g(x)=0. \tag{2.4}$$

The quadratic equation given by Equation (2.4) has the unique solution $x_+\in(0,1)$ given by $x_+=\sqrt{\left(\frac{m_2-1}{2}\right)^2+1}-\left(\frac{m_2-1}{2}\right)$. It follows that $h'(x)>0$ for $0\leq x<x_+$ and $h'(x)<0$ for $x_+<x\leq 1$ and that the minimum value of $h(x)$ on $[0,1]$ equals either $h(0)$ or $h(1)$, with a corresponding statement for $f(x)$.

For $m_1=1$, $\gamma=1$ and $m_2\geq 2$, a calculation gives

$$h(0)-h(1)=\ln(m_2)-\ln(m_2+1)+\frac{1}{m_2}-\frac{1}{m_2+1}+\ln(2)-\frac{1}{2}. \tag{2.5}$$

Viewing $m_2\geq 2$ as a real variable, we see that for $m_2\geq 2$

$$
\begin{aligned}
\Big(\frac{d}{dm_2}\Big)(h(0)-h(1)) &= \Big(\frac{1}{m_2}-\frac{1}{m_2+1}\Big)-\Big(\frac{1}{m_2}-\frac{1}{m_2+1}\Big)\Big(\frac{1}{m_2}+\frac{1}{m_2+1}\Big) \\
&= \Big(\frac{1}{m_2}-\frac{1}{m_2+1}\Big)\Big(1-\frac{1}{m_2}-\frac{1}{m_2+1}\Big)>0.
\end{aligned} \tag{2.6}
$$

It is clear that $\lim_{m_2\to\infty}[h(0)-h(1)]=\ln(2)-\frac{1}{2}>0$, so Equation (2.6) implies that $\ln\left(\frac{m_2}{m_2+1}\right)+\frac{1}{m_2(m_2+1)}<0$ for all $m_2\geq 2$ and if $h(0)-h(1)\geq 0$ for some $m_2\geq 2$, then $h(0)-h(1)\geq 0$ for all positive integers $m_2'\geq m_2$ and

$$\min\{h(x):x\in[0,1]\}=h(1).$$

A calculation, which we leave to the reader, shows that $h(0)-h(1)<0$ for $m_1=1$ and $m_2\in\{2,3\}$ but $h(0)-h(1)>0$ for $m_1=1$ and $m_2=4$. It follows that $h(x)$ achieves its minimum on $[0,1]$ at $x=0$ for $m_1=1$ and $m_2\in\{2,3\}$ and achieves its minimum on $[0,1]$ at $x=1$ for $m_1=1$ and $m_2\geq 4$. □

For the remainder of this section, we shall denote by $F_0$ a finite, nonempty set of positive integers and denote by $m_1$ and $m_2$ positive integers such that $m_1<m_2$ and $m_j\notin F_0$ for $j=1,2$. We shall write $F_1=\{m_1\}\cup F_0$ and $F_2=\{m_2\}\cup F_0$. In the notation of Section 1, $E(F_j)$, $j=1$, 2, denotes the invariant set of the iterated

function system $\{\theta_b : [0,1] \to [0,1] : b \in F_j\}$, where $\theta_b(x) = \frac{1}{x+b}$. Recall that if $s \geq 0, L_{s,F_j}$ (respectively $\Lambda_{s,F_j}$) denotes a bounded linear maps $L_{s,F_j} : C_{\mathbb{R}}([0,1]) \to C_{\mathbb{R}}([0,1])$ (respectively, $\Lambda_{s,F_j} : C_{\mathbb{C}}([0,1]) \to C_{\mathbb{C}}([0,1])$) given by

$$(L_{s,F_j}(f))(x) = \sum_{b \in F_j} |\theta_b'(x)|^s f(\theta_b(x)).$$

As noted in Section 1, $L_{s,F_j}$ has a unique, normalized strictly positive eigenvector $\nu_{s,F_j}$ with eigenvalue $\lambda_{s,F_j} > 0$ and $\lambda_{s,F_j}$ is the functional analysis spectral radius of $L_{s,F_j}$. For each $s \geq 0$, the map $x \to \nu_{s,F_j}(x), x \in [0,1]$, is $C^\infty$, the map $s \to \lambda_{s,F_j}$ is $C^\infty$ and strictly decreasing for $s \geq 0$, and $\dim(F_j)$, the Hausdorff dimension of $F_j$, equals the unique value of $s$ such that $\lambda_{s,F_j} = 1$. If $D_1$ denotes $\frac{d}{dx}$ and $\gamma_j = \min\{m : m \in F_j\}$ we also have (see Equation (1.9), Section 1) that for $s > 0$

$$-\Big(\frac{2s}{\gamma_j}\Big) \leq \frac{D_1 \nu_{s,F_j}(x)}{\nu_{s,F_j}(x)} = D_1 \ln(\nu_{s,F_j}(x)) \leq 0, \; x \in [0,1]. \tag{2.7}$$

THEOREM 2.2. *For all $s > 0$, $\lambda_{s,F_1} > \lambda_{s,F_2}$ and $\dim(F_1) > \dim(F_2)$.*

PROOF. For notational convenience, we define $w_s = \nu_{s,F_2}$, so $L_{s,F_2} w_s = \lambda_{s,F_2} w_s$. If, for $s > 0$, we can prove that there is a constant $\mu_s > 1$ such that $L_{s,F_1} w_s \geq \mu_s \lambda_{s,F_2} w_s$, then the remarks in Section 1 imply that for $s > 0$, $\lambda_{s,F_1} \geq \mu_s \lambda_{s,F_2} > \lambda_{s,F_2}$, which implies Theorem 2.2.

The key step is to compare the terms $\left(\frac{1}{x+m_1}\right)^{2s} w_s\left(\frac{1}{x+m_1}\right)$ and $\left(\frac{1}{x+m_2}\right)^{2s} w_s\left(\frac{1}{x+m_2}\right)$. Since $u \to \ln(w_s(u))$ is a decreasing function and $\frac{1}{x+m_1} \geq \frac{1}{x+m_2}$ for all $x \in [0,1]$ we have, for $\gamma_2$ as above

$$\ln\Big(w_s\Big(\frac{1}{x+m_1}\Big)\Big) - \ln\Big(w_s\Big(\frac{1}{x+m_2}\Big)\Big) \geq -\frac{2s}{\gamma_2}\Big(\frac{1}{x+m_1} - \frac{1}{x+m_2}\Big),$$

which implies that

$$w_s\Big(\frac{1}{x+m_1}\Big) \geq w_s\Big(\frac{1}{x+m_2}\Big) \exp\Big[-\frac{2s}{\gamma_2}\Big(\frac{1}{x+m_1} - \frac{1}{x+m_2}\Big)\Big]. \tag{2.8}$$

We derive from Equation (2.8) that

$$\begin{aligned}&\Big(\frac{1}{x+m_1}\Big)^{2s} w_s\Big(\frac{1}{x+m_1}\Big) \geq \\ &\Big(\frac{x+m_2}{x+m_1}\Big)^{2s} \exp\Big[-\frac{2s}{\gamma_2}\Big(\frac{1}{x+m_1} - \frac{1}{x+m_2}\Big)\Big]\Big(\frac{1}{x+m_2}\Big)^{2s} w_s\Big(\frac{1}{x+m_2}\Big).\end{aligned} \tag{2.9}$$

Since $\gamma_2 \geq 1$ we obtain that

$$\begin{aligned}&\Big(\frac{1}{x+m_1}\Big)^{2s} w_s\Big(\frac{1}{x+m_1}\Big) \geq \\ &\Big(\frac{x+m_2}{x+m_1}\Big)^{2s} \exp\Big[-2s\Big(\frac{1}{x+m_1} - \frac{1}{x+m_2}\Big)\Big]\Big(\frac{1}{x+m_2}\Big)^{2s} w_s\Big(\frac{1}{x+m_1}\Big).\end{aligned} \tag{2.10}$$

If $m_1 \geq 2$ or $m_1 = 1$ and $m_2 \geq 4$, Lemma 2.1 implies that for $0 \leq x \leq 1$,

$$\begin{aligned}&\Big(\frac{x+m_2}{x+m_1}\Big)^{2s} \exp\Big[-2s\Big(\frac{1}{x+m_1} - \frac{1}{x+m_2}\Big)\Big] \\ &\geq \Big[\frac{m_2+1}{m_1+1}\exp\Big(\frac{1}{1+m_2} - \frac{1}{1+m_1}\Big)\Big]^{2s} := \nu^{2s},\end{aligned} \tag{2.11}$$

so we have to prove that $\left(\frac{m_2+1}{m_1+1}\right)\exp\left(\frac{1}{1+m_2}-\frac{1}{1+m_1}\right)>1$ or, equivalently

$$\ln(m_2+1)+\frac{1}{m_2+1}>\ln(m_1+1)+\frac{1}{m_1+1}. \tag{2.12}$$

If we write $g(u):=\ln(u)+\frac{1}{u}$, $g'(u)=\frac{1}{u}-\frac{1}{u^2}>0$ for $u>1$, so $g(m_2+1)>g(m_1+1)$ and Equation (2.12) is satisfied. If $m_1=1$ and $m_2\in\{2,3\}$, Lemma 2.1 implies that for $0\leq x\leq 1$,

(2.13)
$$\left(\frac{x+m_2}{x+m_1}\right)^{2s}\exp\Big[-2s\Big(\frac{1}{x+m_1}-\frac{1}{x+m_2}\Big)\Big]\geq(m_2)^{2s}\exp\Big[-2s\Big(1-\frac{1}{m_2}\Big)\Big].$$

Thus, if $m_1=1$ and $m_2\in\{2,3\}$ it suffices to prove that $m_2\exp\left(\frac{1}{m_2}-1\right)>1$, as the reader can directly verify.

Thus we have that for arbitrary $s>0$ and $m_1,m_2\in\mathbb{N}$ with $m_1<m_2$ there exists a constant $\nu>1$, dependent on $m_1$ and $m_2$, such that

$$\begin{aligned}\Big(\frac{1}{x+m_1}\Big)^{2s}w_s\Big(\frac{1}{x+m_1}\Big)\geq\Big(\frac{1}{x+m_2}\Big)^{2s}w_s\Big(\frac{1}{x+m_2}\Big)+\\ (\nu^{2s}-1)\Big(\frac{1}{x+m_2}\Big)^{2s}w_s\Big(\frac{1}{x+m_2}\Big).\end{aligned} \tag{2.14}$$

Because $w_s(x)$ and $\left(\frac{1}{x+m_2}\right)^{2s}w_s\left(\frac{1}{x+m_2}\right)$ are strictly positive and continuous on $[0,1]$ there is a positive constant $\delta_s$, such that

$$(\nu^{2s}-1)\Big(\frac{1}{x+m_2}\Big)^{2s}w_s\Big(\frac{1}{x+m_2}\Big)\geq\delta_s w_s(x),\ \text{for all } x\in[0,1]. \tag{2.15}$$

Using Equation (2.14) and Equation (2.15) we see that

$$\begin{aligned}((L_{s,F_1})(w_s))(x)&\geq((L_{s,F_2})(w_s))(x)+\delta_s w_s(x)\\ &=(\lambda_{s,F_2}+\delta_s)w_s(x).\end{aligned} \tag{2.16}$$

Using Equation (2.16) and the results of Section 1 we conclude for all $s>0$

$$\lambda_{s,F_1}>\lambda_{s,F_2}+\delta_s, \tag{2.17}$$

which implies that $\dim(F_1)>\dim(F_2)$. □

The same ideas used in the proof of Theorem 2.2 can be used to prove a more general theorem.

THEOREM 2.3. *Let* $\{a_j\in\mathbb{N}:1\leq j\leq n\}:=S_1$ *and* $\{b_j\in\mathbb{N}:1\leq j\leq n\}:=S_2$ *be such that* $a_j<a_{j+1}$ *and* $b_j<b_{j+1}$ *for* $j<n$ *and* $a_j<b_j$ *for* $1\leq j\leq n$. *Assume that* $F_0\subset\mathbb{N}$ *is nonempty if* $n=1$ *and that* $S_1$ *is not contained in* $F_0$ *and* $S_2$ *is not contained in* $F_0$. *Assume that* $a_j\notin F_0$ *whenever* $b_j\notin F_0, 1\leq j\leq n$, *and define* $J=\{j:b_j\notin F_0\}$. *Define* $S_1'=\{a_j:j\in J\}$ *and* $S_2'=\{b_j:j\in J\}$, *so* $F_2':=S_2'\cup F_0=S_2\cup F_0=F_2$ *and* $S_1'\cup F_0:=F_1'\subset S_1\cup F_0:=F_1$. *Then we have that*

$$\dim(F_2)=\dim(F_2')<\dim(F_1')\leq\dim(F_1).$$

PROOF. As in the proof of Theorem 2.2, for $s>0$ we let $w_s=\nu_{s,F_2'}$ be the normalized strictly positive eigenvector for the positive linear operator $L_{s,F_2}=L_{s,F_0}+L_{s,S_2'}$, so $L_{s,F_2}w_s=\lambda_{s,F_2}w_s$ and

$$L_{s,F_2}w_s=L_{s,F_0}w_s+L_{s,S_2'}w_s. \tag{2.18}$$

We have to compare $L_{s,F_1'}w_s$ to Equation (2.18), where

$$L_{s,F_1'}w_s = L_{s,F_0}w_s + L_{s,S_1'}w_s. \tag{2.19}$$

For $j \in J$, we have

$$(L_{s,S_2'}w_s)(x) = \sum_{j\in J}\Big(\frac{1}{x+b_j}\Big)^{2s} w_s\Big(\frac{1}{x+b_j}\Big) \text{ and} \tag{2.20a}$$

$$(L_{s,S_1'}w_s)(x) = \sum_{j\in J}\Big(\frac{1}{x+a_j}\Big)^{2s} w_s\Big(\frac{1}{x+a_j}\Big) \tag{2.20b}$$

as in the proof of Theorem 2.2 (see Equation (2.14)) there is a positive constant, $\nu_j > 1$ such that for $0 \le x \le 1$

$$\begin{aligned} &\Big(\frac{1}{x+a_j}\Big)^{2s} w_s\Big(\frac{1}{x+a_j}\Big) \\ &\ge \Big(\frac{1}{x+b_j}\Big)^{2s} w_s\Big(\frac{1}{x+b_j}\Big) + (\nu_j^{2s}-1)\Big(\frac{1}{x+b_j}\Big)^{2s} w_s\Big(\frac{1}{x+b_j}\Big). \end{aligned} \tag{2.21}$$

Because $J$ is finite, we can replace $\nu_j$ by $\nu = \min\{\nu_j : j \in F\} > 1$ in Equation (2.21). Again because $J$ is finite we can argue as in Theorem 2.2 (see Equation (2.15)) that there is a positive constant $\delta_s$, independent of $j$, such that for all $x \in [0,1]$

$$(\nu^{2s}-1)\Big(\frac{1}{x+b_j}\Big)^{2s} w_s\Big(\frac{1}{x+b_j}\Big) \ge \delta_s w_s(x). \tag{2.22}$$

Using Equation (2.22), we see that

$$\begin{aligned} (L_{s,F_1'})(w_s) &\ge (L_{s,F_2'}w_s)(x) + |J|\delta_s w_s(x) \\ &= (\lambda_{s,F_2'} + |J|\delta_s)w_s(x). \end{aligned} \tag{2.23}$$

As in the proof of Theorem 2.2, Equation (2.23) implies that for all $s > 0$, $\lambda_{s,F_1} \ge \lambda_{s,F_1'} > \lambda_{s,F_2}$ and $\dim(F_1) > \dim(F_2)$. □

## 3. A general question and a suggestive example

Suppose that $G_1 \subset \mathbb{N}$ and $G_2 \subset \mathbb{N}$ are disjoint sets with $|G_1| = |G_2|$ and, in the notation of Section 1, $\dim(G_1) > \dim(G_2)$. Assume that $F_0$ is a non-empty finite subset of $\mathbb{N}$ and $G_1 \cap F_0 = \emptyset$ and $G_2 \cap F_0 = \emptyset$. We write $F_1 = F_0 \cup G_1$ and $F_2 = F_0 \cup G_2$ and we note that we can allow $F_0 = \emptyset$ if $|G_1| > 1$. Under what further condition, if any, is it true that $\dim(F_1) > \dim(F_2)$? For $s \ge 0$, suppose that $L_{s,F_2} : X = C_{\mathbb{R}}([0,1]) \to X$ is the bounded linear operator we have previously defined and $w_s := v_{s,F_2}$ denotes the unique normalized, strictly positive eigenvector of $L_{s,F_2}$ with eigenvalue $\lambda_{s,F_2}$. Recall that we know that $\lambda_{s,F_2}$ equals the spectral radius of $L_{s,F_2}$ and $s \to \lambda_{s,F_2}$ is strictly decreasing, real analytic and $w_s$ is defined and $C^\infty$ on $[0,\infty)$. Also, if $b \ge \gamma$ for all $b \in F_2$ and $s > 0$, then we know that for all $x \ge 0$,

$$-\frac{2s}{\gamma} \le \frac{D_1 w_s(x)}{w_s(x)} < 0. \tag{3.1}$$

If $m_1$ and $m_2$ are any positive integers with $m_1 < m_2$ and $b \geq \gamma \geq 2$ for all $b \in F_2$, then it follows from Equation (2.9) in Section 2 that for $x \geq 0$

$$\begin{aligned}&\left(\frac{1}{x+m_1}\right)^{2s} w_s\left(\frac{1}{x+m_1}\right)\\ (3.2)\quad &\geq \left(\frac{x+m_2}{x+m_1}\right)^{2s} \exp\left[-s\left(\frac{1}{x+m_1}-\frac{1}{x+m_2}\right)\right]\left(\frac{1}{x+m_2}\right)^{2s} w_s\left(\frac{1}{x+m_2}\right).\end{aligned}$$

Since $\gamma \geq 2$, Lemma 2.1 implies that for $0 \leq x \leq 1$,

$$\begin{aligned}&\left(\frac{1}{x+m_1}\right)^{2s} w_s\left(\frac{1}{x+m_1}\right)\\ (3.3)\quad &\geq \left(\frac{m_2+1}{m_1+1}\right)^{2s} \exp\left[-s\left(\frac{1}{m_1+1}-\frac{1}{m_2+1}\right)\right]\left(\frac{1}{x+m_2}\right)^{2s} w_s\left(\frac{1}{x+m_2}\right).\end{aligned}$$

Suppose we can prove that for all $s \geq s_0 = \dim(G_2)$, there exists a positive constant $\delta_s$ such that, for $0 \leq x \leq 1$,

$$\sum_{b\in G_1}\left(\frac{1}{x+b}\right)^{2s} w_s\left(\frac{1}{x+b}\right) \geq \sum_{b\in G_2}\left(\frac{1}{x+b}\right)^{2s} w_s\left(\frac{1}{x+b}\right) + \delta_s w_s(x).$$

Then the same argument used in Section 2 shows that for $s \geq s_0$, $\lambda_{s,F_1} > \lambda_{s,F_2}$. If, for some $s > s_0$, $\lambda_{s,F_2} = 1$, then $\dim(F_2) = s$ and $\lambda_{s,F_1} > 1$. Since $s \to \lambda_{s,F_1}$ is strictly decreasing for $s > 0$, we know that $\lambda_{s,F_1} > 1$ for $0 \leq s \leq s_0$, so $\dim(F_1) > \dim_( F_2)$.

We believe that the argument suggested above can be used to provide a positive answer to Question 3 for a variety of interesting examples. However, rather than consider the general case, we shall restrict attention here to the case that $G_1 = \{1,4\}$, $G_2 = \{2,3\}$ and $F_0 \subset \mathbb{N}$ is a finite set, possibly empty, such that $F_0 \cap G_j = \emptyset$ for $j = 1,2$. We define $F_j = F_0 \cup G_j$ for $j = 1,2$. Our goal is to prove

THEOREM 3.1. *Under the above assumptions, we have*

$$\dim(F_1) > \dim(F_2). \tag{3.4}$$

*Also, it is true that* $\lambda_{s,F_1} > \lambda_{s,F_2}$ *for all* $s > 0$.

We note that it is proved in [**1**] (see Table 3.1 on page 295 of [**1**]) that

$$0.33743678079 \leq \dim(E[2,3]) := \dim(\{2,3\}) \leq 0.33743678082, \tag{3.5}$$

and

$$0.4111827246 \leq \dim(E[1,4]) := \dim(\{1,4\}) \leq 0.41118272481. \tag{3.6}$$

However, we shall not use Equation (3.5) and Equation (3.6) here.

LEMMA 3.2. $\dim(F_2) \geq \dim(G_2) \geq 0.31$.

PROOF. Given $s \geq 0$, let us denote the unique strictly positive eigenvector for $L_{s,G_2}$ with eigenvalue $\lambda_{s,G_2}$ and $u_s(0) = 1$, so, for $0 \leq x \leq 1$,

$$\lambda_{s,G_2} u_s(x) = \left(\frac{1}{x+2}\right)^{2s} u_s\left(\frac{1}{x+2}\right) + \left(\frac{1}{x+3}\right)^{2s} u_s\left(\frac{1}{x+3}\right). \tag{3.7}$$

Because $b \geq 2$ for all $b \in G_2$ and for all $b \in F_2$, we know that

$$-s \leq \frac{u_s'(x)}{u_s(x)} < 0 \quad \forall x > 0.$$

Thus, we see that

$$u_s\left(\frac{1}{2}\right) \geq u_s(0)\exp\left(-\frac{1}{2}s\right) = \exp\left(-\frac{1}{2}s\right) \text{ and } u_s\left(\frac{1}{3}\right) \geq \exp\left(-\frac{1}{3}s\right).$$

It follows that

$$\begin{aligned}\lambda_{s,G_2}u_s(0) = \lambda_{s,G_2} &= \left(\frac{1}{2}\right)^{2s} u_s\left(\frac{1}{2}\right) + \left(\frac{1}{3}\right)^{2s} u_s\left(\frac{1}{3}\right)\\ &\geq \left(\frac{1}{4}\right)^{s} \exp\left(-\frac{1}{2}s\right) + \left(\frac{1}{9}\right)^{s} \exp\left(-\frac{1}{3}s\right). \qquad (3.8)\end{aligned}$$

If $s_* = \dim(G_2)$, so $\lambda_{s_*,G_2} = 1$, then because $s \to \lambda_{s,G_2}$ is strictly decreasing, we see from Equation (3.8) that, for $0 \leq s \leq s_*$,

$$\left(\frac{1}{2}\right)^{2s} u_s\left(\frac{1}{2}\right) + \left(\frac{1}{3}\right)^{2s} u_s\left(\frac{1}{3}\right) > 1. \qquad (3.9)$$

Equation (3.8) then implies that if

$$\left(\frac{1}{4}\right)^{s} \exp\left(-\frac{1}{2}s\right) + \left(\frac{1}{9}\right)^{s} \exp\left(-\frac{1}{3}s\right) = h(s) > 1, \qquad (3.10)$$

then $0 \leq s < s_*$. The map $s \to h(s)$ is strictly decreasing on $[0, \infty)$, so if $h(s_0) > 1$ we see that $h(s) > 1$ for $0 \leq s < s_0$. A calculation shows that $h(0.31) > 1$, which implies that $\dim(G_2) > 0.31$. Since $G_2 \subset F_2$, we see that $\dim(F_2) > 0.31$ □

PROOF OF THEOREM 3.1. For $s \geq 0$, let $w_s$ be the unique strictly positive eigenvector for $L_{s,F_2}$, normalized by $w_s(0) = 1$ and with eigenvalue $\lambda_{s,F_2}$. If we are only interested in proving that $\dim(F_1) > \dim(F_2)$, Lemma 3.2 shows that we can restrict attention to $s \geq 0.31$. However, we shall allow general $s > 0$ and prove that $r(L_{s,F_1}) = \lambda_{s,F_1} > \lambda_{s,F_2} > 0$ for all $s > 0$. For different examples, it may be necessary to restrict $s$ to prove that $\dim(F_1) > \dim(F_2)$. Because $\gamma \geq 2$, we see, using Lemma 2.1, that

$$\begin{aligned}&\left(\frac{1}{x+1}\right)^{2s} w_s\left(\frac{1}{x+1}\right)\\ &\quad\geq \left(\frac{x+2}{x+1}\right)^{2s} \exp\left(-s\left(\frac{1}{x+1} - \frac{1}{x+2}\right)\right)\left(\frac{1}{x+2}\right)^{2s} w_s\left(\frac{1}{x+2}\right)\\ &\quad\geq \left[\left(\frac{3}{2}\right)^{2s} \exp\left(-s\left(\frac{1}{2} - \frac{1}{3}\right)\right)\right]\left(\frac{1}{x+2}\right)^{2s} w_s\left(\frac{1}{x+2}\right) \qquad (3.11)\\ &\quad= \left[\left(\frac{9}{4}\right)\exp\left(-\frac{1}{6}\right)\right]^{s}\left(\frac{1}{x+2}\right)^{2s} w_s\left(\frac{1}{x+2}\right).\end{aligned}$$

A calculation shows that $\frac{9}{4}\exp\left(-\frac{1}{6}\right) \geq 1.9045388$, so

$$\left(\frac{1}{x+1}\right)^{2s} w_s\left(\frac{1}{x+1}\right) \geq \left(\frac{1}{x+2}\right)^{2s} w_s\left(\frac{1}{x+2}\right) + \left(\left[\left(\frac{9}{4}\right)\exp\left(-\frac{1}{6}\right)\right]^s - 1\right)\left(\frac{1}{x+2}\right)^{2s} w_s\left(\frac{1}{x+2}\right). \tag{3.12}$$

Applying Lemma 2.1 and Equation (3.3), we see that

$$\begin{aligned}\left(\frac{1}{x+2}\right)^{2s} w_s\left(\frac{1}{x+2}\right) &\geq \left(\frac{x+3}{x+2}\right)^{2s}\exp\left(-s\left(\frac{1}{x+2}-\frac{1}{x+3}\right)\right)\left(\frac{1}{x+3}\right)^{2s} w_s\left(\frac{1}{x+3}\right)\\ &\geq \left[\left(\frac{4}{3}\right)^{2s}\exp\left(-s\left(\frac{1}{3}-\frac{1}{4}\right)\right)\right]\left(\frac{1}{x+3}\right)^{2s} w_s\left(\frac{1}{x+3}\right)\\ &= \left[\left(\frac{16}{9}\right)\exp\left(-\frac{1}{12}\right)\right]^s\left(\frac{1}{x+3}\right)^{2s} w_s\left(\frac{1}{x+3}\right).\end{aligned} \tag{3.13}$$

Combining Equation (3.12) and Equation (3.13), we obtain:

$$\begin{aligned}&\left(\frac{1}{x+1}\right)^{2s} w_s\left(\frac{1}{x+1}\right)\\ &\geq \left(\frac{1}{x+2}\right)^{2s} w_s\left(\frac{1}{x+2}\right)\\ &\quad + \left(\left[\left(\frac{9}{4}\right)\exp\left(-\frac{1}{6}\right)\right]^s - 1\right)\left[\left(\frac{16}{9}\right)\exp\left(-\frac{1}{12}\right)\right]^s\left(\frac{1}{x+3}\right)^{2s} w_s\left(\frac{1}{x+3}\right).\end{aligned} \tag{3.14}$$

Since, we have

$$\begin{aligned}\left(\frac{1}{x+4}\right)^{2s} w_s\left(\frac{1}{x+4}\right) &\geq \left(\frac{1}{x+4}\right)^{2s} w_s\left(\frac{1}{x+3}\right)\\ &\geq \left(\frac{x+3}{x+4}\right)^{2s}\left(\frac{1}{x+3}\right)^{2s} w_s\left(\frac{1}{x+3}\right)\\ &\geq \left(\frac{3}{4}\right)^{2s}\left(\frac{1}{x+3}\right)^{2s} w_s\left(\frac{1}{x+3}\right),\end{aligned} \tag{3.15}$$

we conclude from Equation (3.14) and Equation (3.15) that

$$\begin{aligned}&\left(\frac{1}{x+1}\right)^{2s} w_s\left(\frac{1}{x+1}\right) + \left(\frac{1}{x+4}\right)^{2s} w_s\left(\frac{1}{x+4}\right)\\ &\qquad \geq \left(\frac{1}{x+2}\right)^{2s} w_s\left(\frac{1}{x+2}\right) + f(s)\left(\frac{1}{x+3}\right)^{2s} w_s\left(\frac{1}{x+3}\right),\end{aligned} \tag{3.16}$$

$$f(s) := \left(\left[\left(\frac{9}{4}\right)\exp\left(-\frac{1}{6}\right)\right]^s - 1\right)\left(\left[\left(\frac{16}{9}\right)\exp\left(-\frac{1}{12}\right)\right]^s\right) + \left(\frac{3}{4}\right)^{2s}, \tag{3.17}$$

as we have remarked before, it suffices to prove that $f(s) > 1$ for all $s > 0$. It is clear that $\frac{16}{9}\exp\left(-\frac{1}{12}\right) > 1$ and $\frac{9}{4}\exp\left(-\frac{1}{6}\right) > 1$, so to prove that $f(s) > 1$ for all $s > 0$ it suffices to prove that for all $s > 0$,

$$\left[\frac{9}{4}\exp\left(-\frac{1}{6}\right)\right]^s - 1 + \left(\frac{3}{4}\right)^{2s} > 1. \tag{3.18}$$

Because $\ln\left[\frac{9}{4}\exp\left(-\frac{1}{6}\right)\right] \geq 0.6442635$, we see that

$$\left[\frac{9}{4}\exp\left(-\frac{1}{6}\right)\right]^s - 1 \geq \exp((0.6442635)s) - 1 = \sum_{j=1}^{\infty}\frac{(0.6442635)^j s^j}{j!}.$$

Similarly, $\left(\frac{3}{4}\right)^{2s} = \exp\left(s\ln\left(\frac{3}{4}\right)\right)$ and $\ln\left(\frac{3}{4}\right) \geq -0.2876821$, so

$$\left(\frac{3}{4}\right)^{2s} \geq \exp((-0.2876821)(2s)) = 1 + \sum_{j=1}^{\infty}\frac{(-1)^j(0.5753642)^j s^j}{j!}.$$

Obviously, we have that for all $j \geq 1$ and $s > 0$

$$(0.6442635)^j s^j + (-1)^j(0.5753642)^j s^j > 0.$$

So we can conclude that for all $s > 0$ Equation (3.18) is true, which proves Theorem 3.1.

## References

[1] R. S. Falk and R. D. Nussbaum, *$C^m$ eigenfunctions of Perron-Frobenius operators and a new approach to numerical computation of Hausdorff dimension: applications in $\mathbb{R}^1$*, J. Fractal Geom. **5** (2018), no. 3, 279–337, DOI 10.4171/JFG/62. MR3827801

[2] R. S. Falk and R. D. Nussbaum, *A new approach to numerical computation of Hausdorff dimension of iterated function systems: applications to complex continued fractions*, Integral Equations Operator Theory **90** (2018), no. 5, Paper No. 61, 46, DOI 10.1007/s00020-018-2485-z. MR3851775

[3] R. S. Falk and R. D. Nussbaum, *Hidden positivity and a new approach to numerical computation of Hausdorff dimension: higher order methods*, J. Fractal Geom. **9** (2022), no. 1-2, 23–72, DOI 10.4171/jfg/111. MR4461209

[4] D. Hensley, *Continued fractions*, World Scientific Publishing Co. Pte. Ltd., Hackensack, NJ, 2006, DOI 10.1142/9789812774682. MR2351741

[5] B. Lemmens and R. Nussbaum, *Birkhoff's version of Hilbert's metric and its applications in analysis*, Handbook of Hilbert geometry, IRMA Lect. Math. Theor. Phys., vol. 22, Eur. Math. Soc., Zürich, 2014, pp. 275–303. MR3329884

[6] R. D. Nussbaum, *$C^m$ positive eigenvectors for linear operators arising in the computation of Hausdorff dimension*, Integral Equations Operator Theory **84** (2016), no. 3, 357–393, DOI 10.1007/s00020-015-2274-x. MR3463454

[7] R. D. Nussbaum, A. Priyadarshi, and S. Verduyn Lunel, *Positive operators and Hausdorff dimension of invariant sets*, Trans. Amer. Math. Soc. **364** (2012), no. 2, 1029–1066, DOI 10.1090/S0002-9947-2011-05484-X. MR2846362

Department of Mathematics, Rutgers University, New Jersey 08854
*Email address*: nussbaum@emeritus.rutgers.edu

Contemporary Mathematics
Volume **825**, 2025
https://doi.org/10.1090/conm/825/16513

# Multifractal zeta-functions and multifractal prime counting formulas

Lars Olsen

Abstract. Prime counting formulas count the number of appropriately defined "prime elements' less than a given number, e.g. the number of prime numbers less than $x$ or the number of prime geodesics whose length is less than $x$.

In this paper we survey the work from [**Ol1**]. For Graph-Directed Self-Conformal Iterated Function Systems, we introduce a multifractal transfer operator $\mathbf{L}^{q,s}$, where $q \in \mathbb{R}$ and $s \in \mathbb{C}$, and define the multifractal zeta-function $\zeta^q(s)$ by

$$\zeta^q(s) = \det(\mathbf{1} - \mathbf{L}^{q,s}) .$$

We provide a detailed study of the growth of the eigenvalues of $\mathbf{L}^{q,s}$ significantly improving the current state of the art estimates by Bandtlow & Jenkinson from 2008.

Using growth of the eigenvalues of $\mathbf{L}^{q,s}$ and a Dirichlet the series expansion of the zeta function $\zeta^q(s)$ obtained from Atiyah-Bott's fixed point theorem, we develop a general theory of fractal and multifractal prime counting formulas for Graph-Directed Self-Conformal Iterated Function Systems. In particular, we show that asymptotic and explicit formulas for the $q$-th moment of self-conformal measures can be obtained in terms of the zeros of the zeta function $\zeta^q$. As a further application we obtain a multifractal prime geodesic counting formula for classical Fuchsian Schottky groups.

## Introduction

For $r > 0$, the Minkowski volume $V_r(E)$ of a Borel subset $E$ of $\mathbb{R}^d$ is defined by

$$\begin{aligned} V_r(E) &= \frac{1}{r^d}\mathcal{L}^d(B(E,r)) \\ &= \frac{1}{r^d}\int_{B(E,r)} d\mathcal{L}^d(x) \end{aligned}$$

where $B(E,r)$ denotes the $r$ neighbourhood of $E$. Tube formulas refer to the study of the Minkowski volume $V_r(E)$ of a Borel subset $E$ of $\mathbb{R}^d$ . For sets $E$ satisfying some (possible very weak) convexity conditions, this has a long history going back to Steiner in the early 19'th century. However, initiated and pioneered independently by Lapidus and Zähle, the systematic study of tube formulas for fractal sets has

2020 *Mathematics Subject Classification.* Primary 28A78.

*Key words and phrases.* Self-similar measures multifractals, tube formulas, zeta-functions, prime counting formulas.

received enormous interest during the past 20 years. Following this, very recently there has been attempts to extent these ideas and develop a theory of multifractal tube formulas for multifractal measures. There are two approaches for investigating these problems.

(1) *Renewal theory.* Using techniques from renewal theory introduced by Lalley [**La1**, **La2**] in the 1990's, Winter [**Wi**] gave a complete description of the asymptotic behaviour (as $r$ tends to 0) of the Minkowski volume $V_r(K)$ of self-similar subsets $K$ of $\mathbb{R}^d$ satisfying the Open Set Condition (OSC).
(2) *Zeta-functions.* Unfortunately, renewal theory techniques do not yield "explicit" expressions for the functions describing the asymptotic behaviour of the Minkowski volume $V_r(K)$ of self-similar subsets $K$ of $\mathbb{R}^d$ satisfying the OSC. This is clearly undesirable. For this reason, Lapidus et al [Lap-vF1] introduceed and developed a second theoretical framework for studying tube formulas of self-similar sets. This approach is based on fractal zeta-functions and provides "explicit" expressions for the tube formulas of self-similar subsets of $\mathbb{R}$. Indeed, using the Mellin transform and the residue theorem, one can express the tube formulas as series involving the residues of an appropriate zeta-function.

The difficult problem of finding "explicit" expressions for the functions describing the asymptotic behaviour (as $r$ tends to 0) of the Minkowski volume $V_r(K)$ of self-similar sets $K$ satisfying the OSC forms the main motivation for the work that we will survey in this paper. In fact, we will investigate this problem in a much broader multifractal context involving the multifractal Minkowski volume of self-conformal measures. The multifractal Minkowski volume is a natural multifractal generalization of the usual Minkowski volume and is defined as follows: for a probability measure $\mu$ on $\mathbb{R}^d$ and $r > 0$, we define the $q$'th multifractal Minkowski volume $V^q_{\mu,r}(E)$ of a Borel subset $E$ of $\mathbb{R}^d$ with respect to $\mu$ by

$$V^q_{\mu,r}(E) = \frac{1}{r^d}\int_{B(E,r)} \mu(B(x,r))^q \, d\mathcal{L}^d(x)\,.$$

It is of interest to investigate the behaviour of the multifractal Minkowski volume for small values of $r$ and for this reason we define the lower and upper $q$'th Renyi dimension (also sometimes called the lower and upper $L^q$-dimension) of $\mu$ by

$$\underline{T}_\mu(q) = \liminf_{r\searrow 0} \frac{\log V^q_{\mu,r}(K)}{-\log r}$$

and

$$\overline{T}_\mu(q) = \limsup_{r\searrow 0} \frac{\log V^q_{\mu,r}(K)}{-\log r}\,,$$

respectively, where $K$ denotes the support of $\mu$. The Renyi dimensions are one of the most fundamental and important objects in multifractal analysis, and the the reader is referred to Section 2 for a comprehensive discussion of this definition and, in particular, the relationship between the multifractal Minkowski volume and multifractal analysis. In this survey we will study and obtain explicit formulas for $q$'th multifractal Minkowski volume of graph-directed self-conformal measures satisfying the Strong Separation Condition (SSC). Indeed, using powerful techniques from functional analysis and, in particular, from the theory of trace-class operators we introduce a multifractal zeta function $\zeta^q$ designed to study $q$'th multifractal

Minkowski volume of graph-directed self-conformal measures satisfying the SSC. We first fix a graph-directed conformal iterated function system

$$\Big(\mathsf{V},\,\mathsf{E},\,(V_{\mathsf{i}})_{\mathsf{i}\in\mathsf{V}},\,(X_{\mathsf{i}})_{\mathsf{i}\in\mathsf{V}},\,(S_{\mathsf{e}})_{\mathsf{e}\in\mathsf{E}},\,(p_{\mathsf{e}})_{\mathsf{e}\in\mathsf{E}}\Big)$$

in $\mathbb{R}^d$ satisfying the SSC; see Section 4.2 for the precise definitions. Next, we consider an "analytic continuation"

$$\Big(\mathsf{V},\,\mathsf{E},\,(\mathbb{V}_{\mathsf{i}})_{\mathsf{i}\in\mathsf{V}},\,(\mathbb{X}_{\mathsf{i}})_{\mathsf{i}\in\mathsf{V}},\,(\mathcal{S}_{\mathsf{e}})_{\mathsf{e}\in\mathsf{E}},\,(p_{\mathsf{e}})_{\mathsf{e}\in\mathsf{E}}\Big)$$

to $\mathbb{C}^d$ of original the graph-directed conformal iterated function system $(\mathsf{V},\mathsf{E},(V_{\mathsf{i}})_{\mathsf{i}\in\mathsf{V}},(X_{\mathsf{i}})_{\mathsf{i}\in\mathsf{V}},(S_{\mathsf{e}})_{\mathsf{e}\in\mathsf{E}},(p_{\mathsf{e}})_{\mathsf{e}\in\mathsf{E}})$; see Section 4.3 for the precise definitions. Using the analytically continued graph-directed conformal iterated function system, we now define the multifractal transfer operator

$$\mathbf{L}^{q,s} : A^2(\mathbb{X}^\circ) \to A^2(\mathbb{X}^\circ)$$

for $s \in \mathbb{C}$ where $\mathbb{X} = \cup_{\mathsf{i}}\mathbb{X}_{\mathsf{i}}$ and $A^2(\mathbb{X}^\circ)$ denotes the Bergman space on $\mathbb{X}^\circ$, i.e. the space of holomorphic square integrable functions on $\mathbb{X}^\circ$; the precise definitions will be given in Section 5. We note that usually transfer operators associated with conformal iterated function systems are defined on $\oplus_{\mathsf{i}\in\mathsf{V}}L^2(K_{\mathsf{i}},\mu_{\mathsf{i}})$ where $(K_{\mathsf{i}})_{\mathsf{i}\in\mathsf{V}}$ and $(\mu_{\mathsf{i}})_{\mathsf{i}\in\mathsf{V}}$ denotes the self-conformal sets and self-conformal measures associated with $(\mathsf{V},\mathsf{E},(V_{\mathsf{i}})_{\mathsf{i}\in\mathsf{V}},(X_{\mathsf{i}})_{\mathsf{i}\in\mathsf{V}},(S_{\mathsf{e}})_{\mathsf{e}\in\mathsf{E}},(p_{\mathsf{e}})_{\mathsf{e}\in\mathsf{E}})$, see, for example, [**PaPo2**]. However, since we will use powerful techniques based on Atiyah-Bott's fixed point theorem introduced by Guillopé, Lin & Zworski in [**GuLiZw**] (but already implicit in [**JePo**]) to study the asymptotic behaviour of $\mathbf{L}^{q,s}$, it is essential that $\mathbf{L}^{q,s}$ is defined on a space of holomorphic square integrable functions on an open subset of $\mathbb{C}^d$, and this explains why we extend the graph-directed conformal iterated function system from $\mathbb{R}^d$ to $\mathbb{C}^d$; the reader is referred to the discussion below for more details of this. It follows from Atiyah-Bott's fixed point theorem that the operator $\mathbf{L}^{q,s}$ is of trace-class and we can therefore define the multifractal zeta function $\zeta^q$ by

$$\zeta^q(s) = \det(\mathbf{1} - \mathbf{L}^{q,s})\,,$$

where det denotes the Fredholm determinant. The zeta function $\zeta^q(s) = \det(\mathbf{1} - \mathbf{L}^{q,s})$ is very general and includes many other zeta functions as special cases. Indeed, by choosing the graph-directed conformal iterated function system $(\mathsf{V},\mathsf{E},(V_{\mathsf{i}})_{\mathsf{i}\in\mathsf{V}},(X_{\mathsf{i}})_{\mathsf{i}\in\mathsf{V}},(S_{\mathsf{e}})_{\mathsf{e}\in\mathsf{E}},(p_{\mathsf{e}})_{\mathsf{e}\in\mathsf{E}})$ appropriately and putting $q = 0$, we obtain, for example, Selberg's zeta function and Ruelle's dynamical zeta-function.

Our first main result concerns bounds for the growth of the eigenvalues of $\mathbf{L}^{q,s}$. If $\lambda_n(\mathbf{L}^{q,s})$ denotes the $n$'th eigenvalue of $\mathbf{L}^{q,s}$ ordered such that $\lambda_1(\mathbf{L}^{q,s}) \geq \lambda_2(\mathbf{L}^{q,s}) \geq \lambda_3(\mathbf{L}^{q,s}) \geq \ldots$, then the current state of the art bounds for the eigenvalues of $\mathbf{L}^{q,s}$ for $q = 0$ are given by [**BaJe3**] and says that there are constants $C_0, c_0, c > 0$ such that

$$|\lambda_n(\mathbf{L}^{q,s})| \leq C_0\,|\mathbf{p}^q|_\infty\, e^{c|s| - c_0 n^{1/d}}$$

for all $n$ where $|\mathbf{p}^q|_\infty = \max_{\mathsf{e}} p_{\mathsf{e}}^q$. We improve this bound substantially. In particular, in Theorem 6.7 we show that for all $c, u > 0$, there are constants $C_0, c_0 > 0$ such that for all $s \in \mathbb{C}$ with $\operatorname{Re}(s) > -c$ and $|\operatorname{Im}(s)| \geq u$, we have

$$|\lambda_n(\mathbf{L}^{q,s})| \leq C_0\,|\mathbf{p}^q|_\infty\, e^{-c_0 \frac{1}{|\operatorname{Im}(s)|^{\dim_{\mathsf{H}} K_{\mathsf{i}}/d}} n^{1/d}}$$

for all $n$ and all $\mathsf{i} \in \mathsf{V}$, and where $\dim_{\mathsf{H}}$ denotes the Hausdorff dimension; recall, that $(K_{\mathsf{i}})_{\mathsf{i}\in\mathsf{V}}$ denotes the self-conformal sets associated with $(\mathsf{V}, \mathsf{E}, (V_{\mathsf{i}})_{\mathsf{i}\in\mathsf{V}}, (X_{\mathsf{i}})_{\mathsf{i}\in\mathsf{V}}, (S_{\mathsf{e}})_{\mathsf{e}\in\mathsf{E}}, (p_{\mathsf{e}})_{\mathsf{e}\in\mathsf{E}})$. The reader is referred to Theorem 6.7 for a more refined version of this result. The reader may also note that this upper bound significantly improves the existing upper bound in [**BaJe3**].

Our second main result utilises the above bound for $|\lambda_n(\mathbf{L}^{q,s})|$ to obtain explicit formulas for a suitable symbolic version of the multifractal Minkowski volume of graph-directed self-conformal measures satisfying the SSC. Recall, that if $(K_{\mathsf{i}})_{\mathsf{i}\in\mathsf{V}}$ and $(\mu_{\mathsf{i}})_{\mathsf{i}\in\mathsf{V}}$ denote the self-conformal sets and self-conformal measures associated with the graph-directed conformal iterated function system $(\mathsf{V}, \mathsf{E}, (V_{\mathsf{i}})_{\mathsf{i}\in\mathsf{V}}, (X_{\mathsf{i}})_{\mathsf{i}\in\mathsf{V}}, (S_{\mathsf{e}})_{\mathsf{e}\in\mathsf{E}}, (p_{\mathsf{e}})_{\mathsf{e}\in\mathsf{E}})$, respectively, then the $q$'th multifractal Minkowski volume of $K_{\mathsf{i}}$ with respect to $\mu_{\mathsf{i}}$ is defined by

$$V^q_{\mu_{\mathsf{i}},r}(K_{\mathsf{i}}) = \frac{1}{r^d}\int_{B(K_{\mathsf{i}},r)} \mu_{\mathsf{i}}(B(x,r))^q \, d\mathcal{L}^d(x)\,.$$

Frustratingly, the problem of finding "explicit" formulas for $V^q_{\mu_{\mathsf{i}},r}(K_{\mathsf{i}})$ is currently completely intractable. Because of this difficulty we will study a natural "symbolic" version of $V^q_{\mu_{\mathsf{i}},r}(K_{\mathsf{i}})$. We immediately note that the problem of finding an explicit formula for the symbolic multifractal Minkowski volume can be viewed as "prime counting". Indeed, if $\Pi_{\mathsf{G}}$ denotes the family of "prime" strings, and $\ell_{\mathbf{I}}$ and $p_{\mathbf{I}}$ denote the length and probability associated with a "prime" string $\mathbf{I} \in \Pi_{\mathsf{G}}$ (the reader is referred to Section 6.1 for the precise definitions), then we define the multifractal prime counting function $\pi^q_{\mathsf{G}}(t)$ by

$$\pi^q_{\mathsf{G}}(t) = \sum_{\substack{\mathbf{I}\in\Pi_{\mathsf{G}}\\ \ell_{\mathbf{I}}\le t}} p^q_{\mathbf{I}}\,,$$

Routine arguments now show that there is a constant $c$ such that

$$\frac{1}{c}\,\pi^q_{\mathsf{G}}(\log\tfrac{1}{r}) \le V^q_{\mu_{\mathsf{i}},r}(K_{\mathsf{i}}) \le c\,\pi^q_{\mathsf{G}}(\log\tfrac{1}{r})$$

for all $r > 0$, and the prime counting function $\pi^q_{\mathsf{G}}(t)$ may therefore be viewed as a natural symbolic version of the multifractal Minkowski volume $V^q_{\mu_{\mathsf{i}},r}(K_{\mathsf{i}})$. Using Wiener-Ikehara's tauberian theorem and the upper bounds for $|\lambda_n(\mathbf{L}^{q,s})|$, we derive various multifractal prime counting formulas for Graph-directed Self-Conformal Iterated Function Systems; this is the content of Theorem 7.1 To formulate this, we use the following notation, namely, if $f, g : (0,\infty) \to (0,\infty)$ are real valued positive functions, then we write $f \sim g$ (or $f(t) \sim g(t)$) provided $\frac{f(t)}{g(t)} \to 1$ as $t \to \infty$. Also, if $\omega \in \mathbb{C}$ and $f$ is a meromorphic function with $f(\omega) = 0$ and $f$ is not constantly equal to 0 in a neighbourhood of $\omega$, then we write $\operatorname{ord}(f;\omega)$ for the order of the zero $\omega$ of $f$. It is well-known that $\operatorname{supp}\mu_{\mathsf{i}} = K_{\mathsf{i}}$, and the Renyi dimensions of $\mu_{\mathsf{i}}$ are therefore given by

$$\underline{T}_{\mu_{\mathsf{i}}}(q) = \liminf_{r\searrow 0} \frac{\log V^q_{\mu_{\mathsf{i}},r}(K_{\mathsf{i}})}{-\log r}$$

and

$$\overline{T}_{\mu_{\mathsf{i}}}(q) = \limsup_{r\searrow 0} \frac{\log V^q_{\mu_{\mathsf{i}},r}(K_{\mathsf{i}})}{-\log r}\,,$$

respectively. A fundamental result in multifractal analysis of self-conformal measures says that $\underline{T}_{\mu_{\mathsf{i}}}(q) = \overline{T}_{\mu_{\mathsf{i}}}(q)$ and that the common value of $\underline{T}_{\mu_{\mathsf{i}}}(q)$ and $\overline{T}_{\mu_{\mathsf{i}}}(q)$ is

of $\mathsf{i}$. Following tradition, we will denote this value by $\beta(q)$, i.e. we will write

$$\beta(q) = \lim_{r\searrow 0} \frac{\log V^q_{\mu_{\mathsf{i}},r}(K_{\mathsf{i}})}{-\log r}.$$

Assuming that the zeta function $\zeta^q$ does not have any other zeros $s$ with $\operatorname{Re}(s) = \beta(q)$ besides $\beta(q)$ and $0 \leq q < 1$, then Theorem 7.1 says that

$$\pi^q_{\mathsf{G}}(t) \sim \operatorname{ord}(\zeta^q; \beta(q)) \frac{e^{\beta(q)t}}{\beta(q)t}.$$

This result promotes the view that:

> *Multifractal analysis of self-conformal measures is prime counting.*

The reader is referred to Section 7 for a discussion of this and for more refined multifractal prime counting formulas.

## 1. Fractal tubes

Let $E$ be a subset of $\mathbb{R}^d$ and $r > 0$. We will write $B(E, r)$ for the open $r$ neighbourhood of $E$, i.e.

$$B(E, r) = \left\{ x \in \mathbb{R}^d \,\middle|\, \operatorname{dist}(x, E) < r \right\}. \tag{1.1}$$

Intuitively we think of the set $B(E, r)$ as consisting of the set $E$ surrounded by a "tube" of width $r$. Our main interest is to compute the volume of the "tube" of width $r$ surrounding $E$. To make this formal, one defines the Minkowski volume $V_r(E)$ of $E$ by

$$V_r(E) = \frac{1}{r^d} \mathcal{L}^d(B(E, r)); \tag{1.2}$$

here and below $\mathcal{L}^d$ denotes the $d$-dimensional Lebesgue measure in $\mathbb{R}^d$ and the normalising factor $\frac{1}{r^d}$ is included to make the subsequent results simpler - we note that different authors use different normalising factors. The Minkowski volume $V_r(E)$ is also important because of its intimate relation to two of the most fundamental notions from Fractal Geometry, namely, the lower and upper box dimension. Indeed, using the Minkowski volume $V_r(E)$, we define the lower and upper Minkowski dimension of $E$ by

$$\underline{\dim}_{\mathsf{M}}(E) = \liminf_{r\searrow 0} \frac{\log V_r(E)}{-\log r},$$

and

$$\overline{\dim}_{\mathsf{M}}(E) = \limsup_{r\searrow 0} \frac{\log V_r(E)}{-\log r},$$

respectively. The link with Fractal Geometry is now explained as follows. Namely, it is not difficult to see that the lower Minkowski dimension equals the lower box dimension and that the upper Minkowski dimension equals the upper box dimension; for the definition of the box dimensions the reader is referred to Falconer's text book [**Fa1**].

After having defined the Minkowski volume, we now return to tube formulas. Tube formulas refers to formulas for computing the Minkowski volume $V_r(E)$ as a function of the width $r$ of the "tube" surrounding $E$. In particular, one is typically interested in the following two types of results:

- Asymptotic behaviour: finding a formula for the asymptotic behaviour of $V_r(E)$ as $r \searrow 0$;
- Explicit formulas: finding an explicit formula for $V_r(E)$ valid for all sufficiently small $r$.

For convex sets $E$, this problem has a rich and fascinating history starting with the work of Steiner in the early 19'th century. Indeed, Steiner showed that if $C$ is a bounded convex subset of $\mathbb{R}^d$, then there are constants $\kappa_0(C), \kappa_1(C), \ldots, \kappa_d(C)$ such that

$$\mathcal{L}^d(B(C,r)) = \sum_l \kappa_l(C)\, r^{d-l} \tag{1.3}$$

for $r > 0$. The coefficients $\kappa_l(C)$ are called the Quermassintegrale or mixed volumes, and the polynomial $\sum_l \kappa_l(C)\, r^{d-l}$ is called the Steiner polynomial. The coefficients have clear geometric interpretations. For example, $\kappa_d(C)$ equals the volume of $C$ and $\kappa_{d-1}(C)$ is equal to the surface area of $C$. Steiner's formula has subsequently been extended to more general classes of sets. For example, in the late 1930's Weyl proved that a similar result holds for compact oriented $d$-dimensional Riemannian manifolds $C$ (with or without boundary) isometrically embedded into Euclidean space. This theory reached its mature form in the 1960's where Federer [**Fe1**, **Fe2**] unified the tube formulas of Steiner for convex bodies and of Weyl for smooth submanifolds, as described in [**BeGo**, **Gr**, **We**], and extended these results to sets of positive reach. Federer's tube formula has since been extended in various directions by a number of researchers in integral geometry and geometric measure theory, including [**Fu1**, **Fu2**, **Sc1**, **Sc2**, **St**, **Zä1**, **Zä2**] and most recently (and most generally) in [**HuLaWe**]. The books [**Gr**, **Mo**, **Sc2**] contain extensive endnotes with further information and many other references. While the above references investigate tube formulas for sets that satisfy some (possibly very weak) convexity and/or smoothness conditions, it is a notoriously difficult problem to find explicit formulas for the Minkowski volume $V_r(E)$ of fractal sets $E$. However, despite of these difficulties very recently there has been significant interest in developing a theory of tube formulas for fractal sets and a number of exciting works have appeared. Indeed, in the early 1990's Lapidus introduced the notion of "complex dimensions" associated with fractal zeta functions and has during the past 20 years very successfully pioneered the use of fractal zeta functions to obtain explicit tube formulas for certain classes of fractal subsets of (mainly) the real line; this exciting theory is described in detail in Lapidus & van Frankenhuysen's intriguing books [Lap-vF1,Lap-vF2] and the more recent book [**LapRaZu**] by Lapidus, Radunovic & DŻubrinic. In a parallel development, and building on earlier work by Lalley [**La1**, **La2**, **La3**] and Gatzouras [**Ga**] (see also [**Fa3**]), Winter [**Wi**] has initiated the systematic study of curvatures of fractal sets and applied this theory to study the asymptotic behaviour of the Minkowski volume $V_r(E)$ of fractal sets $E$ using methods from renewal theory. The work surveyed in this paper may be viewed as a natural higher dimensional multifractal development of this line of research. More precisely, it is our intention to extend the notion of Minkowski volume $V_r(E)$ to multifractals and introduce an appropriate zeta function (defined as the Fredholm determinant of a suitable and natural multifractal transfer operator). This zeta

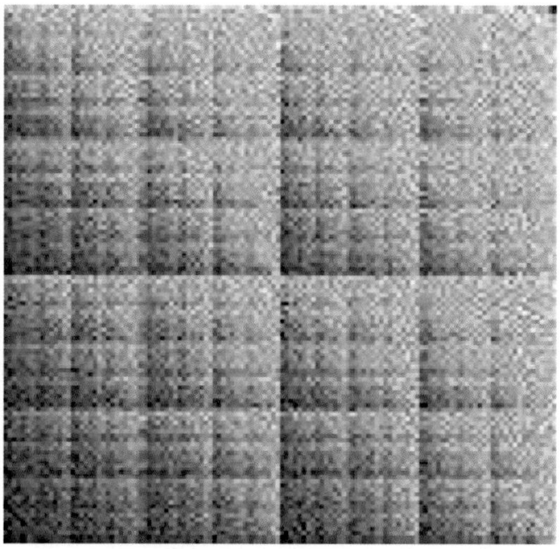

FIGURE 2.1. A typical multifractal, i.e. a measure with widely varying intensity. Dark regions have high concentration of mass and light regions have low concentration of mass.

functions will allow us to obtain explicit formulas for the "symbolic" multifractal Minkowski volume of graph directed self-conformal multifractals. In order to motivate our definitions we will now explain what the term "multifractal" covers.

## 2. Multifractals

**2.1. Multifractal spectra.** Distributions with widely varying intensity occur often in the physical sciences, e.g. the spatial-temporal distribution of rainfall, the spatial distribution of oil and gas in the underground, the distribution of galaxies in the universe, the dissipation of energy in a highly turbulent fluid flow, and the occupation measure on strange attractors. Such distributions are called multifractals and have recently been the focus of much attention in the physics literature.

Figure 2.1 shows a typical multifractal, i.e. a measure with widely varying intensity. Dark regions have high concentration of mass and light regions have low concentration of mass. For a Borel measure $\mu$ on a $\mathbb{R}^d$ and a positive number $\alpha$, let us consider the set $\Delta_\mu(\alpha)$ of those points $x$ in $\mathbb{R}^d$ for which the measure $\mu(B(x,r))$ of the ball $B(x,r)$ with center $x$ and radius $r$ behaves like $r^\alpha$ for small $r$, i.e. the set

$$\Delta_\mu(\alpha) = \left\{ x \in \operatorname{supp}\mu \;\middle|\; \lim_{r \searrow 0} \frac{\log \mu(B(x,r))}{\log r} = \alpha \right\},$$

where $\operatorname{supp}\mu$ denotes the support of the measure. If the intensity of the measure $\mu$ varies very widely, it may happen that the sets $\Delta_\mu(\alpha)$ display a fractal-like character for a range of values of $\alpha$. If this is the case, then the measure is called a multifractal measure or simply a multifractal, and it is natural to study the sizes of the sets $\Delta_\mu(\alpha)$ as $\alpha$ varies. To do this we study the Hausdorff dimension,

$$f_\mu(\alpha) = \dim \Delta_\mu(\alpha)\,, \tag{2.1}$$

of the sets $\Delta_\mu(\alpha)$ as a function of $\alpha$ where dim denotes the Hausdorff dimension. The function in (2.1) and similar functions are generically known as "the multifractal spectrum of $\mu$", "the singularity spectrum of $\mu$" or "the spectrum of scaling indices", and one of the main problems in multifractal analysis is to study these and related functions.

**2.2. Renyi dimensions.** Based on a remarkable insight together with a clever heuristic argument theoretical physicists Halsey et al. [**HaJeKaPrSh**] suggest that the multifractal spectrum $f_\mu(\alpha)$ can be computed in the following way – known as the so-called "Multifractal Formalism" in the physics literature. The "Multifractal Formalism" involves the so-called Renyi dimensions which we will now define. Let $\mu$ be a Borel measure on $\mathbb{R}^d$. For $q \in \mathbb{R}$ and $r > 0$, we define the $q$-th moment $I^q_{\mu,r}(E)$ of a subset $E$ of $\mathbb{R}^d$ with respect to $\mu$ at scale $r$ by

$$I^q_{\mu,r}(E) = \int_E \mu(B(x,r))^{q-1}\, d\mu(x) \tag{2.2}$$

Next, the lower and upper Renyi dimensions of $E$ with respect to $\mu$ are defined by

$$\begin{aligned} \underline{\dim}^q_{\mathsf{R},\mu}(E) &= \liminf_{r\searrow 0} \frac{\log I^q_{\mu,r}(E)}{-\log r}\,, \\ \overline{\dim}^q_{\mathsf{R},\mu}(E) &= \limsup_{r\searrow 0} \frac{\log I^q_{\mu,r}(E)}{-\log r}\,. \end{aligned} \tag{2.3}$$

In particular, the Renyi dimensions of the support of $\mu$ play an important role in the statement of the "Multifractal Formalism". For this reason it is useful to denote these dimensions by separate notation, and we therefore define the lower and upper Renyi spectra $\underline{\tau}_\mu(q), \overline{\tau}_\mu(q) : \mathbb{R} \to [-\infty, \infty]$ of $\mu$ by

$$\begin{aligned} \underline{\tau}_\mu(q) &= \underline{\dim}^q_{\mathsf{R},\mu}(\operatorname{supp}\mu) \\ &= \liminf_{r\searrow 0} \frac{\log I^q_{\mu,r}(\operatorname{supp}\mu)}{-\log r}\,, \\ \overline{\tau}_\mu(q) &= \overline{\dim}^q_{\mathsf{R},\mu}(\operatorname{supp}\mu) \\ &= \limsup_{r\searrow 0} \frac{\log I^q_{\mu,r}(\operatorname{supp}\mu)}{-\log r}\,. \end{aligned} \tag{2.4}$$

**2.3. The Multifractal Formalism.** We can now state the "Multifractal Formalism". Loosely speaking the "Multifractal Formalism" says the the multifractal spectrum $f_\mu$ and the Renyi dimensions carry the same information. More precisely, the "Multifractal Formalism" says the multifractal spectrum equals the Legendre transform of the Renyi dimensions. Before stating this formally, we remind the reader that if $\varphi : \mathbb{R} \to \mathbb{R}$ is a real valued function, then the Legendre transform $\varphi^* : \mathbb{R} \to [-\infty, \infty]$ of $\varphi$ is defined by

$$\varphi^*(x) = \inf_y (xy + \varphi(y))\,.$$

THE MULTIFRACTAL FORMALISM – A PHYSICS FOLKLORE THEOREM. *The multifractal spectrum $f_\mu$ of $\mu$ equals the Legendre transforms, $\underline{\tau}^*_\mu$ and $\overline{\tau}^*_\mu$, of the Renyi dimensions, i.e.*

$$f_\mu(\alpha) = \underline{\tau}^*_\mu(\alpha) = \overline{\tau}^*_\mu(\alpha)$$

*for all $\alpha \geq 0$.*

During the past 20 years there has been an enormous interest in verifying the Multifractal Formalism and computing the multifractal spectra of measures in the mathematical literature. In the mid 1990's Cawley & Mauldin [**CaMa**] and Arbeiter & Patzschke [**ArPa**] verified the Multifractal Formalism for self-similar measures satisfying the Open Set Condition, and within the last 10 years the multifractal spectra of various classes of measures in Euclidean space $\mathbb{R}^d$ exhibiting some degree of self-similarity have been computed rigorously, cfthe text [**Fa2, Pi**] and the references therein.

## 3. Multifractal tubes

**3.1. Multifractal tubes.** Motivated by Lapidus & van Frankenhuysen investigations [Lap-vF1,Lap-vF2] of tube formulas for fractal sets, it it natural to seek to develop a theory of multifractal tube formulas for multifractal measures. In this section we will present a framework attempting to do this. As an example, we will also give a complete description of the asymptotic behaviour of the multifractal tube formulas for self-similar measures satisfying the Open Set Condition.

Multufractal tube formulas are defined as follows. First note that if $r > 0$ and $E$ is a subset of $\mathbb{R}^d$, then the Minkowski volume $V_r(E)$ is given by

$$\begin{aligned} V_r(E) &= \frac{1}{r^d}\,\mathcal{L}^d(B(E,r)) \\ &= \frac{1}{r^d}\int_{B(E,r)} d\mathcal{L}^d(x)\,, \end{aligned}$$

where we have rewritten the Lebesgue measure $\mathcal{L}^d(B(E,r))$ of $B(E,r)$ as the integral $\int_{B(E,r)} d\mathcal{L}^d(x)$. Motivated by the Renyi dimensions (i.e. (2.2) and (2.3)) and the above expression for $V_r(E)$, we now define the multifractal Minkowski volume as follows. Namely, let $r > 0$ and $E$ be a subset of $\mathbb{R}^d$. For real number $q$ and a Borel measure $\mu$ on $\mathbb{R}^d$, we now define the multifractal $q$ Minkowski volume $V^q_{\mu,r}(E)$ of $E$ with respect to the measure $\mu$ by

$$V^q_{\mu,r}(E) = \frac{1}{r^d}\int_{B(E,r)} \mu(B(x,r))^q\, d\mathcal{L}^d(x)\,. \tag{3.1}$$

Note, that if $q = 0$, then the $q$ multifractal Minkowski volume $V^q_{\mu,r}(E)$ reduces to the usual Minkowski volume, iė

$$V^0_{\mu,r}(E) = V_r(E)\,.$$

The importance of the Renyi dimensions in multifractal analysis together with the formal resemblance between the multifractal Minkowski volume $V^q_{\mu,r}(E)$ and the moments $I^q_{\mu,r}(E)$ used in the definition the Renyi dimensions may be seen as a justification for calling the quantity $V^q_{\mu,r}(E)$ for the *multifractal* Minkowski volume. However, it should be noted that the resemblance between the multifractal Minkowski volume $V^q_{\mu,r}(E)$ and the moments $I^q_{\mu,r}(E)$ is not merely formal providing further justification for calling the quantity $V^q_{\mu,r}(E)$ for the *multifractal* Minkowski volume. Indeed, if we define the lower and upper multifractal $q$ Minkowski dimension of $E$, by

$$\underline{\dim}^q_{\mathsf{M},\mu}(E) = \liminf_{r\searrow 0} \frac{\log V^q_{\mu,r}(E)}{-\log r}\,,$$

and

$$\overline{\dim}^q_{\mathsf{M},\mu}(E) = \limsup_{r\searrow 0} \frac{\log V^q_{\mu,r}(E)}{-\log r},$$

respectively, then the proposition below shows thatfor $q \geq 0$, the multifractal Minkowski dimensions and the Renyi dimensions coincide. This provides further justification for calling the quantity $V^q_{\mu,r}(E)$ for the *multifractal* Minkowski volume.

PROPOSITION 3.1. *Let $\mu$ be a Borel measure on $\mathbb{R}^d$ and $E \subseteq \mathbb{R}^d$. If $q \geq 0$, then*

$$\underline{\dim}^q_{\mathsf{M},\mu}(E) = \underline{\dim}^q_{\mathsf{R},\mu}(E),$$
$$\overline{\dim}^q_{\mathsf{M},\mu}(E) = \overline{\dim}^q_{\mathsf{R},\mu}(E).$$

*In particular, if $q \geq 0$, then*

$$\underline{\dim}^q_{\mathsf{M},\mu}(\operatorname{supp}\mu) = \underline{\tau}_\mu(q),$$
$$\overline{\dim}^q_{\mathsf{M},\mu}(\operatorname{supp}\mu) = \overline{\tau}_\mu(q).$$

PROOF. This follows from the definitions using routine arguments and we have therefore decided to omit the proof. □

We will introduce a multifractal zeta function (defined in as the Fredholm determinant of a suitable and natural multifractal transfer operator). This zeta functions will allow us to obtain explicit formulas for the "symbolic" multifractal Minkowski volume of graph directed self-conformal multifractals. However, before doing so, we first recall the definition of self-conformal constructions.

## 4. Graph-directed self-conformal sets and measures

**4.1. Notation from symbolic dynamics.** The purpose of this section is to recall notation and terminology from symbolic dynamics that we will use.

We start be recalling the definition of a finite directed multi-graph. A finite directed multi-graph is a quadruple $\mathsf{G} = (\mathsf{V}, \mathsf{E}, \mathrm{i}, \mathrm{t})$, also sometimes simply written as $\mathsf{G} = (\mathsf{V}, \mathsf{E})$, where $\mathsf{V}$ and $\mathsf{E}$ are finite sets and $\mathrm{i}, \mathrm{t} : \mathsf{E} \to \mathsf{V}$ are functions; the elements in $\mathsf{V}$ are called vertices, the elements in $\mathsf{E}$ are called edges, and for an edge $\mathsf{e} \in \mathsf{E}$, the vertex $\mathrm{i}(\mathsf{e})$ is called the initial vertex of $\mathsf{e}$, the vertex $\mathrm{t}(\mathsf{e})$ is called the terminal vertex of $\mathsf{e}$, and the edge $\mathsf{e}$ is called an edge from $\mathrm{i}(\mathsf{e})$ to $\mathrm{t}(\mathsf{e})$. For $\mathsf{i}, \mathsf{j} \in \mathsf{V}$, write

$$\begin{aligned} \mathsf{E}_{\mathsf{i}} &= \left\{ \mathsf{e} \in \mathsf{E} \,\middle|\, \mathrm{i}(\mathsf{e}) = \mathsf{i} \right\}, \\ \mathsf{E}_{\mathsf{i},\mathsf{j}} &= \left\{ \mathsf{e} \in \mathsf{E} \,\middle|\, \mathrm{i}(\mathsf{e}) = \mathsf{i} \text{ and } \mathrm{t}(\mathsf{e}) = \mathsf{j} \right\}; \end{aligned} \tag{4.1}$$

i.e. $\mathsf{E}_{\mathsf{i}}$ is the family of all edges starting at $\mathsf{i}$ and $\mathsf{E}_{\mathsf{i},\mathsf{j}}$ is the is the family of all edges starting at $\mathsf{i}$ and ending at $\mathsf{j}$. Also, for a positive integer $n$, we write

$$\begin{aligned}
\Sigma_{\mathsf{G}}^{n} = \Big\{\mathsf{e}_1\dots\mathsf{e}_n \,\Big|\, & \mathsf{e}_i\in\mathsf{E} \text{ for } 1\le i\le n,\\
& \mathrm{t}(\mathsf{e}_1)=\mathrm{i}(e_2)\,,\\
& \mathrm{t}(e_{i-1})=\mathrm{i}(\mathsf{e}_i) \text{ and } \mathrm{t}(\mathsf{e}_i)=\mathrm{i}(e_{i+1}) \text{ for } 1<i<n,\\
& \mathrm{t}(e_{n-1})=\mathrm{i}(\mathsf{e}_n)\Big\}
\end{aligned}$$

$$\Sigma_{\mathsf{G}}^{*}=\bigcup_n \Sigma_{\mathsf{G}}^{n}\,, \tag{4.2}$$

$$\begin{aligned}
\Sigma_{\mathsf{G}}^{\mathbb{N}} = \Big\{\mathsf{e}_1\mathsf{e}_2\dots \,\Big|\, & \mathsf{e}_i\in\mathsf{E} \text{ for } 1\le i,\\
& \mathrm{t}(\mathsf{e}_1)=\mathrm{i}(e_2)\,,\\
& \mathrm{t}(e_{i-1})=\mathrm{i}(\mathsf{e}_i) \text{ and } \mathrm{t}(\mathsf{e}_i)=\mathrm{i}(e_{i+1}) \text{ for } 1<i\Big\}\,;
\end{aligned}$$

i.e. $\Sigma_{\mathsf{G}}^{n}$ is the family of all finite strings $\mathbf{i}=\mathsf{e}_1\dots\mathsf{e}_n$ consisting of finite paths in $\mathsf{G}$ of length $n$, $\Sigma_{\mathsf{G}}^{*}$ is the family of all finite strings $\mathbf{i}=\mathsf{e}_1\dots\mathsf{e}_m$ with $m\in\mathbb{N}$ consisting of finite paths in $\mathsf{G}$ and $\Sigma_{\mathsf{G}}^{\mathbb{N}}$ is the family of all infinite strings $\mathbf{i}=\mathsf{e}_1\mathsf{e}_2\dots$ consisting of infinite paths in $\mathsf{G}$. For a finite string $\mathbf{i}=\mathsf{e}_1\dots\mathsf{e}_n\in\Sigma_{\mathsf{G}}^{n}$, we write

$$\mathrm{i}(\mathbf{i})=\mathrm{i}(\mathsf{e}_1)\,,\quad \mathrm{t}(\mathbf{i})=\mathrm{t}(\mathsf{e}_n)\,, \tag{4.3}$$

and for an infinite string $\mathbf{i}=\mathsf{e}_1\mathsf{e}_2\dots\in\Sigma_{\mathsf{G}}^{\mathbb{N}}$, we write

$$\mathrm{i}(\mathbf{i})=\mathrm{i}(\mathsf{e}_1)\,. \tag{4.4}$$

Also, for an infinite string $\mathbf{i}=\mathsf{e}_1\mathsf{e}_2\dots\in\Sigma_{\mathsf{G}}^{\mathbb{N}}$ and a positive integer $n$, we will write $\mathbf{i}|n=\mathsf{e}_1\dots\mathsf{e}_n$. Finally, for a positive integer $n$ and a finite string $\mathbf{i}=\mathsf{e}_1\dots\mathsf{e}_n\in\Sigma_{\mathsf{G}}^{n}$ with length equal to $n$, we will write $|\mathbf{i}|=n$.

**4.2. Graph-directed Self-Conformal Iterated Function Systems.** We now recall the definition of graph-directed self-conformal sets and measures. A Graph-directed Conformal Iterated Function System is a list

$$\Big(\mathsf{V},\,\mathsf{E},\,(V_{\mathsf{i}})_{\mathsf{i}\in\mathsf{V}},\,(X_{\mathsf{i}})_{\mathsf{i}\in\mathsf{V}},\,(S_{\mathsf{e}})_{\mathsf{e}\in\mathsf{E}},\,(p_{\mathsf{e}})_{\mathsf{e}\in\mathsf{E}}\Big) \tag{4.5}$$

satisfying the conditions listed below. Below we will write $C^{\omega}$ for the family of real analytic functions and if $\mathbb{F}=\mathbb{R}$ or $\mathbb{F}=\mathbb{C}$ and $f:U\to V$ is a differentiable function where $U$ and $V$ are open subsets of $\mathbb{F}^n$ and $\mathbb{F}^m$, respectively, then we will write

$$(Df)(x):\mathbb{F}^n\to\mathbb{F}^m \tag{4.6}$$

for the derivative of $f$ at $x\in U$; also, if $M$ is a $n\times m$ matrix with entries from $\mathbb{F}$, then $|M|$ denotes the operator norm of the linear map $M:\mathbb{F}^m\to\mathbb{F}^n$ induced by $M$, i.e.

$$|M|=\sup_{u\in\mathbb{F}^m\,,\,|u|=1}|Mu|\,.$$

Basic Conditions:

- $\mathsf{G}=(\mathsf{V},\mathsf{E})$ is a finite directed multi-graph.
- For each $\mathsf{i}\in\mathsf{V}$ we have: $V_{\mathsf{i}}$ is an open, connected subset of $\mathbb{R}^d$.
- For each $\mathsf{i}\in\mathsf{V}$ we have: $X_{\mathsf{i}}$ is a compact subset of $\mathbb{R}^d$ with $X_{\mathsf{i}}\subseteq V_{\mathsf{i}}$ and $X_{\mathsf{i}}^{\circ\,-}=X_{\mathsf{i}}$.

- For each $\mathsf{i},\mathsf{j}\in\mathsf{V}$ and $\mathsf{e}\in E_{\mathsf{i},\mathsf{j}}$ we have: for $d=1$, the function $S_{\mathsf{e}}:V_{\mathsf{j}}\to V_{\mathsf{i}}$ is a contractive $C^{\omega}$ diffeomorphism such that $S_{\mathsf{e}}(X_{\mathsf{j}})\subseteq X_{\mathsf{i}}$, and for $d>1$, the function $S_{\mathsf{e}}:V_{\mathsf{j}}\to V_{\mathsf{i}}$ is a contractive $C^{1+\gamma}$ diffeomorphism with $0<\gamma<1$ such that $S_{\mathsf{e}}(X_{\mathsf{j}})\subseteq X_{\mathsf{i}}$.
- The Conformality Condition. For each $\mathsf{i},\mathsf{j}\in\mathsf{V}$, $\mathsf{e}\in E_{\mathsf{i},\mathsf{j}}$ and $x\in V_{\mathsf{j}}$ we have: there exists $r_{\mathsf{e}}(x)\in(0,1)$ such that

$$|(DS_{\mathsf{e}})(x)u| = r_{\mathsf{e}}(x)|u| \tag{4.7}$$

  for all $u\in\mathbb{R}^d$, i.e. $(DS_{\mathsf{e}})(x)$ is a similarity matrix with similarity ratio equal to $r_{\mathsf{e}}(x)$.
- For each $\mathsf{i}\in\mathsf{V}$ we have: $(p_{\mathsf{e}})_{\mathsf{e}\in\mathsf{E}_{\mathsf{i}}}$ is a probability vector.

Usually the definition of a Graph-directed Self-Conformal Iterated Function System assumes that the maps $S_{\mathsf{e}}$ are $C^{1+\gamma}$ for all values of $d$. However, in the definition above we assume that $S_{\mathsf{e}}$ is $C^{\omega}$ for $d=1$. The reason for this is that we will use Atiyah-Bott's fixed point theorem to study the multifractal transfer operator $\mathbf{L}^{q,s}$ introduced in Section 5. However, Atiyah-Bott's fixed point theorem is a statement about operators defined on spaces of holomorphic functions on subdomains of $\mathbb{C}^d$ and for this reason it is necessary to extend the Self-Conformal Iterated Function System in (4.5) holomorphically to a domain in $\mathbb{C}^d$ and this can only be done if the maps $S_{\mathsf{e}}$ are real analytic. For $d>1$ it is well known that any conformal $C^{1+\gamma}$ diffeomorphism between domains in $\mathbb{R}^d$ is real analytic (see, for example, [**BePe**] or Observation 4.1 in Section 4.3). However, this is clearly false for $d=1$, and for this reason we have to assume that the maps $S_{\mathsf{e}}$ are real analytic for $d=1$.

We will also assume the well-known and commonly used Strong Separation Condition (SSC) and a further related separation condition. The SSC and the related separation condition are stated below.

Separation Conditions:

- The Strong Separation Condition. For all $\mathsf{i}\in\mathsf{V}$ and all $\mathsf{e}_1,\mathsf{e}_2\in\mathsf{E}_{\mathsf{i}}$ with $\mathsf{e}_1\neq\mathsf{e}_2$ we have: $S_{\mathsf{e}_1}\big(X_{\mathsf{t}(\mathsf{e}_1)}\big)\cap S_{\mathsf{e}_2}\big(X_{\mathsf{t}(\mathsf{e}_2)}\big)=\varnothing$.
- For all $\mathsf{i},\mathsf{j}\in\mathsf{V}$ with $\mathsf{i}\neq\mathsf{j}$ we have: $X_{\mathsf{i}}\cap X_{\mathsf{j}}\neq\varnothing$.

Finally, we will assume two topological conditions that are usually not part of the definition of a Graph-directed Conformal Iterated Function System. The first of those conditions says that the set $X_{\mathsf{i}}^{\circ}$ is simply connected, and the second condition says that not only is $S_{\mathsf{e}}(X_{\mathsf{j}})$ a subset of $X_{\mathsf{i}}^{\circ}$ for all $\mathsf{i},\mathsf{j}\in\mathsf{V}$ and $\mathsf{e}\in E_{\mathsf{i},\mathsf{j}}$, but the set $S_{\mathsf{e}}(X_{\mathsf{j}})$ is "well separated" from the boundary of $X_{\mathsf{i}}^{\circ}$; more precisely, we will assume that $S_{\mathsf{e}}(X_{\mathsf{j}})\subseteq X_{\mathsf{i}}^{\circ}$. These assumptions are needed in order to prove that the trace class norms of certain operators are uniformly bounded. Below we formally state the assumptions.

Topological Conditions:

- The Simply Connected Condition. For each $\mathsf{i}\in\mathsf{V}$ we have: $X_{\mathsf{i}}^{\circ}$ is simply connected.
- The Interior Condition. For each $\mathsf{i},\mathsf{j}\in\mathsf{V}$ and $\mathsf{e}\in E_{\mathsf{i},\mathsf{j}}$ we have: $S_{\mathsf{e}}(X_{\mathsf{j}})\subseteq X_{\mathsf{i}}^{\circ}$.

It follows from [**Hu**] that there exists a unique list $(K_{\mathsf{i}})_{\mathsf{i}\in\mathsf{V}}$ of non-empty compact sets $K_{\mathsf{i}} \subseteq X_{\mathsf{i}}$ such that

$$K_{\mathsf{i}} = \bigcup_{\mathsf{e}\in\mathsf{E}_{\mathsf{i}}} S_{\mathsf{e}} K_{\mathsf{t}(\mathsf{e})}\,, \tag{4.8}$$

and a unique list $(\mu_{\mathsf{i}})_{\mathsf{i}\in\mathsf{V}}$ of probability measures with $\operatorname{supp}\mu_{\mathsf{i}} = K_{\mathsf{i}}$ such that

$$\mu_{\mathsf{i}} = \sum_{\mathsf{e}\in E_{\mathsf{i}}} p_{\mathsf{e}}\, \mu_{\mathsf{t}(\mathsf{e})} \circ S_{\mathsf{e}}^{-1}\,. \tag{4.9}$$

The sets $(K_{\mathsf{i}})_{\mathsf{i}\in\mathsf{V}}$ and measures $(\mu_{\mathsf{i}})_{\mathsf{i}\in\mathsf{V}}$ are called the self-conformal sets and self-conformal measures associated with the list (4.5), respectively. Finally, for $\mathbf{i} = \mathsf{e}_1 \dots \mathsf{e}_n \in \Sigma_{\mathsf{G}}^n$, we write

$$\begin{aligned} S_{\mathbf{i}} &= S_{\mathsf{e}_1} \cdots S_{\mathsf{e}_n}\,, \\ K_{\mathbf{i}} &= S_{\mathsf{e}_1} \cdots S_{\mathsf{e}_n}\big(K_{\mathsf{t}(\mathsf{e}_n)}\big)\,, \\ p_{\mathbf{i}} &= p_{\mathsf{e}_1} \cdots p_{\mathsf{e}_n}\,. \end{aligned} \tag{4.10}$$

**4.3. Complex analytic continuation of Graph-directed Self-Conformal Iterated Functions Systems.** We will use Atiyah-Bott's fixed point theorem to study the multifractal transfer operator $\mathbf{L}^{q,s}$ introduced in Section 5. However, Atiyah-Bott's fixed point theorem is a statement about operators defined on spaces of holomorphic functions on subdomains of $\mathbb{C}^d$. Hence, in order to apply Atiyah-Bott's fixed point theorem it is necessary to extend the maps $S_{\mathsf{e}}$ in the Graph-directed Self-Conformal Iterated Function System (4.5) to subdomains of $\mathbb{C}^d$. In this section we will describe how to do this. We first make the following definition and Observation 4.1 below. For $\mathsf{i},\mathsf{j} \in \mathsf{V}$ and $\mathsf{e} \in E_{\mathsf{i},\mathsf{j}}$ define the function $Q_{\mathsf{e}} : V_{\mathsf{j}} \to \mathbb{R}$ by

$$Q_{\mathsf{e}}(x) = |(DS_{\mathsf{e}})(x)|\,. \tag{4.11}$$

We now have the following observation.

OBSERVATION 4.1. *For each* $\mathsf{e} \in \mathsf{E}$*, the function* $Q_{\mathsf{e}}$ *is real analytic.*

PROOF. For a square matrix $M$, write $\lambda_{\max}(M) = \max\{|\lambda| \mid \lambda \text{ is an eigenvalue}$ of $M$ and let $M^{\mathsf{T}}$ denote the transpose of $M$. Next, we note that since $Q_{\mathsf{e}}(x) = |(DS_{\mathsf{e}})(x)| = \sqrt{\lambda_{\max}((DS_{\mathsf{e}})(x)^{\mathsf{T}}(DS_{\mathsf{e}})(x))}$ and $\lambda_{\max}((DS_{\mathsf{e}})(x)^{\mathsf{T}}(DS_{\mathsf{e}})(x)) > 0$ for all $x$ (because $Q_{\mathsf{e}}(x) > 0$ for all $x$), it suffices to show that for each $\mathsf{e} \in \mathsf{E}$, the function $S_{\mathsf{e}}$ is real analytic. For $d = 1$, it is part of the assumptions in Section 2.2 that the functions $S_{\mathsf{e}}$ are real analytic for all $\mathsf{e} \in \mathsf{E}$. For $d = 2$, it follows from [**BePe**, Theorem A.3.2] that the conformal diffeomorphism $S_{\mathsf{e}}$ is holomorphic or antiholomorphic and therefore real analytic. For $d \geq 3$, it follows from Liouville's theorem [**BePe**, Theorem A.3.7] that the conformal diffeomorphism $S_{\mathsf{e}}$ is a Möbius transformation and therefore real analytic. □

It follows from the Basic Conditions in Section 4.2 and Observation 4.1 that we can choose a list

$$\Big( (\mathbb{V}_{\mathsf{i}})_{\mathsf{i}\in\mathsf{V}},\, (\mathbb{X}_{\mathsf{i}})_{\mathsf{i}\in\mathsf{V}},\, (\mathcal{S}_{\mathsf{e}})_{\mathsf{e}\in\mathsf{E}},\, (\mathcal{Q}_{\mathsf{e}})_{\mathsf{e}\in\mathsf{E}} \Big) \tag{4.12}$$

satisfying the Basic Conditions below.

Basic Conditions:

- For each $\mathsf{i} \in \mathsf{V}$ we have: $\mathbb{V}_{\mathsf{i}}$ is an open, connected subset of $\mathbb{C}^d$ with $V_{\mathsf{i}} \subseteq \mathbb{V}_{\mathsf{i}}$.

- For each $\mathsf{i} \in \mathsf{V}$ we have: $\mathbb{X}_\mathsf{i}$ is a compact subset of $\mathbb{C}^d$ with $X_\mathsf{i} \subseteq \mathbb{X}_\mathsf{i} \subseteq \mathbb{V}_\mathsf{i}$.
- For each $\mathsf{i}, \mathsf{j} \in \mathsf{V}$ and $\mathsf{e} \in E_{\mathsf{i},\mathsf{j}}$ we have: $\mathcal{S}_\mathsf{e} : \mathbb{V}_\mathsf{j} \to \mathbb{V}_\mathsf{i}$ is a holomorphic map with $\mathcal{S}_\mathsf{e}(\mathbb{X}_\mathsf{j}) \subseteq \mathbb{X}_\mathsf{i}$ and
$$\mathcal{S}_\mathsf{e}|_{X_\mathsf{j}} = S_\mathsf{e}$$
and there exists $s_\mathsf{e} \in (0, 1)$ such that
$$|D\mathcal{S}_\mathsf{e}(z)| \leq s_\mathsf{e} \tag{4.13}$$
for all $z \in \mathbb{X}_\mathsf{j}$.
- For each $\mathsf{i}, \mathsf{j} \in \mathsf{V}$ and $\mathsf{e} \in E_{\mathsf{i},\mathsf{j}}$ we have: $\mathcal{Q}_\mathsf{e} : \mathbb{V}_\mathsf{j} \to \mathbb{C}$ is a holomorphic map with
$$\mathcal{Q}_\mathsf{e}|_{V_\mathsf{j}} = Q_\mathsf{e}$$
and there exists a lower bound $q_\mathsf{e} > 0$ such that
$$q_\mathsf{e} \leq |\mathcal{Q}_\mathsf{e}(z)| \leq 1 \tag{4.14}$$
for all $z \in \mathbb{X}_\mathsf{j}$.

Also, for each $\mathsf{i}, \mathsf{j} \in \mathsf{V}$, $\mathsf{e} \in E_{\mathsf{i},\mathsf{j}}$ and $s \in \mathbb{C}$, we define the function $\mathcal{Q}_\mathsf{e}^s : \mathbb{V}_\mathsf{j} \to \mathbb{C}$ by
$$\mathcal{Q}_\mathsf{e}^s(z) = \mathcal{Q}_\mathsf{e}(z)^s\,, \tag{4.15}$$
where we define $u^v$ by $u^v = e^{v \log |u|} e^{iv \operatorname{Arg} u}$ for $u, v \in \mathbb{C}$ with $u \neq 0$. Next, we note the following non-trivial property of the maps $S_\mathsf{e}$.

OBSERVATION 4.2. *For each* $\mathsf{e} \in \mathsf{E}$, *the function* $x \to Q_\mathsf{e}(x) = |(DS_\mathsf{e})(x)|$ *is Lipschitz on* $X_{\mathsf{t}(\mathsf{e})}$.

PROOF. For $d = 1$, the statement follows from the mean value theorem since $X^\circ_{\mathsf{t}(\mathsf{i})}$ is assumed to be simply connected and therefore is an interval, and for $d \geq 2$, the statement follows from Koebe's theorem, see [**MaUr**, Proposition 4.2.1] for more details. □

It follows from Observation 4.2 that if we choose the sets $(\mathbb{X}_\mathsf{i})_\mathsf{i}$ sufficiently small that we may assume that the maps $\mathcal{Q}_\mathsf{e}$ are Lipschitz in $\mathbb{X}_\mathsf{e}$ for all $\mathsf{e}$, i.e. we can assume that the following Lipschitz Condition is satisfied.

Lipschitz Condition:

- For each $\mathsf{i}, \mathsf{j} \in \mathsf{V}$ and $\mathsf{e} \in E_{\mathsf{i},\mathsf{j}}$ we have: there is a number $C_\mathsf{e} \geq 1$ such that
$$|\mathcal{Q}_\mathsf{e}(z) - \mathcal{Q}_\mathsf{e}(w)| \leq C_\mathsf{e}|z - w| \tag{4.16}$$
for all $z, w \in \mathbb{X}_\mathsf{j}$.

Since the list (4.5) satisfied the Topological Conditions and the Separation Conditions in Section 4.2, we can choose the sets $(\mathbb{X}_\mathsf{i})_{\mathsf{i} \in \mathsf{V}}$ such that the list (4.12) satisfies the following conditions.

Topological Conditions:

- The Simply Connected Condition. For each $\mathsf{i} \in \mathsf{V}$ we have: $\mathbb{X}_\mathsf{i}^\circ$ is simply connected and has connected, smooth boundary that is spherical (i.e. locally biholomorphically equivalent to a piece of the boundary of a ball).
- The Interior Condition. For each $\mathsf{i}, \mathsf{j} \in \mathsf{V}$ and $\mathsf{e} \in E_{\mathsf{i},\mathsf{j}}$ we have: $\mathcal{S}_\mathsf{e}(\mathbb{X}_\mathsf{j}) \subseteq \mathbb{X}_\mathsf{i}^\circ$.

Separation Conditions:

- The Strong Separation Condition. For all $\mathsf{i} \in \mathsf{V}$ and all $\mathsf{e}_1, \mathsf{e}_2 \in \mathsf{E}_\mathsf{i}$ with $\mathsf{e}_1 \neq \mathsf{e}_2$ we have: $\mathcal{S}_{\mathsf{e}_1}\big(\mathbb{X}_{\mathsf{t}(\mathsf{e}_1)}\big) \cap \mathcal{S}_{\mathsf{e}_2}\big(\mathbb{X}_{\mathsf{t}(\mathsf{e}_2)}\big) = \varnothing$.
- For all $\mathsf{i}, \mathsf{j} \in \mathsf{V}$ with $\mathsf{i} \neq \mathsf{j}$ we have: $\mathbb{X}_\mathsf{i} \cap \mathbb{X}_\mathsf{j} \neq \varnothing$.

Finally, by choosing the sets $(\mathbb{X}_\mathsf{i})_{\mathsf{i}\in\mathsf{V}}$ sufficiently small we can also guarantee that the following neighbourhood condition is satisfied. To state this condition we first introduce the neighbourhood radius $r_{\text{nbh}}$ defined by

$$r_{\text{nbh}} = \tfrac{1}{4\sqrt{2}}\big(\min_{\mathsf{e}} \tfrac{q_\mathsf{e}}{C_\mathsf{e}}\big)(1 - s_{\max}) \tag{4.17}$$

where $s_{\max} = \max_\mathsf{e} s_\mathsf{e} < 1$.

Neighbourhood Condition:

- For each $\mathsf{i} \in \mathsf{V}$ we have: for each $z \in \mathbb{X}_\mathsf{i}$ there is $x \in X_\mathsf{i}$ with $|x - z| < r_{\text{nbh}}$ such that $[x, z] \subseteq \mathbb{X}_\mathsf{i}$ where $[x, z]$ denotes the line-segment from $x$ to $z$.

Finally, for $\mathbf{i} = \mathsf{e}_1 \dots \mathsf{e}_n \in \Sigma_\mathsf{G}^*$, we write

$$\mathcal{S}_\mathbf{i} = \mathcal{S}_{\mathsf{e}_1} \cdots \mathcal{S}_{\mathsf{e}_n} \,. \tag{4.18}$$

## 5. The Multifractal Zeta Function $\zeta^q(s)$

We can now define the multifractal transfer operator and the multifractal zeta function $\zeta^q(s)$. We start by recalling the definition of the Bergman space. For an open subset $\mathbb{G}$ of $\mathbb{C}^d$, the Bergman space $A^2(\mathbb{G})$ is defined by

$$A^2(\mathbb{G}) = \left\{ f : \mathbb{G} \to \mathbb{C} \,\middle|\, f \text{ is holomorphic}\,, \quad \int |f(z)|^2 \, dz < \infty \right\}.$$

The Bergman space $A^2(\mathbb{G})$ will always be equipped with the canonical inner product $\langle\cdot|\cdot\rangle$ defined by

$$\langle f|g\rangle = \int f(z)\,\overline{g(z)}\,dz\,.$$

It is well-known that the Bergman space $(A^2(\mathbb{G}), \langle\cdot|\cdot\rangle)$ is a Hilbert space; the reader is referred to the texts [**HeKoZh**] for a proof of this and for further discussion of Bergman spaces. We can now define the $(q, s)$-multifractal transfer operator.

DEFINITION (The $(q, s)$-multifractal transfer operator). *Let*

$$\mathbb{X} = \bigcup_{\mathsf{i}} \mathbb{X}_\mathsf{i}\,.$$

*For $q \in \mathbb{R}$ and $s \in \mathbb{C}$, the $(q, s)$-multifractal transfer operator*

$$\mathbf{L}^{q,s} : A^2(\mathbb{X}^\circ) \to A^2(\mathbb{X}^\circ)$$

*is defined by*

$$(\mathbf{L}^{q,s} f)(z) = \sum_{\substack{\mathsf{e}\in\mathsf{E}\\ z\in\mathbb{X}^\circ_{\mathsf{t}(\mathsf{e})}}} p_\mathsf{e}^q\, \mathcal{Q}_\mathsf{e}^s(z)\, f(\mathcal{S}_\mathsf{e}(z))$$

*for $f \in A^2(\mathbb{X}^\circ)$.*

REMARK. The reader may wonder why the transfer operator $\mathbf{L}^{q,s}$ is not defined by the following alternative (and perhaps more natural) expression $\mathbf{L}_{\text{alt}}^{q,s}$ given by

$$(\mathbf{L}_{\text{alt}}^{q,s} f)(z) = \sum_{\substack{\mathsf{e}\in\mathsf{E}\\ z\in\mathbb{X}^\circ_{\mathsf{t}(\mathsf{e})}}} p_\mathsf{e}^q\, |(D\mathcal{S}_\mathsf{e})(z)|^s\, f(\mathcal{S}_\mathsf{e}(z))$$

for $f \in A^2(\mathbb{X}^\circ)$. However, the function $z \to |(D\mathcal{S}_{\mathsf{e}})(z)|$ is, in general, not holomorphic and the operator $\mathbf{L}^{q,s}_{\text{alt}}$ does therefore, in general, not map $A^2(\mathbb{X}^\circ)$ into $A^2(\mathbb{X}^\circ)$. This is the reason why we introduce the holomorphic functions $\mathcal{Q}_{\mathsf{e}}$ and the reason why $|(D\mathcal{S}_{\mathsf{e}})(z)|^s$ has been replaced by $\mathcal{Q}^s_{\mathsf{e}}(z)$ in the definition of $\mathbf{L}^{q,s}$.

It follows immediately from Atiyah-Bott's fixed point theorem and the Interior Condition in Section 4.3 that the operator $\mathbf{L}^{q,s}$ is of trace-class, this is the content of Theorem 5.1.

THEOREM 5.1. *The operator* $\mathbf{L}^{q,s}$ *is of trace-class.*

Since the operator $\mathbf{L}^{q,s}$ is of trace-class, the Fredholm determinant $\det(\mathbf{1}-\mathbf{L}^{q,s})$ of $\mathbf{1}-\mathbf{L}^{q,s}$ is well-defined and we can therefore define the $q$'th multifractal zeta function $\zeta^q$ as follows.

DEFINITION (The $q$'th multifractal zeta function). *For* $q \in \mathbb{R}$, *the* $q$*'th multifractal zeta function* $\zeta^q : \mathbb{C} \to \mathbb{C}$ *is defined by*

$$\zeta^q(s) = \det(\mathbf{1} - \mathbf{L}^{q,s}).$$

## 6. Properties of the Multifractal Zeta Function $\zeta^q(s)$

We will present the key properties of the multifractal zeta function $\zeta^q(s)$. In particular, we will relate the zeros of the $q$'th multifractal zeta function $\zeta^q(s)$ to the Renyi dimensions $\underline{\tau}_{\mu_{\mathsf{i}}}(q)$ and $\overline{\tau}_{\mu_{\mathsf{i}}}(q)$ of the measures $\mu_{\mathsf{i}}$. As noted in the Introduction, a fundamental result in multifractal analysis says that the lower and upper $L^q$ dimensions of a graph directed self-conformal measure $\mu_{\mathsf{i}}$ associated with a Graph-directed Self-Conformal Iterated Function System satisfying the Open Set Condition conincide and that the common value of $\underline{\tau}_{\mu_{\mathsf{i}}}(q)$ and $\overline{\tau}_{\mu_{\mathsf{i}}}(q)$ is independent of $\mathsf{i}$, see, for example, [**MaUr**] for a proof and discussion of this. Also, recall from the Introduction that the common value of $\underline{\tau}^q(\mu_{\mathsf{i}})$ and $\overline{\tau}^q(\mu_{\mathsf{i}})$ is traditionally denoted by $\beta(q)$ and that we will follow this convention, i.e. we will write

$$\underline{\tau}_{\mu_{\mathsf{i}}}(q) = \overline{\tau}_{\mu_{\mathsf{i}}}(q) = \beta(q) \tag{6.1}$$

for all $\mathsf{i}$. We note that the number $\beta(q)$ is the solution to an equation involving the topological pressure (see, for example, [**MaUr**]), but since this result is not needed we have decided to omit the details. Below we present the main properties of $\zeta^q(s)$.

**6.1. Dirichlet series expansion.** We first show that the zeta function $\zeta^q(s)$ has a Dirichlet series expansion.

THEOREM 6.1 (Dirichlet series expansion). *For* $\mathbf{i} \in \Sigma^*_{\mathsf{G}}$ *with* $\mathsf{i}(\mathbf{i}) = \mathsf{t}(\mathbf{i})$, *let* $x_{\mathbf{i}} \in X_{\mathsf{t}(\mathbf{i})}$ *be the unique fixed point of* $S_{\mathbf{i}} : X_{\mathsf{t}(\mathbf{i})} \to X_{\mathsf{i}(\mathbf{i})}$. *For all* $s \in \mathbb{C}$ *with* $\operatorname{Re}(s) > \beta(q)$, *we have*

$$\zeta^q(s) = \exp\left( -\sum_n \frac{1}{n} \sum_{\substack{\mathbf{i}\in\Sigma^n_{\mathsf{G}} \\ \mathsf{i}(\mathbf{i})=\mathsf{t}(\mathbf{i})}} \frac{p^q_{\mathbf{i}}\, |(DS_{\mathbf{i}})(x_{\mathbf{i}})|^s}{\det(1-(DS_{\mathbf{i}})(x_{\mathbf{i}}))} \right).$$

For the proof of Theorem 6.1 the reader is referred to [**Ol1**]. The proof is based on Atiyah-Bott's fixed point theorem and the Interior Condition from Section 4.2 together with a number of delicate multifractal estimates. Theorem 6.1 plays a crucial role in obtaining the lower bound for the zeta function $\zeta^q(s)$ in Section 6.2, and for obtaining functional equations for the zeta function $\zeta^q(s)$ in [**Ol1**].

The Dirichlet series expansion for $\zeta^q(s)$ in Theorem 6.1 can be written as a sum of "primes"; this is the content of Theorem 6.2 below. The fact that the Dirichlet series of $\zeta^q(s)$ in Theorem 6.1 can be written as a sum of "primes" plays an important role when using the Wiener-Ikehara's tauberian theorem and the fact that the Dirichlet series for $\zeta^q(s)$ has an analytic continuation to derive various multifractal prime counting formulas for Graph-directed Self-Conformal Iterated Function Systems, and, in particular, obtain explicit formulas for the "symbolic" multifractal Minkowski volume of graph directed self-conformal multifractals; this will be discussed in Section 7. We will now explain how the series in Theorem 6.1 can be written as a sum of "primes". In order to so we introduce the following notation. A string $\mathbf{i} \in \Sigma_{\mathsf{G}}^*$ is called periodic if there is a string $\mathbf{j} \in \Sigma_{\mathsf{G}}^*$ and a positive integer $n > 1$ such that $\mathbf{i} = \mathbf{j}^n$, and a string is called prime if it is not periodic. We denote the set of prime loop strings by $\Pi_{\mathsf{G}}$, i.e.

$$\Pi_{\mathsf{G}} = \left\{ \mathbf{i} \in \Sigma_{\mathsf{G}}^* \,\middle|\, \text{the string } \mathbf{i} \text{ is prime with } \mathrm{i}(\mathbf{i}) = \mathrm{t}(\mathbf{i}) \right\}.$$

Next, for $\mathbf{i} = \mathsf{e}_1 \dots \mathsf{e}_n \in \Sigma_{\mathsf{G}}^n$ and a permutation $\sigma \in S_n$, we write

$$\sigma(\mathbf{i}) = \mathsf{e}_{\sigma(1)} \dots \mathsf{e}_{\sigma(n)},$$

and define the equivalence relation $\approx$ in $\Pi_{\mathsf{G}}$ by:

$$\mathbf{i} \approx \mathbf{j} \quad \Leftrightarrow \quad \begin{array}{l} \text{there is a positive integer } n \text{ and} \\ \text{a cyclic permutation } \sigma \in S_n \text{ such that} \\ |\mathbf{i}| = |\mathbf{j}| = n \text{ and } \mathbf{i} = \sigma(\mathbf{j}) \end{array}$$

Finally, let

$$\mathit{\Pi}_{\mathsf{G}} = \Pi_{\mathsf{G}} / \approx$$

denote the family of equivalence classes. We first note that if $\mathbf{i}, \mathbf{j} \in \Sigma_{\mathsf{G}}^*$ satisfy $\mathbf{i} \approx \mathbf{j}$, then clearly $p_{\mathbf{i}} = p_{\mathbf{j}}$, and for $\mathbf{I} \in \mathit{\Pi}_{\mathsf{G}}$, we can therefore define $p_{\mathbf{I}}$ by letting

$$p_{\mathbf{I}} = p_{\mathbf{i}}$$

for any $\mathbf{i} \in \mathbf{I}$. We also introduce the following notation, namely, for $\mathbf{i} \in \Sigma_{\mathsf{G}}^*$ with $\mathrm{i}(\mathbf{i}) = \mathrm{t}(\mathbf{i})$, we write

$$\begin{aligned} \ell_{\mathbf{i}} &= -\log |(DS_{\mathbf{i}})(x_{\mathbf{i}})|, \\ \Delta_{\mathbf{i}} &= \det(1 - (DS_{\mathbf{i}})(x_{\mathbf{i}})); \end{aligned}$$

recall that $x_{\mathbf{i}} \in X_{\mathrm{t}(\mathbf{i})}$ denotes the unique fixed point of $S_{\mathbf{i}} : X_{\mathrm{t}(\mathbf{i})} \to X_{\mathrm{i}(\mathbf{i})}$. Next, we note that it is not difficult to see that if $\mathbf{i}, \mathbf{j} \in \Sigma_{\mathsf{G}}^*$ satisfy $\mathbf{i} \approx \mathbf{j}$, then $\ell_{\mathbf{i}} = \ell_{\mathbf{j}}$ and $\Delta_{\mathbf{i}} = \Delta_{\mathbf{j}}$, and for $\mathbf{I} \in \mathit{\Pi}_{\mathsf{G}}$, we can therefore define $\ell_{\mathbf{I}}$ by letting

$$\begin{aligned} \ell_{\mathbf{I}} &= \ell_{\mathbf{i}}, \\ \Delta_{\mathbf{I}} &= \Delta_{\mathbf{i}}, \end{aligned}$$

for any $\mathbf{i} \in \mathbf{I}$. Using the above notation Theorem 6.1 can now be reformulated in the following way.

THEOREM 6.2 (Dirichlet series expansion). *For $\mathbf{I} \in \mathit{\Pi}_{\mathsf{G}}$, let $p_{\mathbf{I}}$, $\ell_{\mathbf{I}}$ and $\Delta_{\mathbf{I}}$ be defined as above. For all $s \in \mathbb{C}$ with* $\mathrm{Re}(s) > \beta(q)$, *we have*

$$\zeta^q(s) = \exp\left( - \sum_{\mathbf{I} \in \mathit{\Pi}_{\mathsf{G}}} \frac{p_{\mathbf{I}}^q}{\Delta_{\mathbf{I}}} e^{-\ell_{\mathbf{I}} s} \right).$$

PROOF. This follows immediately from Theorem 6.1 since $\operatorname{card}\mathbf{I} = |\mathbf{i}|$ for all $\mathbf{I} \in \Pi_{\mathsf{G}}$ and all $\mathbf{i} \in \mathbf{I}$. □

**6.2. Lower bound and zero free half plane.** Our first main result provides lower bounds for the zeta function $\zeta^q(s)$. In particular, for each $t > \beta(q)$, we obtain a *uniform* lower bound for $\zeta^q(s)$ for $s$ with $\operatorname{Re}(s) \geq t$ and, as an application of this, we show that $\zeta^q(s)$ does not have any zeros for $s \in \mathbb{C}$ with $\operatorname{Re}(s) > \beta(q)$. This is the contents of Theorem 6.3 and Theorem 6.4 below.

THEOREM 6.3 (Lower bound). *Let $q \in \mathbb{R}$ and $t > \beta(q)$. Then there is a (uniform) constant $c > 0$ such that*

$$|\zeta^q(s)| \geq c$$

*for all $s \in \mathbb{C}$ with $\operatorname{Re}(s) \geq t$.*

The proof of Theorem 6.3 is based on the Dirichlet series expansion of $\zeta^q(s)$ in Theorem 6.1. Indeed, the terms in the Dirichlet series in Theorem 6.1 can be estimated and this leads to the lower bound in the statement of Theorem 6.3; the reader is referred to [**Ol1**] for more details.

It follows immediate from Theorem 6.3 that $\zeta^q(s)$ does not have any zeros for $s \in \mathbb{C}$ with $\operatorname{Re}(s) > \beta(q)$. This is the content of Theorem 6.4 below.

THEOREM 6.4 (Zero free half plane). *Let $q \in \mathbb{R}$. Then $\zeta^q(s) \neq 0$ for all $s \in \mathbb{C}$ with $\operatorname{Re}(s) > \beta(q)$.*

PROOF. This follows immediately from Theorem 6.3. □

Theorem 6.4 says that the zeta function $\zeta^q(s)$ does not have any zeros for $\operatorname{Re}(s) > \beta(q)$. This statement is optimal since it follows from Theorem 6.5 below that the real number $\beta(q)$ is always a zero of $\zeta^q(s)$. Theorem 6.5 also provides information about the non-real zeros of $\zeta^q(s)$ on the "critical line" $\operatorname{Re}(s) = \beta(q)$. Indeed, Theorem 6.5 shows that if the numbers $\log |(DS_{\mathsf{e}_i})(S_{\mathsf{e}_{i+1}} \cdots S_{\mathsf{e}_n} x_{\mathbf{i}})|$ are rationally independent for all $n \in \mathbb{N}$, $1 \leq i \leq n$ and $\mathbf{i} = \mathsf{e}_1 \ldots \mathsf{e}_n \in \Sigma_{\mathsf{G}}^n$ (recall, that for $\mathbf{i} \in \Sigma_{\mathsf{G}}^n$ with $\mathrm{i}(\mathbf{i}) = \mathrm{t}(\mathbf{i})$, we let $x_{\mathbf{i}} \in X_{\mathrm{t}(\mathbf{i})}$ denote the unique fixed point of $S_{\mathbf{i}} : X_{\mathrm{t}(\mathbf{i})} \to X_{\mathrm{i}(\mathbf{i})}$), then the zeta function $\zeta^q$ has an infinite lattice of zeros whose real parts equal $\beta(q)$. In the statement of Theorem 6.5 we use the following notation, namely, if $f$ is an entire function, then we denote the set of zeros of $f$ by $Z(f)$, and if $(t_i)_{i\in I}$ is a family of real numbers, then $\langle t_i | i \in I\rangle$ denotes the additive subgroup of $\mathbb{R}$ generated by the family $(t_i)_{i\in I}$.

THEOREM 6.5 (Critical line). *Let $q \in \mathbb{R}$.*

(1) *We have*

$$\beta(q) \in Z(\zeta^q)\,.$$

(2) *For $\mathbf{i} \in \Sigma_{\mathsf{G}}^n$ with $\mathrm{i}(\mathbf{i}) = \mathrm{t}(\mathbf{i})$, let $x_{\mathbf{i}} \in X_{\mathrm{t}(\mathbf{i})}$ be the unique fixed point of $S_{\mathbf{i}} : X_{\mathrm{t}(\mathbf{i})} \to X_{\mathrm{i}(\mathbf{i})}$, and write*

$$\mathcal{G} = \left\langle \log |(DS_{\mathsf{e}_i})(S_{\mathsf{e}_{i+1}} \cdots S_{\mathsf{e}_n} x_{\mathbf{i}})| \,\middle|\, n \in \mathbb{N}\,,\ 1 \leq i \leq n\,,\ \mathbf{i} = \mathsf{e}_1 \ldots \mathsf{e}_n \in \Sigma_{\mathsf{G}}^n \right\rangle. \tag{6.2}$$

*If $\mathcal{G} = u\mathbb{Z}$ for some $u > 0$, then we have*

$$\beta(q) + i\tfrac{2\pi}{u}\mathbb{Z} \subseteq Z(\zeta^q)\,.$$

For the proof of Theorem 6.5, the reader is referred to [**Ol1**]. We believe that the statement in Theorem 6.5 can be strengthened as follows: the zeros of the zeta function $\zeta^q$ equals an infinite lattice of points with real parts equal to $\beta(q)$ if and only of the subgroup $\mathcal{G}$ in (6.2) is discrete, and that $\beta(q)$ is the only zero of the zeta function $\zeta^q$ with real part equal to $\beta(q)$ if and only of the subgroup $\mathcal{G}$ in (6.2) is non-discrete. We formulate this formally in Conjecture 6.6 below.

CONJECTURE 6.6 (Critical line lattice dichotomy). *Let $q \in \mathbb{R}$ and let $\mathcal{G}$ be the subgroup in* (6.2).

(1) *If $\mathcal{G} \neq u\mathbb{Z}$ for all $u > 0$ (i.e. if $\mathcal{G}$ is dense in $\mathbb{R}$), then*

$$\{\beta(q)\} = Z(\zeta^q) \cap \{s \in \mathbb{C} \mid \mathrm{Re}(s) = \beta(q)\}\,.$$

(2) *If $\mathcal{G} = u\mathbb{Z}$ for some $u > 0$, then*

$$\beta(q) + i\tfrac{2\pi}{u}\mathbb{Z} = Z(\zeta^q) \cap \{s \in \mathbb{C} \mid \mathrm{Re}(s) = \beta(q)\}\,.$$

While we have not been able to prove Conjecture 6.6 in full generality, we show in [**Ol1**] (see also Section 8) that Conjecture 6.6 is true for the most studied (and, arguably, most important) class of Graph-directed Self-Conformal Iterated Functions Systems, namely, the Graph-directed Self-Similar Iterated Functions Systems, i.e. those Graph-directed Self-Conformal Iterated Functions Systems for which all the maps $S_{\mathsf{e}}$ are contracting similarities. Indeed, in Section 4 we show that if $d = 1$, then Conjecture 6.6 is true for all Graph-directed Self-Similar Iterated Functions Systems and if $d > 1$, then Conjecture 6.6 is true for those Graph-directed Self-Similar Iterated Functions Systems for which the linear parts of the maps $S_{\mathsf{e}}$ commute.

**6.3. Upper bound.** For a Hilbert space $H$ and a compact operator $\mathbf{A} : H \to H$, we let $(\lambda_n(\mathbf{A}))_n$ denote the sequence of eigenvalues of $\mathbf{A}$ ordered such that $|\lambda_1(\mathbf{A})| \geq |\lambda_2(\mathbf{A})| \geq \ldots$. Our second and third main results, namely Theorem 6.7 and Theorem 6.8 below, provide upper bounds for the eigenvalues of $\mathbf{L}^{q,s}$ and an upper bound for the zeta function $\zeta^q$, respectively. In particular, we show that eigenvalues $\lambda_n(\mathbf{L}^{q,s})$ tend to 0 exponentially fast and that the decay rate is proportional to $\frac{1}{|\mathrm{Im}(s)|^{\dim_{\mathsf{H}} K_{\mathsf{i}}/d}}$ (this is the contents of Theorem 6.7) and that the growth rate of $\log|\zeta^q(s)|$ as $|s|$ tends to infinity is bounded above by $|\mathrm{Im}(s)|^{\dim_{\mathsf{H}} K_{\mathsf{i}}}$ (this is the contents of Theorem 6.8). To state Theorem 6.7 and Theorem 6.8 we first introduce the embeddebility radius $r_{\mathrm{emb}}$. To define $r_{\mathrm{emb}}$ we note that if $\mathsf{i} \in \mathsf{V}$ is a vertex, then it follows from the Simply Connected Condition in Section 4.3 and Riemann's mapping Theorem that there is a surjective bi-holomorphic map

$$\varphi_{\mathsf{i}} : \mathbb{X}_{\mathsf{i}}^{\circ} \to B(0,1)\,,$$

and so

$$\mathbb{X}_{\mathsf{i}}^{\circ} = \varphi_{\mathsf{i}}^{-1}(B(0,1))\,.$$

We now define the embeddebility radius $r_{\mathrm{emb}}$ by

$$r_{\mathrm{emb}} = \max_{\mathsf{i},\mathsf{j}\in\mathsf{E}}\ \max_{\mathsf{e}\in\mathsf{E}_{\mathsf{i},\mathsf{j}}}\ \inf\left\{r > 0 \,\middle|\, \mathcal{S}_{\mathsf{e}}(\mathbb{X}_{\mathsf{j}}) \subseteq \varphi_{\mathsf{i}}^{-1}(B(0,r)) \subseteq \mathbb{X}_{\mathsf{i}}^{\circ}\right\};$$

note that since $\mathcal{S}_{\mathsf{e}}(\mathbb{X}_{\mathsf{j}})$ is a compact subset of $\mathbb{X}_{\mathsf{i}}^{\circ}$ and $\mathbb{X}_{\mathsf{i}}^{\circ} = \varphi_{\mathsf{i}}^{-1}(B(0,1))$, we have $r_{\mathrm{emb}} < 1$. It is clear that the embeddability radius $r_{\mathrm{emb}}$ provides a quantitative measure for how well $\mathcal{S}_{\mathsf{e}}(\mathbb{X}_{\mathsf{j}})$ embeds into $\varphi_{\mathsf{i}}^{-1}(B(0,1))$ for $\mathsf{e}\in\mathsf{E}_{\mathsf{i},\mathsf{j}}$. We also define the logarithmic embeddability index $l_{\mathrm{emb}}$ by

$$l_{\mathrm{emb}}=\log\frac{1}{r_{\mathrm{emb}}}\,.$$

Next, for a vector $\mathbf{x}=(x_{\mathsf{e}})_{\mathsf{e}\in\mathsf{E}}\in\mathbb{R}^{\mathsf{E}}$, we write $|\mathbf{x}|_\infty$ for the uniform norm of $\mathbf{x}$, i.e. $|\mathbf{x}|_\infty=\max_{\mathsf{e}\in\mathsf{E}}|x_{\mathsf{e}}|$, and for a vector $\mathbf{x}=(x_{\mathsf{e}})_{\mathsf{e}\in\mathsf{E}}\in\mathbb{R}^{\mathsf{E}}$ with $x_{\mathsf{e}}>0$ for all $\mathsf{e}$ and $t\in\mathbb{R}$, we write $\mathbf{x}^t=(x_{\mathsf{e}}^t)_{\mathsf{e}\in\mathsf{E}}$. We now define the vector $\mathbf{p}$ by $\mathbf{p}=(p_{\mathsf{e}})_{\mathsf{e}\in\mathsf{E}}$ and consider the uniform norm $|\mathbf{p}^q|_\infty$ of the vector $\mathbf{p}^q=(p_{\mathsf{e}}^q)_{\mathsf{e}\in\mathsf{E}}$, i.e.

$$|\mathbf{p}^q|_\infty=\max_{\mathsf{e}\in\mathsf{E}}p_{\mathsf{e}}^q\,.$$

Finally, for $\varepsilon>0$, we define $L_\varepsilon$ by

$$L_\varepsilon=\inf_{\substack{r_{\mathrm{emb}}<u<v<1\\ \frac{v}{u}\le\frac{1}{r_{\mathrm{emb}}}\le e^{\varepsilon}\frac{v}{u}}}\ \max_{\mathsf{i}}\ \sup_{\substack{z\in\varphi_{\mathsf{i}}^{-1}(\,\overline{B(0,v)}\,)\\ w\in\varphi_{\mathsf{i}}^{-1}(\,\overline{B(0,u)}\,)}}\frac{|\det((D\varphi_{\mathsf{i}})(z))|}{|\det((D\varphi_{\mathsf{i}})(w))|}\,,$$

and we define the ambient dimension index by

$$\mathrm{ind}_d=\frac{1}{4\,c_{\mathrm{bd}}^2\,(4\,c_{\mathrm{bv}}^2\,c_{\max}^2\,c_{\mathsf{G}}\,|\mathsf{E}|)^{1/d}}\,\frac{d}{1+d}\,d$$

where $c_{\mathrm{bd}}$, $c_{\mathrm{bv}}$, $c_{\max}$ and $c_{\mathsf{G}}$ are four universal constants associated with the Graph-directed Self-Conformal Iterated Function System in (4.5); the reader is referred to [**Ol1**] for the precise definition of the constants $c_{\mathrm{bd}}$, $c_{\mathrm{bv}}$, $c_{\max}$ and $c_{\mathsf{G}}$.

THEOREM 6.7 (Upper bounds for the eigenvalues of $\mathbf{L}^{q,s}$).

(1) ***Upper bounds for "large" values of*** $s$. *Let $c,u>0$. Then there is a constant $C>0$ such that for all $0<\varepsilon<l_{\mathrm{emb}}$, $\rho>0$, $q\in\mathbb{R}$ and $s\in\mathbb{C}$ with* $\mathrm{Re}(s)>-c$ *and* $|\mathrm{Im}(s)|\ge u$, *we have*

$$|\lambda_n(\mathbf{L}^{q,s})|\le C\,\frac{1}{\rho^{d-1}}\,\frac{L_\varepsilon}{((l_{\mathrm{emb}}-\varepsilon)e^{(l_{\mathrm{emb}}-\varepsilon)})^d}\,|\mathbf{p}^q|_\infty\,e^{-(1-\rho)\,(l_{\mathrm{emb}}-\varepsilon)\,\mathrm{ind}_d\,\frac{1}{|\mathrm{Im}(s)|^{\dim_{\mathsf{H}}K_{\mathsf{i}}/d}}\,n^{1/d}}$$

*for all $n$ and all $\mathsf{i}\in\mathsf{V}$.*

(2) ***Upper bounds for "small" values of*** $s$. *Let $c,U>0$. Then there is a constant $C>0$ such that for all $0<\varepsilon<l_{\mathrm{emb}}$, $\rho>0$, $q\in\mathbb{R}$ and $s\in\mathbb{C}$ with* $\mathrm{Re}(s)>-c$ *and* $|\mathrm{Im}(s)|\le U$, *we have*

$$|\lambda_n(\mathbf{L}^{q,s})|\le C\,\frac{1}{\rho^{d-1}}\,\frac{L_\varepsilon}{((l_{\mathrm{emb}}-\varepsilon)e^{(l_{\mathrm{emb}}-\varepsilon)})^d}\,|\mathbf{p}^q|_\infty\,e^{-(1-\rho)\,(l_{\mathrm{emb}}-\varepsilon)\,\mathrm{ind}_d\,n^{1/d}}$$

*for all $n$.*

The proof of Theorem 6.7 in presented in [**Ol1**]. Theorem 6.7 immediately implies our third main result, namely Theorem 6.8 below. Theorem 6.8 provides an upper bound for the growth rate of $\log|\zeta^q(s)|$ as $|s|$ tends to infinity. This growth rate is also determined by the Hausdorff dimension $\dim_{\mathsf{H}}K_{\mathsf{i}}$ of the self-conformal sets $K_{\mathsf{i}}$.

THEOREM 6.8 (Upper bound for the zeta-function $\zeta^q(s)$).

(1) ***Upper bound for "large" values of*** $s$. *Let $c,u>0$. Then there is a constant $C>0$ such that for all $0<\varepsilon<l_{\mathrm{emb}}$, $q\in\mathbb{R}$ and $s\in\mathbb{C}$ with* $\mathrm{Re}(s)>-c$ *and* $|\mathrm{Im}(s)|\ge u$, *we have*

$$\log|\zeta^q(s)|\le C\,\frac{L_\varepsilon}{((l_{\mathrm{emb}}-\varepsilon)^2e^{(l_{\mathrm{emb}}-\varepsilon)})^d}\,|\mathbf{p}^q|_\infty\,|\mathrm{Im}(s)|^{\dim_{\mathsf{H}}K_{\mathsf{i}}}$$

*for all $\mathsf{i}\in\mathsf{V}$.*

(2) ***Upper bound for "small" values of*** $s$. *Let $c, U > 0$. Then there is a constant $C > 0$ such that for all $0 < \varepsilon < l_{\text{emb}}$, $q \in \mathbb{R}$ and $s \in \mathbb{C}$ with $\operatorname{Re}(s) > -c$ and $|\operatorname{Im}(s)| \le U$, we have*

$$\log|\zeta^q(s)| \le C \frac{L_\varepsilon}{((l_{\text{emb}}-\varepsilon)^2 e^{(l_{\text{emb}}-\varepsilon)})^d} |\mathbf{p}^q|_\infty .$$

PROOF. (1) For $s \in \mathbb{C}$, we have

$$\begin{aligned} \log|\zeta^q(s)| &= \log|\det(\mathbf{1} - \mathbf{L}^{q,s})| \\ &= \log\left|\prod_n (1 + \lambda_n(-\mathbf{L}^{q,s}))\right| \\ &\le \log \prod_n (1 + |\lambda_n(\mathbf{L}^{q,s})|) \qquad (6.3) \\ &= \sum_n \log(1 + |\lambda_n(\mathbf{L}^{q,s})|) \\ &\le \sum_n |\lambda_n(\mathbf{L}^{q,s})| . \end{aligned}$$

It follows from Theorem 6.7 that there is a constant $C > 0$ such that for all $0 < \varepsilon < l_{\text{emb}}$, $\rho > 0$, $q \in \mathbb{R}$, $s \in \mathbb{C}$ with $\operatorname{Re}(s) > -c$ and $|\operatorname{Im}(s)| \ge u$, we have
(6.4)

$$|\lambda_n(\mathbf{L}^{q,s})| \le C \frac{1}{\rho^{d-1}} \frac{L_\varepsilon}{((l_{\text{emb}}-\varepsilon) e^{(l_{\text{emb}}-\varepsilon)})^d} |\mathbf{p}^q|_\infty \, e^{-(1-\rho)\,(l_{\text{emb}}-\varepsilon)\,\text{ind}_d \frac{1}{|\operatorname{Im}(s)|^{\dim_{\mathsf{H}} K_{\mathsf{i}}/d}} n^{1/d}}$$

for all $n$ and all $\mathsf{i} \in \mathsf{V}$. Combining (6.3) and (6.4) we conclude that for all $0 < \varepsilon < l_{\text{emb}}$, $\rho > 0$, $q \in \mathbb{R}$, $s \in \mathbb{C}$ with $\operatorname{Re}(s) > -c$ and $|\operatorname{Im}(s)| \ge u$, we have
(6.5)

$$\log|\zeta^q(s)| \le C \frac{1}{\rho^{d-1}} \frac{L_\varepsilon}{((l_{\text{emb}}-\varepsilon) e^{(l_{\text{emb}}-\varepsilon)})^d} |\mathbf{p}^q|_\infty \sum_n e^{-(1-\rho)\,(l_{\text{emb}}-\varepsilon)\,\text{ind}_d \frac{1}{|\operatorname{Im}(s)|^{\dim_{\mathsf{H}} K_{\mathsf{i}}/d}} n^{1/d}} .$$

However, for $x > 0$, we have $\sum_n e^{-xn^{1/d}} \le \int_0^\infty e^{-xt^{1/d}}\, dt = x^{-d} \int_0^\infty e^{-t^{1/d}}\, dt = x^{-d} I$ where $I = \int_0^\infty e^{-t^{1/d}}\, dt < \infty$, and it now follows from this and (6.5) that for all $0 < \varepsilon < l_{\text{emb}}$, $q \in \mathbb{R}$, $s \in \mathbb{C}$ with $\operatorname{Re}(s) > -c$ and $|\operatorname{Im}(s)| \ge u$, we have

$$\begin{aligned} \log|\zeta^q(s)| &\le C\,I \frac{1}{\rho^{d-1}} \frac{L_\varepsilon}{((l_{\text{emb}}-\varepsilon) e^{(l_{\text{emb}}-\varepsilon)})^d} |\mathbf{p}^q|_\infty \left( (1-\rho)\,(l_{\text{emb}}-\varepsilon)\,\text{ind}_d \frac{1}{|\operatorname{Im}(s)|^{\dim_{\mathsf{H}} K_{\mathsf{i}}/d}} \right)^{-d} \\ &= C\,I \frac{1}{\text{ind}_d^d} \frac{1}{\rho^{d-1}(1-\rho)^d} \frac{L_\varepsilon}{((l_{\text{emb}}-\varepsilon)^2 e^{(l_{\text{emb}}-\varepsilon)})^d} |\mathbf{p}^q|_\infty \, |\operatorname{Im}(s)|^{\dim_{\mathsf{H}} K_{\mathsf{i}}} \end{aligned}$$

for all $0 < \rho < 1$, and so

$$\begin{aligned} \log|\zeta^q(s)| &\le C\,I \frac{1}{\text{ind}_d^d} \left( \inf_{0<\rho<1} \frac{1}{\rho^{d-1}(1-\rho)^d} \right) \frac{L_\varepsilon}{((l_{\text{emb}}-\varepsilon)^2 e^{(l_{\text{emb}}-\varepsilon)})^d} |\mathbf{p}^q|_\infty \, |\operatorname{Im}(s)|^{\dim_{\mathsf{H}} K_{\mathsf{i}}} \\ &= C\,I \frac{1}{\text{ind}_d^d} \frac{(d-1)^{d-1} d^d}{(2d-1)^{2d-1}} \frac{L_\varepsilon}{((l_{\text{emb}}-\varepsilon)^2 e^{(l_{\text{emb}}-\varepsilon)})^d} |\mathbf{p}^q|_\infty \, |\operatorname{Im}(s)|^{\dim_{\mathsf{H}} K_{\mathsf{i}}} \end{aligned}$$

since an easy calculus argument shows that $\inf_{0<\rho<1} \frac{1}{\rho^{d-1}(1-\rho)^d} = \frac{(d-1)^{d-1} d^d}{(2d-1)^{2d-1}}$.
(2) The proof of part (2) is similar to the proof of part (1) and is therefore omitted. □

REMARK (Comparing Theorem 6.7 with previous results). The study of upper bounds for eigenvalues of transfer operators defined on Bergman spaces has a rich and interesting history and the reader is referred to [**Pi**] for an account of this. The state-of-the-art upper bounds for eigenvalues of transfer operators defined on

Bergman spaces consisting of holomorphic $\mathcal{L}^2$ functions defined on general open subsets $\mathbb{G}$ of $\mathbb{C}^d$ are given by [**BaJe3**] . Bounds similar to those in [**BaJe3**] appeared in [FaRo, Thm. 4] in the setting of dynamical systems on the torus, and in [**BaJe1**, **BaJe2**] in the special case where $\mathbb{G}$ is an Euclidean ball. Bounds similar to those in [**BaJe3**] in the special case where $\mathbb{G}$ is an Euclidean ball also appeared in [**GuLiZw**, (3.6), p. 157]). The setting in [**BaJe3**] is somewhat more general that the setting in this paper. While we consider transfer operators defined on the Bergman space $A^2(\mathbb{X}^\circ)$ on the set $\mathbb{X}^\circ$ associated with the Graph-directed Self- Conformal Iterated Function Systems in Section 4, [**BaJe3**] considers transfer operators defined on the Bergman space $A^2(\mathbb{G})$ on a general open subset $\mathbb{G}$ of $\mathbb{C}^d$. Also, while the transfer operators considered in this work are weighted by $p_{\mathsf{e}}^q$, [**BaJe3**] considers transfer operators weighted be arbitrary holomorphic weight functions. However, when adapted to the setting in this work, the main result in [**BaJe3**] says that there are constants $c_0, C_0 > 0$ such that for all $q \in \mathbb{R}$ and $s \in \mathbb{C}$, we have

$$|\lambda_n(\mathbf{L}^{q,s})| \leq C_0 \, |\mathbf{p}^q|_\infty \left( \max_{\mathsf{e}\in\mathsf{E}} \sup_{z\in\mathbb{X}^\circ_{\mathsf{t}(\mathsf{e})}} |\mathcal{Q}^s_{\mathsf{e}}(z)| \right) e^{-c_0 n^{1/d}}$$

for all $n$. Since it is easily seen that there is a constant $c$ such that

$$\max_{\mathsf{e}\in\mathsf{E}} \sup_{z\in\mathbb{X}^\circ_{\mathsf{t}(\mathsf{e})}} |\mathcal{Q}^s_{\mathsf{e}}(z)| \leq e^{c|s|}$$

for all $s$, we now conclude that the main result in [**BaJe3**] says that there are constants $c, c_0, C_0 > 0$ such that for all $q \in \mathbb{R}$ and $s \in \mathbb{C}$, we have

$$|\lambda_n(\mathbf{L}^{q,s})| \leq C_0 \, |\mathbf{p}^q|_\infty \, e^{c|s| - c_0 n^{1/d}} \tag{6.6}$$

for all $n$. We immediately note that the upper bound in (6.8) is substantially weaker than the bound in Theorem 6.7.

Using arguments similar to those in the proof of Theorem 6.8, the following upper bound for $\zeta^q(s)$ follows immediately from inequality (6.8), namely, there are constants $c, C > 0$ such that for all $q \in \mathbb{R}$ and $s \in \mathbb{C}$, we have

$$\log|\zeta^q(s)| \leq C \, |\mathbf{p}^q|_\infty \, e^{c|s|} \, . \tag{6.7}$$

We note that this bound is substantially weaker than the statement in Theorem 6.8.

It is not difficult to improve the upper bound in (6.7) somewhat using only slightly more careful estimates. Indeed, for $s, q \in \mathbb{C}$, we have

$$\begin{aligned} \log|\zeta^q(s)| &= \log|\det(\mathbf{1} - \mathbf{L}^{q,s})| \\ &= \log\left|\prod_n (1 + \lambda_n(-\mathbf{L}^{q,s}))\right| \\ &\leq \log\prod_n (1 + |\lambda_n(\mathbf{L}^{q,s})|) = \sum_n \log(1 + |\lambda_n(\mathbf{L}^{q,s})|) \, . \end{aligned} \tag{6.8}$$

Combining (6.6) and (6.8) we now conclude that for all $q \in \mathbb{R}$ and $s \in \mathbb{C}$, we have we have

$$\log|\zeta^q(s)| \leq \sum_n \log\left(1 + C_0 \, |\mathbf{p}^q|_\infty \, e^{c|s| - c_0 \, n^{1/d}}\right) . \tag{6.9}$$

To proceed further we use the following estimate from real analysis: if $h$, $k$ and $\Delta$ are positive constants, then there are constants $K, t_0 > 0$ such that

$$\sum_n \log(1 + xe^{ht - kn^{\frac{1}{\Delta}}}) \leq K\,(1+x)\,t^{\Delta+1} \tag{6.10}$$

for all $x > 0$ and all $t \geq t_0$. Combining (6.9) and (6.10) it follows that we can find constants $C, u > 0$ such that for all $q \in \mathbb{R}$ and $s \in \mathbb{C}$ with $|s| \geq u$, we have

$$\log|\zeta^q(s)| \leq C\left(|\mathbf{p}^q|_\infty + 1\right)|s|^{d+1}\,. \tag{6.11}$$

While the upper bound for $\zeta^q(s)$ in (6.11) improves (6.7), we note that since $\dim_{\mathsf{H}} K_{\mathsf{i}} \leq d < d+1$, it is still substantially weaker than the statement in Theorem 6.8.

Remark (A brief and direct proof of a weak version of Theorem 6.7). In order to prove Theorem 6.7 and improve the existing upper bound (6.6) from [**BaJe3**], a new idea is needed; namely, we use the "self-similarity" encoded in the definition of a Graph-directed Self-Conformal Iterated Function System in (4.5) to decompose the operator $\mathbf{L}^{q,s}$ into a sum "smaller but similar" operators. In order to illustrate the need for this decomposition and to highlight the key technical difficulty that we encounter when implementing this idea, it is instructive to present the brief and direct proof of the substantially weaker upper bound in (6.6). Recall that (6.6) says the following: there are constants $c, c_0, C_0 > 0$ such that for all $q \in \mathbb{R}$ and $s \in \mathbb{C}$, we have

$$|\lambda_n(\mathbf{L}^{q,s})| \leq C_0\,|\mathbf{p}^q|_\infty\, e^{c|s| - c_0 n^{1/d}}$$

for all $n \in \mathbb{N}$. To prove this inequality we start be defining the following operators

$$\begin{aligned}
\mathbf{D} &: A^2(\mathbb{X}^\circ) \to \bigoplus_{\mathsf{e}\in\mathsf{E}} A^2(\mathbb{X}^\circ_{\mathsf{i}(\mathsf{e})})\,,\\
\mathbf{S}_{\mathsf{e}} &: A^2(\mathbb{X}^\circ_{\mathsf{i}(\mathsf{e})}) \to A^2(\mathbb{X}^\circ_{\mathsf{t}(\mathsf{e})})\,,\\
\mathbf{Q}^s_{\mathsf{e}} &: A^2(\mathbb{X}^\circ_{\mathsf{t}(\mathsf{e})}) \to A^2(\mathbb{X}^\circ_{\mathsf{t}(\mathsf{e})})\,,\\
\mathbf{L}^{q,s}_{\mathsf{e}} &: A^2(\mathbb{X}^\circ_{\mathsf{i}(\mathsf{e})}) \to A^2(\mathbb{X}^\circ_{\mathsf{t}(\mathsf{e})})\,,\\
\mathbf{A} &: \bigoplus_{\mathsf{e}\in\mathsf{E}} A^2(\mathbb{X}^\circ_{\mathsf{t}(\mathsf{e})}) \to A^2(\mathbb{X}^\circ)
\end{aligned}$$

by

$$\begin{aligned}
\mathbf{D}f &= \left(f|_{X^\circ_{\mathsf{i}(\mathsf{e})}}\right)_{\mathsf{e}\in\mathsf{E}}\,,\\
(\mathbf{S}_{\mathsf{e}}f)(z) &= f(\mathcal{S}_{\mathsf{e}}(z))\,,\\
(\mathbf{Q}^s_{\mathsf{e}}f)(z) &= Q^s_{\mathsf{e}}(z)f(z)\,,\\
\mathbf{L}^{q,s}_{\mathsf{e}} &= p^q_{\mathsf{e}}\,\mathbf{Q}^s_{\mathsf{e}}\,\mathbf{S}_{\mathsf{e}}\,,\\
\left(\mathbf{A}\left((f_{\mathsf{e}})_{\mathsf{e}\in\mathsf{E}}\right)\right)(z) &= \sum\nolimits^{\mathsf{e}\in\mathsf{E}}_{z\in\mathbb{X}^\circ_{\mathsf{t}(\mathsf{e})}} f_{\mathsf{e}}(z)\,.
\end{aligned}$$

It clear that $\mathbf{L}^{q,s}$ can be written as a sum of the $\mathbf{L}^{q,s}_{\mathsf{e}}$'s, more precisely we have

$$\mathbf{L}^{q,s} = \mathbf{A}\left(\bigoplus_{\mathsf{e}\in\mathsf{E}} \mathbf{L}^{q,s}_{\mathsf{e}}\right)\mathbf{D}\,. \tag{6.12}$$

For Hilbert spaces $H$ and $K$ and a compact operator $\mathbf{A}: H \to K$, we let $(\nu_n(\mathbf{A}))_{n\in\mathbb{N}}$ denote the sequence of singular values of $\mathbf{A}$ ordered such that $\nu_1(\mathbf{A}) \geq \nu_2(\mathbf{A}) \geq \ldots$. Using (6.12) and Weyl's inequality, we now have

$$
\begin{aligned}
|\lambda_n(\mathbf{L}^{q,s})| &\leq \left(\prod_{i=1}^{n} |\lambda_i(\mathbf{L}^{q,s})|\right)^{\frac{1}{n}} \\
&\leq \left(\prod_{i=1}^{n} \nu_i(\mathbf{L}^{q,s})\right)^{\frac{1}{n}} && \text{[by Weyl's inequality]} \\
&\leq \left(\prod_{i=1}^{n} \nu_i\left(\mathbf{A}\left(\bigoplus_{\mathsf{e}\in\mathsf{E}} \mathbf{L}_{\mathsf{e}}^{q,s}\right)\mathbf{D}\right)\right)^{\frac{1}{n}} && \text{[by (6.12)]} \\
&\leq \left(\prod_{i=1}^{n} \|\mathbf{A}\|\,\|\mathbf{D}\|\,\nu_i\left(\bigoplus_{\mathsf{e}\in\mathsf{E}} \mathbf{L}_{\mathsf{e}}^{q,s}\right)\right)^{\frac{1}{n}} \\
&= \|\mathbf{A}\|\,\|\mathbf{D}\| \left(\prod_{i=1}^{n} \nu_i\left(\bigoplus_{\mathsf{e}\in\mathsf{E}} \mathbf{L}_{\mathsf{e}}^{q,s}\right)\right)^{\frac{1}{n}}.
\end{aligned}
\tag{6.13}
$$

To proceed further we note the following basic and easily established inequality: if $I$ is a finite set and $\mathbf{A}_i : H_i \to K_i$ is a compact operator between Hilbert spaces $H_i$ and $K_i$ for $i \in I$, then

$$
\nu_n\left(\bigoplus_{i\in I} \mathbf{A}_i\right) \leq \max_{i\in I} \nu_{[\frac{n}{|I|}]}(\mathbf{A}_i)\,;
\tag{6.14}
$$

see [**Ol1**] for a proof of this inequality. Combining (6.14) and inequality (6.13), we have

$$
\begin{aligned}
|\lambda_n(\mathbf{L}^{q,s})| &\leq \|\mathbf{A}\|\,\|\mathbf{D}\| \left(\prod_{i=1}^{n} \max_{\mathsf{e}} \nu_{[\frac{i}{|\mathsf{E}|}]}(\mathbf{L}_{\mathsf{e}}^{q,s})\right)^{\frac{1}{n}} \\
&= \|\mathbf{A}\|\,\|\mathbf{D}\| \left(\prod_{i=1}^{n} \max_{\mathsf{e}} \nu_{[\frac{i}{|\mathsf{E}|}]}(p_{\mathsf{e}}^{q}\,\mathbf{Q}_{\mathsf{e}}^{s}\,\mathbf{S}_{\mathsf{e}})\right)^{\frac{1}{n}} \\
&\leq \|\mathbf{A}\|\,\|\mathbf{D}\| \left(\prod_{i=1}^{n} \max_{\mathsf{e}} p_{\mathsf{e}}^{q}\,\|\mathbf{Q}_{\mathsf{e}}^{s}\| \nu_{[\frac{i}{|\mathsf{E}|}]}(\mathbf{S}_{\mathsf{e}})\right)^{\frac{1}{n}} \\
&\leq \|\mathbf{A}\|\,\|\mathbf{D}\|\,|\mathbf{p}^{q}|_{\infty} \left(\max_{\mathsf{e}} \|\mathbf{Q}_{\mathsf{e}}^{s}\|\right) \left(\prod_{i=1}^{n} \max_{\mathsf{e}} \nu_{[\frac{i}{|\mathsf{E}|}]}(\mathbf{S}_{\mathsf{e}})\right)^{\frac{1}{n}}.
\end{aligned}
\tag{6.15}
$$

However, as noted above, it is not difficult to see that there is a constant $c$ such that $\|\mathbf{Q}_{\mathsf{e}}^{s}\| = \max_{\mathsf{e}\in\mathsf{E}} \sup_{z\in\mathbb{X}_{\mathsf{t}(\mathsf{e})}^{\circ}} |\mathcal{Q}_{\mathsf{e}}^{s}(z)| \leq e^{c|s|}$ for all $\mathsf{e} \in \mathsf{E}$ and all $s$, and we deduce from this and (6.15) that

$$
|\lambda_n(\mathbf{L}^{q,s})| \leq \|\mathbf{A}\|\,\|\mathbf{D}\|\,|\mathbf{p}^{q}|_{\infty}\, e^{c|s|} \left(\prod_{i=1}^{n} \max_{\mathsf{e}} \nu_{[\frac{i}{|\mathsf{E}|}]}(\mathbf{S}_{\mathsf{e}})\right)^{\frac{1}{n}}.
\tag{6.16}
$$

To proceed further we need an estimate for $\nu_n(\mathbf{S}_{\mathsf{e}})$. This estimate is the only non-trivial part in the proof of (6.6). As part of the proof of Theorem 6.7, we obtain estimates for the singular values of a certain canonical restriction operator. This is done in [**Ol1**, Proposition 10.1], and it follows from the estimate in [**Ol1**, Proposition 10.1] that for each fixed $\mathsf{e} \in \mathsf{E}$ there are constants $a_{\mathsf{e}}, A_{\mathsf{e}} > 0$ such that

$$\nu_n(\mathbf{S}_{\mathsf{e}}) \leq A_{\mathsf{e}} e^{-a_{\mathsf{e}} n^{1/d}} \tag{6.17}$$

for all $n$. Letting

$$A = \max_{\mathsf{e}} A_{\mathsf{e}}$$

and

$$a = \min_{\mathsf{e}} a_{\mathsf{e}} ,$$

we now have the following uniform estimate, namely,

$$\nu_n(\mathbf{S}_{\mathsf{e}}) \leq A e^{-a n^{1/d}} \tag{6.18}$$

for all $n$ and *all* edges $\mathsf{e}$. Combining (6.18) and the previous inequality, we see that

$$\begin{aligned}
|\lambda_n(\mathbf{L}^{q,s})| &\leq \|\mathbf{A}\| \, \|\mathbf{D}\| \, |\mathbf{p}^q|_\infty \, e^{c|s|} \left( \prod_{i=1}^{n} e^{-a[\frac{i}{|\mathsf{E}|}]^{1/d}} \right)^{\frac{1}{n}} \\
&\leq \|\mathbf{A}\| \, \|\mathbf{D}\| \, |\mathbf{p}^q|_\infty \, e^{c|s|} \, e^{-a\frac{1}{n} \sum_{i=1}^{n} [\frac{i}{|\mathsf{E}|}]^{1/d}} \\
&\leq \|\mathbf{A}\| \, \|\mathbf{D}\| \, |\mathbf{p}^q|_\infty \, e^{c|s|} \, e^{-a\frac{1}{n} \int_0^n [\frac{x}{|\mathsf{E}|}]^{1/d} \, dx} .
\end{aligned}$$

We conclude immediately from the above inequality that there are universal constants $c_0, C_0 > 0$ such that for all $q \in \mathbb{R}$ and $s \in \mathbb{C}$, we have

$$|\lambda_n(\mathbf{L}^{q,s})| \leq C_0 \, |\mathbf{p}^q|_\infty \, e^{c|s| - c_0 n^{1/d}}$$

for all $n \in \mathbb{N}$; this proves (6.6).

The reader will note that the above argument only involves *global* estimates, i.e. the argument only involves estimating the eigenvalues $\lambda_n(\mathbf{L}^{q,s})$ of the transfer operator $\mathbf{L}^{q,s}$ itself. To improve the upper bound in (6.6), we apply an idea that permeates almost all studies of fractal geometry of self-similar constructions, namely, we use the "self-similarity" encoded in the definition of a Graph-directed Self-Conformal Iterated Function System in (4.5) to decompose the operator $\mathbf{L}^{q,s}$ into a sum of "smaller but similar" operators $\mathbf{L}^{q,s}_{\mathsf{e},\mathbf{i},\mathbf{j}}$ and then estimate each of the "smaller" operators $\mathbf{L}^{q,s}_{\mathsf{e},\mathbf{i},\mathbf{j}}$ using delicate *local* estimates. This approach leads to the substantially improved upper bounds for $\lambda_n(\mathbf{L}^{q,s})$ and $\zeta^q(s)$ in Theorem 6.7 and Theorem 6.8, respectively.

**6.4. Counting zeros.** We can apply Jensen's theorem and the estimates in Theorem 6.8 to obtain upper bounds for the number of zeros of the zeta function $\zeta^q(s)$ in balls centred at points near (but to the right of) the line $\mathrm{Re}(s) = \beta(q)$; this is the content of Theorem 6.9 below. For an entire function $f : \mathbb{C} \to \mathbb{C}$, write

$$Z(f) = \{ s \in \mathbb{C} \mid f(s) = 0 \} .$$

We can now state Theorem 6.9.

THEOREM 6.9. *Let $q \in \mathbb{R}$. Fix $R, \varepsilon > 0$. For $t > 0$, write*

$$\Pi^q(t) = \Big| B(\, \beta(q) + \varepsilon + it \, , \, R \,) \cap Z(\zeta^q) \Big| .$$

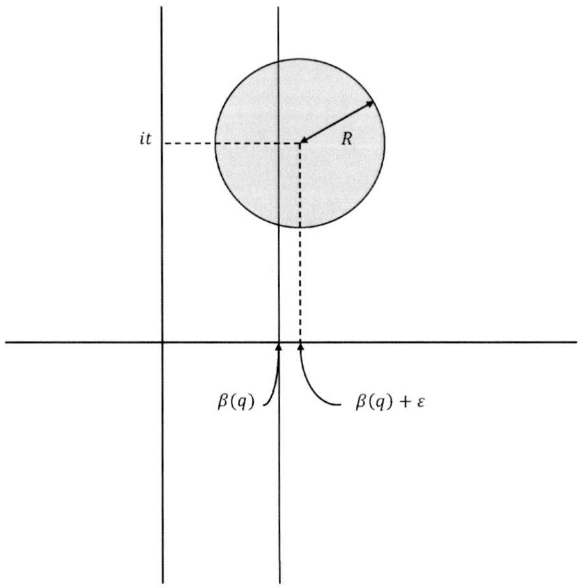


FIGURE 6.1. For fixed positive numbers $R$ and $\varepsilon$, the statement in Theorem 3.9 provides an upper bound for the number of zeros of $\zeta^q(s)$ in the shaded ball for large values of $|t|$.

*Then there are constants $C, t_0 > 0$ such that for all $t \in \mathbb{R}$ with $|t| > t_0$, we have*

$$\Pi^q(t) \le C\, |\mathbf{p}^q|_\infty\, |t|^{\dim_{\mathsf{H}} K_{\mathrm{i}}}$$

*for all* $\mathrm{i}$.

The proof of Theorem 6.9 is given in [**Ol1**]. The Figure 6.1 illustrates the statement in Theorem 6.9.

## 7. Multifracal Prime Counting Formulas

**7.1. Prime counting formulas.** Using Wiener-Ikehara's tauberian theorem and the fact that the series for $\zeta^q$ in Theorem 6.2 has an analytic continuation, we can derive various multifractal prime counting formulas for Graph-directed Self-Conformal Iterated Function Systems; this is the content of Theorem 7.1 below. As a special case of Theorem 7.1 we obtain a multifractal prime geodesic counting theorem of classical Fuchsian Schottky groups which for $q = 0$ reduces to the classical prime geodesic counting theorem of classical Fuchsian Schottky groups, see Theorem 7.2 and the subsequent discussion.

Ideally, we would like to obtain an explicit formula the $q$'th multifractal Minkowski volume

$$V^q_{\mu_{\mathsf{i}},r}(K_{\mathsf{i}}) = \frac{1}{r^d}\int_{B(K_{\mathsf{i}},r)} \mu_{\mathsf{i}}(B(x,r))^q \, d\mathcal{L}^d(x) \,.$$

of $K_{\mathsf{i}}$. Regrettably, the problem of finding a formula for $V^q_{\mu_{\mathsf{i}},r}(K_{\mathsf{i}})$ seems completely out of reach with existing methods and technology. Because of those difficulties we will study a "symbolic" version of $V^q_{\mu_{\mathsf{i}},r}(K_{\mathsf{i}})$. It turns out that the problem of finding an explicit formula for the symbolic multifractal Minkowski volume may be viewed as "prime counting". Indeed, if $\Pi_{\mathsf{G}}$ denotes the family of "prime" strings and $p_{\mathbf{I}}$ denotes the probability associated with a "prime" string $\mathbf{I} \in \Pi_{\mathsf{G}}$ (the reader is referred to Section 6.1 for the precise definitions), then we, as noted in the Introduction, define the multifractal prime counting function $\pi^q_{\mathsf{G}}(t)$ by

$$\pi^q_{\mathsf{G}}(t) = \sum_{\substack{\mathbf{I}\in\Pi_{\mathsf{G}} \\ \ell_{\mathbf{I}}\le t}} p^q_{\mathbf{I}} \,.$$

As also noted in the Introduction, routine arguments show that there is a constant $c > 0$ such that

$$\frac{1}{c}\,\pi^q_{\mathsf{G}}(\log\tfrac{1}{r}) \le V^q_{\mu_{\mathsf{i}},r}(K_{\mathsf{i}}) \le c\,\pi^q_{\mathsf{G}}(\log\tfrac{1}{r}) \tag{7.1}$$

for all $r > 0$, and the prime counting function $\pi^q_{\mathsf{G}}(t)$ may therefore be viewed as a natural symbolic version of the multifractal Minkowski volume $V^q_{\mu_{\mathsf{i}},r}(K_{\mathsf{i}})$. This result and analogy prompt us to promote the view that:

> *Multifractal analysis of self-conformal measures is prime counting.*

We will now present Theorem 7.1 providing an asymptotic formula for the multifractal prime counting function $\pi^q_{\mathsf{G}}(\log\frac{1}{r})$ and hence, because of (7.1), for the multifratcal Minkowski volume $V^q_{\mu_{\mathsf{i}},r}(K_{\mathsf{i}})$. To formulate Theorem 7.1, we first recall that $\Pi_{\mathsf{G}}$ denotes the family of equivalence classes of prime strings defined in Section 6.1, and for $\mathbf{I} \in \Pi_{\mathsf{G}}$, we let $p_{\mathbf{I}}$ and $\ell_{\mathbf{I}}$ be defined as in Section 6.1. In the statement of Theorem 7.1 we also use the following notation from the Introduction, namely, if $f, g : (0,\infty) \to (0,\infty)$ are real valued positive functions, then we write $f \sim g$ (or $f(t) \sim g(t)$) provided $\frac{f(t)}{g(t)} \to 1$ as $t \to \infty$. Finally, recall that if $\omega \in \mathbb{C}$ and $f$ is a meromorphic function with $f(\omega) = 0$ and $f$ is not constantly equal to 0 in a neighbourhood of $\omega$, then we write $\operatorname{ord}(f;\omega)$ for the order of the zero $\omega$ of $f$. Also recall (see (6.1)) that we write $\beta(q)$ for the common value of the of the lower and upper $q$ Renyi dimensions of $\mu_{\mathsf{i}}$, i.e.

$$\beta(q) = \underline{\tau}_{\mu_{\mathsf{i}}}(q) = \overline{\tau}_{\mu_{\mathsf{i}}}(q) \,.$$

We can now state Theorem 7.1.

THEOREM 7.1 (Abstract multifractal prime counting theorem for Graph-directed Self-Conformal Iterated Function Systems). *Assume that the zeta function $\zeta^q$ does not have any other zeros $s$ with $\operatorname{Re}(s) = \beta(q)$ besides $\beta(q)$. If $0 \le q < 1$, then*

$$\pi^q_{\mathsf{G}}(t) \sim \operatorname{ord}(\zeta^q;\beta(q))\,\frac{e^{\beta(q)t}}{\beta(q)t} \,.$$

The proof of Theorem 7.1 is presented in [**Ol1**] and is based on Wiener-Ikehara's Tauberian theorem and the fact that the series for $\zeta^q$ in Theorem 6.2 has an analytic continuation.

We immediately note that Theorem 7.1 implies the following multifractal prime geodesic counting theorem for classical Fuchsian Schottky groups. Let $\Gamma$ be a classical Fuchsian Schottky group and let $\Lambda(\Gamma)$ denote the limit set of $\Gamma$. It follows from [**Ol1**], see also [**MaUr**], that there is a "canonical" Graph-directed Self-Conformal Iterated Function System $(\mathsf{V},\mathsf{E},(V_i)_{i\in\mathsf{V}},(X_i)_{i\in\mathsf{V}},(S_{\mathsf{e}})_{i\in\mathsf{E}},(p_{\mathsf{e}})_{i\in\mathsf{E}})$ with associated self-conformal sets $(K_i)_{i\in\mathsf{V}}$ such that

$$\Lambda(\Gamma)=\bigcup_i K_i\,,$$

and the results in this work can therefore be applied to the study of the limit set $\Lambda(\Gamma)$. We will now utilise this to obtain classical and multifractal prime geodesic counting theorems for classical Fuchsian Schottky groups. We first recall that the length spectrum $\mathcal{L}(\Gamma)$ of the Fuchsian Schottky group $\Gamma$ is defined by

$$\mathcal{L}((\Gamma)=\Big\{\ell\,\Big|\,\ell\text{ is the length of a primitive oriented closed geodesic}\\ \text{on the hyperbolic surface }\Gamma/\mathbb{H}\Big\}\,,$$

and note that it is well-known that there is a bijection

$$\mathbf{\Pi}:\mathcal{L}(\Gamma)\to\mathit{\Pi}_{\mathsf{G}}$$

such that

$$\ell=-\log|S_{\mathbf{i}}'(x_{\mathbf{i}})|$$

for all $\mathbf{i}\in\mathbf{\Pi}(\ell)$, see, for example, [**Bo**] for more details. The following theorem now follows immediately from the above and Theorem 7.1.

THEOREM 7.2 (Multifractal prime geodesic counting theorem for classical Fuchsian Schottky groups). *Let $\Gamma$ be a classical Fuchsian Schottky group and define the multifractal prime geodesic counting function $\pi_\Gamma^q:[0,\infty)\to[0,\infty)$ of $\Gamma$ by*

$$\pi_\Gamma^q(t)=\sum_{\substack{\ell\in\mathcal{L}(\Gamma)\\ \ell\le t}}p_{\mathbf{\Pi}(\ell)}^q\,.$$

*If $0\le q<1$, then*

$$\pi_\Gamma^q(t)\sim\frac{e^{\beta(q)t}}{\beta(q)t}\,.$$

Note that if $\pi_\Gamma:[0,\infty)\to[0,\infty)$ denotes the usual prime geodesic counting function of $\Gamma$, i.e. if

$$\pi_\Gamma(t)=\operatorname{card}\big\{\ell\in\mathcal{L}(\Gamma)\,\big|\,\ell\le t\big\}\,,$$

then $\pi_\Gamma(t)=\pi_\Gamma^0(t)$, and Theorem 7.2 therefore simplifies to the following classical result, namely,

$$\pi_\Gamma(t)\sim\frac{e^{\delta t}}{\delta t}\,,$$

where $\delta=\beta(0)$ equals the Hausdorff dimension of $\Lambda(\Gamma)$.

**7.2. Refined and explicit multifractal prime counting formulas.** Refined and explicit multifractal prime counting theorems for Graph-directed Self-Conformal Iterated Function Systems can also be obtained by a more careful analysis of the zeros of the zeta function $\zeta^q$ and provides the main motivation for the detailed study of the zeta function $\zeta^q$ provided in this work. These ideas will be explored further in [**O12**]. If we define $h : \{s \in \mathbb{C} \mid \operatorname{Re}(s) > \beta(q)\} \to \mathbb{C}$ and $T : (0,\infty) \to (0,\infty)$ by

$$h(s) = \sum_{\mathbf{I}\in\Pi_{\mathsf{G}}} p_{\mathbf{I}}^q\, \ell_{\mathbf{I}}\, e^{-\ell_{\mathbf{I}} s}$$

(it is not difficult to see that the series $\sum_{\mathbf{I}\in\Pi_{\mathsf{G}}} p_{\mathbf{I}}^q\, \ell_{\mathbf{I}}\, e^{-\ell_{\mathbf{I}} s}$ is convergent for $\operatorname{Re}(s) > \beta(q)$, and the function $h$ is therefore well-defined) and

$$T(r) = \begin{cases} \displaystyle\sum_{\substack{\mathbf{I}\in\Pi_{\mathsf{G}}\\ \ell_{\mathbf{I}}\le -\log r}} p_{\mathbf{I}}^q\, \ell_{\mathbf{I}} & \text{for } 0 < r \le 1;\\ 0 & \text{for } 1 < r, \end{cases}$$

then it follows from Mellin's inversion theorem that

$$T(r) = \frac{1}{2\pi i}\int_{c-i\infty}^{c+i\infty} r^{-s}\frac{h(s)}{s}\,ds$$

for all $c > \beta(q)$. Also, since $\zeta^q$ is entire and $\zeta^q(s)$ can be represented by the series in Theorem 6.2 for $s \in \mathbb{C}$ with $\operatorname{Re}(s) > \beta(q)$, it can be shown that $h$ has an analytic continuation to $\{s \in \mathbb{C} \mid \operatorname{Re}(s) > \beta(q) - 1\}$ and we can therefore "move the path of integration in the integral $\int_{c-i\infty}^{c+i\infty} r^{-s}\frac{h(s)}{s}\,ds$ to the left". "Moving the path of integration to the left" and using the estimates in Theorem 6.8 together with a careful analysis of the zeros of $\zeta^q$, we now conclude that there is a continuous function $a : \mathbb{R} \to (0,\infty)$ with $\lim_{t\to\pm\infty} a(t) = 0$ such that if $\gamma$ denotes the path defined by $\gamma = \{\beta(q) - a(t) + it \mid t \in \mathbb{R}\}$, then

$$\frac{1}{2\pi i}\int_{c-i\infty}^{c+i\infty} r^{-s}\frac{h(s)}{s}\,ds + \frac{1}{2\pi i}\int_{\gamma} r^{-s}\frac{h(s)}{s}\,ds = \sum_{\substack{\omega \text{ is a zero of } \zeta^q\\ \operatorname{Re}(\omega)=\beta(q)}} \operatorname{Res}\left(r^{-s}\frac{h(s)}{s};\omega\right),$$

and

$$\frac{1}{2\pi i}\int_{\gamma} r^{\beta(q)-s}\frac{h(s)}{s}\,ds \to 0 \text{ as } r \searrow 0.$$

Hence, writing

$$\varepsilon(r) = \frac{1}{2\pi i}\int_{\gamma} r^{\beta(q)-s}\frac{h(s)}{s}\,ds$$

and noticing that $T(r) = \frac{1}{2\pi i}\int_{c-i\infty}^{c+i\infty} r^{-s}\frac{h(s)}{s}\,ds$ and $\operatorname{Res}(r^{-s}\frac{h(s)}{s};\omega) = \frac{r^{-\omega}}{\omega}\operatorname{ord}(h;\omega)$, the above shows that

$$r^{\beta(q)}T(r) = \sum_{\substack{\omega \text{ is a zero of } \zeta^q\\ \operatorname{Re}(\omega)=\beta(q)}} \frac{r^{\beta(q)-\omega}}{\omega}\operatorname{ord}(h;\omega) + \varepsilon(r)$$

where $\varepsilon(r) \to 0$ as $r \searrow 0$. This is an "explicit" prime counting formula for $T(r)$ with a "refined" error term given by $\varepsilon(r) = \frac{1}{2\pi i}\int_{\gamma} r^{\beta(q)-s}\frac{h(s)}{s}\,ds$. Since it can be shown that

$$\frac{t\pi_{\mathsf{G}}^q(t)}{T(e^{-t})} \sim 1,$$

the "explicit" prime counting formula for $T(r)$ also yields an "explicit" prime counting formula for $\pi^q_{\mathsf{G}}(t)$; the reader is referred to **[Ol1]** for the details.

## 8. An Application: Self-similar constructions

There is a particular important class of Graph-directed Self-Conformal Iterated Function Systems, namely, the Graph-directed Self-Similar Iterated Function Systems. A Graph-directed Self-Similar Iterated Function System is a Graph-directed Self-Conformal Iterated Function System in which all the maps $S_{\mathsf{e}}$ are similarities, i.e. for each $\mathsf{e} \in \mathsf{E}$ there is a number $r_{\mathsf{e}} \in (0,1)$ such that $|S_{\mathsf{e}}(x) - S_{\mathsf{e}}(y)| = r_{\mathsf{e}}|x-y|$ for all $x$ and $y$. The importance of Graph-directed Self-Similar Iterated Function Systems is, in part, due to the fact that many classical fractal sets, including, for example, the Cantor set, the Sierpinski triangle and the Menger cube, are limits sets associated with this type of Iterated Function Systems. As an application of the theory developed in this work, we will now compute the zeta function $\zeta^q(s)$ for Graph-directed Self-Similar Iterated Function System. Because of the simple nature of similarity maps, it turns out that explicit formulas for the zeta function $\zeta^q(s)$ can be obtained for this type of Iterated Function Systems. Using these formulas we can also obtain functional equations for the zeta function $\zeta^q(s)$.

**8.1. The matrix $W^q(s,k,\mathbf{k})$.** Let $(\,\mathsf{V},\ \mathsf{E},\ (V_{\mathsf{i}})_{\mathsf{i}\in\mathsf{V}},\ (X_{\mathsf{i}})_{\mathsf{i}\in\mathsf{V}},\ (S_{\mathsf{e}})_{\mathsf{e}\in\mathsf{E}},\ (p_{\mathsf{e}})_{\mathsf{e}\in\mathsf{E}}\,)$ be a Graph-directed Self-Similar Iterated Function System. Fix $\mathsf{i},\mathsf{j} \in \mathsf{V}$ and $\mathsf{e} \in \mathsf{E}_{\mathsf{i},\mathsf{j}}$. Since $S_{\mathsf{e}}$ is a contractive similarity there is a number $r_{\mathsf{e}} \in (0,1)$ such that $|S_{\mathsf{e}}(x) - S_{\mathsf{e}}(y)| = r_{\mathsf{e}}|x-y|$ for all $x, y \in V_{\mathsf{j}}$. In particular, this implies that there is a vector $a_{\mathsf{e}} \in \mathbb{R}^d$ and a $d \times d$ orthogonal matrix $O_{\mathsf{e}}$ such that

$$S_{\mathsf{e}}(x) = r_{\mathsf{e}} O_{\mathsf{e}} x + a_{\mathsf{e}}$$

for all $x \in V_{\mathsf{j}}$. Let $\lambda_{1,\mathsf{e}},\ \ldots,\ \lambda_{d,\mathsf{e}}$ be the eigenvalues of $O_{\mathsf{e}}$ where each eigenvalue is listed according to its multiplicity and put

$$\boldsymbol{\lambda}_{\mathsf{e}} = (\lambda_{1,\mathsf{e}}, \ldots, \lambda_{d,\mathsf{e}})\,.$$

Below we will use the following notation, namely, for a $d$-tuple $\mathbf{x} = (x_1, \ldots, x_d)$ of positive real numbers and a multi-index $\mathbf{k} = (k_1, \ldots, k_d)$ of integers, we will write $|\mathbf{k}| = k_1 + \cdots + k_d$ and

$$\mathbf{x}^{\mathbf{k}} = x_1^{k_1} \cdots x_d^{k_d}\,.$$

Next, for $q \in \mathbb{R}$, $s \in \mathbb{C}$ and a positive integer $k$ and a multi-index $\mathbf{k} = (k_1, \ldots, k_d)$ of positive integers, we define the matrix

$$W^q(s,k,\mathbf{k}) = (w^q_{\mathsf{i},\mathsf{j}}(s,k,\mathbf{k}))_{\mathsf{i},\mathsf{j}\in\mathsf{V}}$$

by

$$w^q_{\mathsf{i},\mathsf{j}}(s,k,\mathbf{k}) = \sum_{\mathsf{e}\in\mathsf{E}_{\mathsf{i},\mathsf{j}}} p^q_{\mathsf{e}} r_{\mathsf{e}}^{s+k}\, \boldsymbol{\lambda}_{\mathsf{e}}^{\mathbf{k}}\,.$$

In Section 8.2, we will show that the zeta function $\zeta^q(s)$ can be expressed as a product of the determinants of the matrices $W^q(s,k,\mathbf{k})$ and using this result we also obtain a functional equation for $\zeta^q(s)$.

**8.2. The zeta function $\zeta^q(s)$ for $d \in \mathbb{N}$.** We will now show that the zeta function $\zeta^q(s)$ can be expressed as a product of the determinants of the matrices $W^q(s,k,\mathbf{k})$.

THEOREM 8.1. *Let $q \in \mathbb{R}$ and $s \in \mathbb{C}$. Assume that the matrices in the family $(O_{\mathsf{e}})_{\mathsf{e}\in\mathsf{E}}$ commute. Then we have*

$$\zeta^q(s) = \prod_{k=0}^{\infty} \prod_{\substack{\mathbf{k}\in[0,\infty)^d \\ |\mathbf{k}|=k}} \det(1 - W^q(s,k,\mathbf{k}))\,.$$

*In particular, if $(\mathsf{V},\mathsf{E})$ has only one vertex and $N$ edges labelled $1, \ldots, N$, then we have*

$$\zeta^q(s) = \prod_{k=0}^{\infty} \prod_{\substack{\mathbf{k}\in[0,\infty)^d \\ |\mathbf{k}|=k}} \left(1 - \sum_{i=1}^{N} p_i^q r_i^{s+k} \boldsymbol{\lambda}_{\mathsf{e}}^{\mathbf{k}}\right). \tag{8.1}$$

PROOF. For $\mathbf{i} = \mathsf{e}_1 \ldots \mathsf{e}_n \in \Sigma_{\mathsf{G}}^*$, write $r_{\mathbf{i}} = r_{\mathsf{e}_1} \cdots r_{\mathsf{e}_n}$ and $O_{\mathbf{i}} = O_{\mathsf{e}_1} \cdots O_{\mathsf{e}_n}$, and note that since $S_{\mathsf{e}}(x) = r_{\mathsf{e}} O_{\mathsf{e}} x + a_{\mathsf{e}}$ for all $x \in V_{\mathsf{j}}$, it follows that

$$(DS_{\mathbf{i}})(x) = r_{\mathbf{i}} O_{\mathbf{i}} \tag{8.2}$$

for all $\mathbf{i} \in \Sigma_{\mathsf{G}}^*$ and $x \in V_{\mathsf{t}(\mathbf{i})}$. Also, since the matrices in family $(O_{\mathsf{e}})_{\mathsf{e}\in\mathsf{E}}$ commute, we conclude that the family $(O_{\mathsf{e}})_{\mathsf{e}\in\mathsf{E}}$ is simultaneously diagonalisable, i.e. if $D_{\mathsf{e}}$ denotes the $d \times d$ diagonal matrix defined by

$$D_{\mathsf{e}} = \begin{pmatrix} \lambda_{1,\mathsf{e}} & & & \\ & \lambda_{2,\mathsf{e}} & & \\ & & \ddots & \\ & & & \lambda_{d,\mathsf{e}} \end{pmatrix},$$

then there is an invertible $d \times d$ matrix $P$ such that for all $\mathsf{e} \in \mathsf{E}$ we have $O_{\mathsf{e}} = P^{-1} D_{\mathsf{e}} P$.

Noticing that both sides of (8.1) are analytic functions of $s$, we conclude that it suffices to prove (8.1) for $s \in \mathbb{C}$ with $\mathrm{Re}(s) > \beta(q)$. Since $\mathrm{Re}(s) > \beta(q)$, we conclude from Theorem 6.1 that

$$\zeta^q(s) = \exp\left( -\sum_n \frac{1}{n} \sum_{\substack{\mathbf{i}\in\Sigma_{\mathsf{G}}^n \\ \mathsf{i}(\mathbf{i})=\mathsf{t}(\mathbf{i})}} \frac{p_{\mathbf{i}}^q |(DS_{\mathbf{i}})(x_{\mathbf{i}})|^s}{\det(1 - (DS_{\mathbf{i}})(x_{\mathbf{i}}))} \right); \tag{8.3}$$

recall, that for $\mathbf{i} = \mathsf{e}_1 \ldots \mathsf{e}_n \in \Sigma_{\mathsf{G}}^n$, we let $x_{\mathbf{i}} \in X_{\mathsf{t}(\mathbf{i})}$ denote the unique fixed point of the function $S_{\mathbf{i}} : X_{\mathsf{t}(\mathbf{i})} \to X_{\mathsf{i}(\mathbf{i})}$. We will now compute the interior sum in (8.3). For $\mathbf{i} = \mathsf{e}_1 \ldots \mathsf{e}_n \in \Sigma_{\mathsf{G}}^n$ with $\mathsf{i}(\mathbf{i}) = \mathsf{t}(\mathbf{i})$, write $D_{\mathbf{i}} = D_{\mathsf{e}_1} \cdots D_{\mathsf{e}_n}$ and $\lambda_{l,\mathbf{i}} = \lambda_{l,\mathsf{e}_1} \cdots \lambda_{l,\mathsf{e}_n}$, and observe that it follows from (8.2) that $|(DS_{\mathbf{i}})(x_{\mathbf{i}})| = r_{\mathbf{i}}$ and $\det(1-(DS_{\mathbf{i}})(x_{\mathbf{i}})) = \det(1 - r_{\mathbf{i}} O_{\mathbf{i}}) = \det(P^{-1}(1 - r_{\mathbf{i}} D_{\mathbf{i}})P) = \prod_{l=1}^{d}(1 - r_{\mathbf{i}}\lambda_{l,\mathbf{i}})$, whence (by summing a geometric series)

$$\sum_{\substack{\mathbf{i}\in\Sigma_{\mathsf{G}}^n \\ \mathsf{i}(\mathbf{i})=\mathsf{t}(\mathbf{i})}} \frac{p_{\mathbf{i}}^q |(DS_{\mathbf{i}})(x_{\mathbf{i}})|^s}{\det(1 - (DS_{\mathbf{i}})(x_{\mathbf{i}}))} = \sum_{\mathsf{i}\in\mathsf{V}} \sum_{\substack{\mathbf{i}\in\Sigma_{\mathsf{G}}^n \\ \mathsf{i}(\mathbf{i})=\mathsf{t}(\mathbf{i})=\mathsf{i}}} p_{\mathbf{i}}^q r_{\mathbf{i}}^s \prod_{l=1}^{d} \sum_{k_l=0}^{\infty} r_{\mathbf{i}}^{k_l} \lambda_{l,\mathbf{i}}^{k_l}\,. \tag{8.4}$$

Next, writing $\boldsymbol{\lambda}_{\mathbf{i}} = (\lambda_{1,\mathbf{i}}, \ldots, \lambda_{d,\mathbf{i}})$ for $\mathbf{i} \in \Sigma_{\mathsf{G}}^n$, it follows from Cauchy multiplication of series that

$$\prod_{l=1}^{d} \sum_{k_l=0}^{\infty} r_{\mathbf{i}}^{k_l} \lambda_{l,\mathbf{i}}^{k_l} = \sum_{k=0}^{\infty} \sum_{\substack{\mathbf{k} \in [0,\infty)^d \\ |\mathbf{k}|=k}} \boldsymbol{\lambda}_{\mathbf{i}}^{\mathbf{k}} r_{\mathbf{i}}^{k}. \tag{8.5}$$

For brevity we now introduce the following notation, namely, for $\mathsf{i}, \mathsf{j} \in \mathsf{V}$ and a positive integer $n$, we will denote the $(\mathsf{i},\mathsf{j})$'th entry of the matrix $W^q(s,k,\mathbf{k})^n$ by $w_{\mathsf{i},\mathsf{j}}^{q,n}(s,k,\mathbf{k})$. Using this notation we now conclude from (8.4) and (8.5) that

$$\begin{aligned} \sum_{\substack{\mathbf{i} \in \Sigma_{\mathsf{G}}^n \\ \mathrm{i}(\mathbf{i})=\mathrm{t}(\mathbf{i})}} \frac{p_{\mathbf{i}}^q Q_{\mathbf{i}}(x_{\mathbf{i}})^s}{\det(1-(DS_{\mathbf{i}})(x_{\mathbf{i}}))} &= \sum_{k=0}^{\infty} \sum_{\substack{\mathbf{k} \in [0,\infty)^d \\ |\mathbf{k}|=k}} \sum_{\mathsf{i} \in \mathsf{V}} \sum_{\substack{\mathbf{i} \in \Sigma_{\mathsf{G}}^n \\ \mathrm{i}(\mathbf{i})=\mathrm{t}(\mathbf{i})=\mathsf{i}}} p_{\mathbf{i}}^q r_{\mathbf{i}}^{s+k} \boldsymbol{\lambda}_{\mathbf{i}}^{\mathbf{k}} \\ &= \sum_{k=0}^{\infty} \sum_{\substack{\mathbf{k} \in [0,\infty)^d \\ |\mathbf{k}|=k}} \sum_{\mathsf{i} \in \mathsf{V}} w_{\mathsf{i},\mathsf{i}}^{q,n}(s,k,\mathbf{k}) \\ &= \sum_{k=0}^{\infty} \sum_{\substack{\mathbf{k} \in [0,\infty)^d \\ |\mathbf{k}|=k}} \operatorname{tr}(W^q(s,k,\mathbf{k})^n), \end{aligned} \tag{8.6}$$

where we have used the fact that $\sum_{\mathbf{i} \in \Sigma_{\mathsf{G}}^n, \mathrm{i}(\mathbf{i})=\mathrm{t}(\mathbf{i})=\mathsf{i}} p_{\mathbf{i}}^q r_{\mathbf{i}}^{s+k} \boldsymbol{\lambda}_{\mathbf{i}}^{\mathbf{k}} = w_{\mathsf{i},\mathsf{i}}^{q,n}(s,k,\mathbf{k})$ and $\sum_{\mathsf{i} \in \mathsf{V}} w_{\mathsf{i},\mathsf{i}}^{q,n}(s,k,\mathbf{k}) = \operatorname{tr}(W^q(s,k,\mathbf{k})^n)$.

Finally, combining (8.3) and (8.6) and using the fact that $\sum_n \frac{1}{n} \operatorname{tr}(M^n) = -\log\det(1-M)$ for any square matrix $M$, we now conclude that

$$\begin{aligned} \zeta^q(s) &= \exp\left( -\sum_{k=0}^{\infty} \sum_{\substack{\mathbf{k} \in [0,\infty)^d \\ |\mathbf{k}|=k}} \sum_{n} \frac{1}{n} \operatorname{tr}(W^q(s,k,\mathbf{k})^n) \right) \\ &= \exp\left( \sum_{k=0}^{\infty} \sum_{\substack{\mathbf{k} \in [0,\infty)^d \\ |\mathbf{k}|=k}} \log\det(1-W^q(s,k,\mathbf{k})) \right) \\ &= \prod_{k=0}^{\infty} \prod_{\substack{\mathbf{k} \in [0,\infty)^d \\ |\mathbf{k}|=k}} \det(1-W^q(s,k,\mathbf{k})). \end{aligned}$$

This completes the proof. □

**8.3. The zeta function $\zeta^q(s)$ for $d = 1$.** For $d = 1$, the explicit formula in Theorem 8.1 becomes particularly simple and we also obtain a simple functional equation for $\zeta^q(s)$. This is the contents of Theorem 8.2 and Theorem 8.3 below.

THEOREM 8.2. *Let $q \in \mathbb{R}$ and $s \in \mathbb{C}$. Assume that $(\mathsf{V}, \mathsf{E})$ has only one vertex and $N$ edges labelled $1, \ldots, N$ and that $d = 1$. For $i = 1, \ldots, N$, let the map $S_i$ be*

$$S_i(x) = r_i \lambda_i x + a_i$$

*where* $r_i \in (0,1)$, $\lambda_i \in \{-1,1\}$ *and* $a_i \in \mathbb{R}$. *Then we have*

$$\zeta^q(s) = \prod_{k=0}^{\infty}\left(1 - \sum_{i=1}^{N} p_i^q r_i^{s+k}\,\lambda_i^k\right).$$

PROOF. Since $O_i = (\lambda_i)$, the results follows immediately from Theorem 8.1. □

THEOREM 8.3 (Functional Equation). *Let* $q \in \mathbb{R}$ *and* $s \in \mathbb{C}$. *Assume that* $d = 1$, *and for* $\mathsf{e} \in \mathsf{E}$, *let the map* $S_\mathsf{e}$ *be given by*

$$S_\mathsf{e}(x) = r_\mathsf{e}\lambda_\mathsf{e} x + a_\mathsf{e}$$

*where* $r_\mathsf{e} \in (0,1)$, $\lambda_\mathsf{e} \in \{-1,1\}$ *and* $a_\mathsf{e} \in \mathbb{R}$. *Define the matrices* $A^q(s)$ *and* $B^q(s)$ *by*

$$A^q(s) = \left(\sum_{\mathsf{e}\in\mathsf{E}_{\mathsf{i},\mathsf{j}}} p_\mathsf{e}^q r_\mathsf{e}^s\right)_{\mathsf{i},\mathsf{j}\in\mathsf{V}},$$

$$B^q(s) = \left(\sum_{\mathsf{e}\in\mathsf{E}_{\mathsf{i},\mathsf{j}}} p_\mathsf{e}^q r_\mathsf{e}^{s+1}\,\lambda_\mathsf{e}\right)_{\mathsf{i},\mathsf{j}\in\mathsf{V}}.$$

*Then we have*

$$\zeta^q(s) = \det(1 - A^q(s))\,\zeta^q(s+1) \qquad \textit{if } \{\lambda_\mathsf{e} \mid \mathsf{e}\in\mathsf{E}\} = \{1\},$$
$$\zeta^q(s) = \det(1 - A^q(s))\,\det(1 - B^q(s))\,\zeta^q(s+2) \qquad \textit{if } \{\lambda_\mathsf{e} \mid \mathsf{e}\in\mathsf{E}\} \neq \{1\}.$$

*In particular, if* $(\mathsf{V},\mathsf{E})$ *has only one vertex and* $N$ *edges labelled* $1, \ldots, N$, *then we have*

$$\zeta^q(s) = \left(1 - \sum_{i=1}^{N} p_i^q r_i^s\right)\zeta^q(s+1) \qquad \textit{if } \{\lambda_1,\ldots,\lambda_N\} = \{1\},$$

$$\zeta^q(s) = \left(1 - \sum_{i=1}^{N} p_i^q r_i^s\right)\left(1 - \sum_{i=1}^{N} p_i^q r_i^{s+1}\,\lambda_i\right)\zeta^q(s+2) \qquad \textit{if } \{\lambda_1,\ldots,\lambda_N\} \neq \{1\}.$$

We refer the reader to [**Ol1**] for the proof of Theorem 8.3.

## References

[ArPa] M. Arbeiter and N. Patzschke, *Random self-similar multifractals*, Math. Nachr. **181** (1996), 5–42, DOI 10.1002/mana.3211810102. MR1409071
**276** (2007), 901–905.

[BaJe1] O. F. Bandtlow and O. Jenkinson, *Explicit a priori bounds on transfer operator eigenvalues*, Comm. Math. Phys. **276** (2007), no. 3, 901–905, DOI 10.1007/s00220-007-0355-7. MR2350441

[BaJe2] O. F. Bandtlow and O. Jenkinson, *On the Ruelle eigenvalue sequence*, Ergodic Theory Dynam. Systems **28** (2008), no. 6, 1701–1711, DOI 10.1017/S0143385708000059. MR2465596

[BaJe3] O. F. Bandtlow and O. Jenkinson, *Explicit eigenvalue estimates for transfer operators acting on spaces of holomorphic functions*, Adv. Math. **218** (2008), no. 3, 902–925, DOI 10.1016/j.aim.2008.02.005. MR2414325

[BeGo] M. Berger and B. Gostiaux, *Differential Geometry: Manifolds, Curves and Surfaces*, Springer-Verlag, Berlin, 1988.

[BePe] *Lectures on Hyperbolic Geometry*, Springer-Verlag, Berlin, 1992.

[Bo] D. Borthwick, *Spectral theory of infinite-area hyperbolic surfaces*, 2nd ed., Progress in Mathematics, vol. 318, Birkhäuser/Springer, [Cham], 2016, DOI 10.1007/978-3-319-33877-4. MR3497464

[CaMa] R. Cawley and R. D. Mauldin, *Multifractal decompositions of Moran fractals*, Adv. Math. **92** (1992), no. 2, 196–236, DOI 10.1016/0001-8708(92)90064-R. MR1155465

[Fa1] K. Falconer, *Fractal geometry: Mathematical foundations and applications*, John Wiley & Sons, Ltd., Chichester, 1990. MR1102677

[Fa2] K. Falconer, *Techniques in fractal geometry*, John Wiley & Sons, Ltd., Chichester, 1997. MR1449135

[Fa3] K. J. Falconer, *On the Minkowski measurability of fractals*, Proc. Amer. Math. Soc. **123** (1995), no. 4, 1115–1124, DOI 10.2307/2160708. MR1224615

[FaRo] F. Faure and N. Roy, *Ruelle-Pollicott resonances for real analytic hyperbolic maps*, Nonlinearity **19** (2006), no. 6, 1233–1252, DOI 10.1088/0951-7715/19/6/002. MR2229997

[Fe1] H. Federer, *Curvature measures*, Trans. Amer. Math. Soc. **93** (1959), 418–491, DOI 10.2307/1993504. MR110078

[Fe2] H. Federer, *Geometric measure theory*, Die Grundlehren der mathematischen Wissenschaften, Band 153, Springer-Verlag New York, Inc., New York, 1969. MR257325

[Fu1] J. H. G. Fu, *Tubular neighborhoods in Euclidean spaces*, Duke Math. J. **52** (1985), no. 4, 1025–1046, DOI 10.1215/S0012-7094-85-05254-8. MR816398

[Fu2] J. H. G. Fu, *Curvature measures of subanalytic sets*, Amer. J. Math. **116** (1994), no. 4, 819–880, DOI 10.2307/2375003. MR1287941

[Ga] D. Gatzouras, *Lacunarity of self-similar and stochastically self-similar sets*, Trans. Amer. Math. Soc. **352** (2000), no. 5, 1953–1983, DOI 10.1090/S0002-9947-99-02539-8. MR1694290

[Gr] A. Gray, *Tubes*, 2nd ed., Progress in Mathematics, vol. 221, Birkhäuser Verlag, Basel, 2004, DOI 10.1007/978-3-0348-7966-8. MR2024928

[GuLiZw] L. Guillopé, K. K. Lin, and M. Zworski, *The Selberg zeta function for convex co-compact Schottky groups*, Comm. Math. Phys. **245** (2004), no. 1, 149–176, DOI 10.1007/s00220-003-1007-1. MR2036371

[HaJeKaPrSh] T. C. Halsey, M. H. Jensen, L. P. Kadanoff, I. Procaccia, and B. I. Shraiman, *Fractal measures and their singularities: the characterization of strange sets*, Phys. Rev. A (3) **33** (1986), no. 2, 1141–1151, DOI 10.1103/PhysRevA.33.1141. MR823474

[HeKoZh] H. Hedenmalm, B. Korenblum, and K. Zhu, *Theory of Bergman spaces*, Graduate Texts in Mathematics, vol. 199, Springer-Verlag, New York, 2000, DOI 10.1007/978-1-4612-0497-8. MR1758653

[Hu] J. E. Hutchinson, *Fractals and self-similarity*, Indiana Univ. Math. J. **30** (1981), no. 5, 713–747, DOI 10.1512/iumj.1981.30.30055. MR625600

[HuLaWe] D. Hug, G. Last, and W. Weil, *A local Steiner-type formula for general closed sets and applications*, Math. Z. **246** (2004), no. 1-2, 237–272, DOI 10.1007/s00209-003-0597-9. MR2031455

[JePo] O. Jenkinson and M. Pollicott, *Calculating Hausdorff dimensions of Julia sets and Kleinian limit sets*, Amer. J. Math. **124** (2002), no. 3, 495–545. MR1902887

[La1] S. P. Lalley, *The packing and covering functions of some self-similar fractals*, Indiana Univ. Math. J. **37** (1988), no. 3, 699–710, DOI 10.1512/iumj.1988.37.37034. MR962930

[La2] S. P. Lalley, *Probabilistic methods in certain counting problems of ergodic theory*, Ergodic theory, symbolic dynamics, and hyperbolic spaces (Trieste, 1989), Oxford Sci. Publ., Oxford Univ. Press, New York, 1991, pp. 223–258. MR1130178

[La3] S. P. Lalley, *Renewal theorems in symbolic dynamics, with applications to geodesic flows, non-Euclidean tessellations and their fractal limits*, Acta Math. **163** (1989), no. 1-2, 1–55, DOI 10.1007/BF02392732. MR1007619

[LapRaZu] M. L. Lapidus, G. Radunović, and D. Žubrinić, *Zeta functions and complex dimensions of relative fractal drums: theory, examples and applications*, Dissertationes Math. **526** (2017), 105, DOI 10.4064/dm757-4-2017. MR3764036

[LapRo] M. L. Lapidus and J. A. Rock, *Towards zeta functions and complex dimensions of multifractals*, Complex Var. Elliptic Equ. **54** (2009), no. 6, 545–559, DOI 10.1080/17476930802326758. MR2537255

[LapLe-VeRo] M. L. Lapidus, J. Lévy-Véhel, and J. A. Rock, *Fractal strings and multifractal zeta functions*, Lett. Math. Phys. **88** (2009), no. 1-3, 101–129, DOI 10.1007/s11005-009-0302-y. MR2512142

[Lap-vF1] M. L. Lapidus and M. van Frankenhuysen, *Fractal geometry and number theory: Complex dimensions of fractal strings and zeros of zeta functions*, Birkhäuser Boston, Inc., Boston, MA, 2000, DOI 10.1007/978-1-4612-5314-3. MR1726744

[Lap-vF2] M. L. Lapidus and M. van Frankenhuijsen, *Fractal geometry, complex dimensions and zeta functions*, Springer Monographs in Mathematics, Springer, New York, 2006. Geometry and spectra of fractal strings, DOI 10.1007/978-0-387-35208-4. MR2245559

[Le-VeMe] J. Lévy Véhel and F. Mendivil, *Multifractal and higher-dimensional zeta functions*, Nonlinearity **24** (2011), no. 1, 259–276, DOI 10.1088/0951-7715/24/1/013. MR2746147

[MaUr] R. D. Mauldin and M. Urbański, *Graph directed Markov systems: Geometry and dynamics of limit sets*, Cambridge Tracts in Mathematics, vol. 148, Cambridge University Press, Cambridge, 2003, DOI 10.1017/CBO9780511543050. MR2003772

[Mo] J.-M. Morvan, *Generalized curvatures*, Geometry and Computing, vol. 2, Springer-Verlag, Berlin, 2008, DOI 10.1007/978-3-540-73792-6. MR2428231

[Ol1] L. Olsen, *Multifractal zeta-functions and multifractal prime counting formulas, Part 1*, Preprint (2024).

[Ol2] L. Olsen, *Multifractal zeta-functions and multifractal prime counting formulas, Part 2* Preprint (2024).

[PaPo1] W. Parry and M. Pollicott, *An analogue of the prime number theorem for closed orbits of Axiom A flows*, Ann. of Math. (2) **118** (1983), no. 3, 573–591, DOI 10.2307/2006982. MR727704

[PaPo2] W. Parry and M. Pollicott, *Zeta functions and the periodic orbit structure of hyperbolic dynamics* (English, with French summary), Astérisque **187-188** (1990), 268. MR1085356

[Pi] A. Pietsch, *Eigenvalues and s-numbers*, Cambridge Studies in Advanced Mathematics, vol. 13, Cambridge University Press, Cambridge, 1987. MR890520

[Sc1] R. Schneider, *Curvature measures of convex bodies*, Ann. Mat. Pura Appl. (4) **116** (1978), 101–134, DOI 10.1007/BF02413869. MR506976

[Sc2] R. Schneider, *Convex bodies: the Brunn-Minkowski theory*, Encyclopedia of Mathematics and its Applications, vol. 44, Cambridge University Press, Cambridge, 1993, DOI 10.1017/CBO9780511526282. MR1216521

[St] L. L. Stachó, *On curvature measures*, Acta Sci. Math. (Szeged) **41** (1979), no. 1-2, 191–207. MR534512

[Te] G. Tenenbaum, *Introduction to analytic and probabilistic number theory*, Cambridge Studies in Advanced Mathematics, vol. 46, Cambridge University Press, Cambridge, 1995. MR1342300

[We] H. Weyl, *On the Volume of Tubes*, Amer. J. Math. **61** (1939), no. 2, 461–472, DOI 10.2307/2371513. MR1507388

[Wi] S. Winter, *Curvature measures and fractals*, Dissertationes Math. **453** (2008), 66, DOI 10.4064/dm453-0-1. MR2423952

[Zä1] M. Zähle, *Integral and current representation of Federer's curvature measures*, Arch. Math. (Basel) **46** (1986), no. 6, 557–567, DOI 10.1007/BF01195026. MR849863

[Zä2] M. Zähle, *Curvatures and currents for unions of sets with positive reach*, Geom. Dedicata **23** (1987), no. 2, 155–171, DOI 10.1007/BF00181273. MR892398


Department of Mathematics, University of St. Andrews, St. Andrews, Fife KY16 9SS, Scotland

*Email address*: `lo@st-and.ac.uk`

Contemporary Mathematics
Volume **825**, 2025
https://doi.org/10.1090/conm/825/16514

# Spectrality of certain self-affine measures

Megala and Srijanani Anurag Prasad

ABSTRACT. In this paper, different conditions on the matrix $M$ are derived so that the measure $\mu_{M,D}$ is a spectrum for a specific four-element digit set $D$. Further, the measure $\mu_{M,D}$ is shown to be a spectrum for specific $M$ and general four-element digit set $D$.

## 1. Introduction

Let $\mu$ be Borel probability measure on $\mathbb{R}^n$ with compact support and $L^2(\mu)$ be corresponding Hilbert space. There is a long history of research into approximating functions in $L^2(\mu)$ with "good functions". Fourier discovered that the exponential functions $\{e^{2\pi i\langle\lambda,x\rangle} : \lambda \in \mathbb{Z}\}$ form an orthonormal basis for $L^2([0,1]^n)$ concerning Lebesgue measure, and this discovery is considered as one of the cornerstones of modern mathematics. It is natural to wonder what other measures have this property, that their $L^2$-space has an orthonormal basis formed by a family of exponential functions.

Suppose $\mu$ is a Borel probability measure and $\langle\cdot,\cdot\rangle$ is the standard inner product.

DEFINITION 1.1 ([**1**]). If a countable set $\Lambda \subset \mathbb{R}^n$ exists such that $E_\Lambda := \{e^{2\pi i\langle\lambda,\cdot\rangle} : \lambda \in \Lambda\}$ forms an orthonormal basis of $L^2(\mu)$, then $\mu$ is said to be a spectral measure. The set $\Lambda$ is called a spectrum for $\mu$, and the pair $(\mu,\Lambda)$ is known as a spectral pair.

The Fourier transform $\hat{\mu}(\xi)$ of the measure $\mu$ is defined as

$$\hat{\mu}(\xi) = \int e^{-2\pi i\langle\xi,x\rangle} d\mu(x)$$

It is obvious that the measure $\mu$ is a spectral if and only if

(i) $\hat{\mu}(\lambda - \lambda') = 0$ for all distinct $\lambda, \lambda' \in \Lambda$,

(ii) If $f \in L^2(\mu)$ with $\int f(x)e^{-2\pi i\langle\lambda,x\rangle} d\mu(x) = 0$, for all $\lambda \in \Lambda$ then $f = 0$.

A set $\Omega$ is called a spectral set if the normalized Lebesgue measure restricted to $\Omega$ is a spectral measure. Fuglede [**1**] laid the foundation stone for this topic. He conjectured that $\Omega$ is a spectral set if and only if $\Omega$ is a translation tile. The notions connecting the spectrality of a measure and the tile of its support are essential in the analysis of the spectrality of a Borel probability with compact

2020 *Mathematics Subject Classification.* Primary 28A80, 42C05; Secondary 46C05.
*Key words and phrases.* Fractals, self-affine measure, spectrum, spectrality.
The first author is supported by an MHRD grant.

support, even though the conjecture has been proven to fail in $\mathbb{R}^n, n \geq 3$. Following Fuglede's pioneering work, the spectral problem has been researched in various mathematical domains and exciting directions, some of which are prompted by the self-similar/self-affine measure, such as the construction of Fourier series and Fourier transforms on fractals.

A system of affine mappings in $\mathbb{R}^n$ generates the self-affine fractals, a significant class of fractals. The affine mappings can be represented in the following manner:

$$\phi_j(x) = M^{-1}(x+j) \quad j \in D \tag{1.1}$$

where, $M \in M_n(\mathbb{Z})$ is an $n \times n$ expansive matrix and $D \subset \mathbb{Z}^n$ is a finite digit set with cardinality $N$. According to Hutchinson [**2**], there exists a unique non-empty compact subset, denoted by $T := T(M, D)$ such that $MT = \bigcup\limits_{j \in D} (T+j)$ is an attractor for the IFS mentioned above. This attractor $T$ is generally a fractal set, which is self-affine. Also, there exists a unique probability measure, represented by $\mu := \mu_{M,D}$ and supported on $T(M, D)$, such that

$$\mu(\cdot) = \frac{1}{N} \sum_{j \in D} \mu \circ \phi_j^{-1}(\cdot). \tag{1.2}$$

The measure $\mu_{M,D}$ is a self-affine measure corresponding to the IFS (1.1). Let the mask polynomial of $D$ be defined as $m_D(x) = \frac{1}{|D|} \sum\limits_{j \in D} e^{2\pi i \langle j, x \rangle}$, $x \in \mathbb{R}^2$. Then, using Eq.(1.2) and mask polynomial definition, it is easily seen that the Fourier transform satisfies

$$\hat{\mu}(\xi) = \int e^{-2\pi i \langle \xi, x \rangle} d\mu(x) = \prod_{k=1}^{\infty} m_D(M^{*-k}\xi), \qquad \xi \in \mathbb{R}^2. \tag{1.3}$$

A crucial concern is when $\mu_{M,D}$ is spectral. Studying spectrality and non-spectrality for $L^2(\mu_{M,D})$ is a basic problem in Harmonic analysis. It is of most interest when it comes to the fractal measure. Jorgensen and Pedersen [**3**] first observed that for certain $M$ and $D$, the corresponding $\mu_{M,D}$ is a spectral measure. The authors proved that one-fourth Cantor measure is spectral and one-third Cantor measure is not. This means that a sequence of frequencies exists in the fractal measure for which the corresponding Fourier exponential functions form an orthonormal basis in the corresponding Hilbert space of square-integrable functions. This astonishing revelation has opened up new avenues for investigating harmonic analysis on fractals. The research on the spectrality and non-spectrality of self-affine measures corresponding to different $M$ and $D$ has garnered much attention [**4**–**11**] because it allows the well-developed Fourier technique to be used on some fractals. For the planar self-affine measures with three-element digit sets, spectrality and non-spectrality are well known [**12**–**18**]. However, for the four-element digit sets, the finite or infinite $\mu_{M,D}$-orthogonal exponentials and the spectrality of the corresponding self-affine measure differ from that of the three-element digit sets. A few results are obtained in this direction [**19**–**25**]. The studies on spectrality for Cantor measures are all in one dimension [**4**–**6**, **8**, **9**, **26**]. To our knowledge, no results exist for cantor-type measures in higher dimensions. In this paper, we studied spectrality for certain Cantor-type measures in two dimensions. We will derive various conditions on the matrix $M$ such that the family of certain Cantor dust-type measures $\mu_{M,D}$ are spectral. Further, for some unique $M$, we will find some general four-element digit set $D$ for which the measure $\mu_{M,D}$ is spectral.

## 2. Preliminaries

In this section, we state some of the existing results in the literature.

DEFINITION 2.1. Let $M \in M_n(\mathbb{Z})$ be an $n \times n$ expansive matrix. Let $D, S \subset \mathbb{Z}^n$ be a finite set of integer vectors with $N = |D| = |S|$ and $0 \in D \cap S$. The system $(M, D, S)$ is called Hadamard triple or $(M^{-1}D, S)$ is a compatible pair or the pair $(M, D)$ is admissible if the matrix $H = [\frac{1}{\sqrt{N}} e^{2\pi i \langle M^{-1}j, s\rangle}]_{j\in D, s\in S}$ is unitary.

THEOREM 2.2 ([**27**]). *If $(M, D, S)$ is a Hadamard triple, then $\mu_{M,D}$ is a spectral measure.*

PROPOSITION 2.3 ([**28**]). *Let $\mu_{M,D}$ be a self-affine measure, and let $P$ be a non-singular matrix with real entries. Then $\Lambda$ is a spectrum (res. orthonormal set) of $\mu_{M,D}$ if and only if $P^*\Lambda$ is a spectrum(res. orthonormal set) of $\mu_{P^{-1}MP, P^{-1}D}$.*

Jian-Lin Li [**19, 21**] gave a criterion for the existence of infinite families of orthogonal exponentials as follows:

PROPOSITION 2.4 ([**21**]). *Let $D \subset \mathbb{Z}^2$ be a finite digit set such that $(1/2, 0)$, $(0, 1/2)$, $(1/2, 1/2) \in \{x \in \mathbb{R}^2 : m_D(x) = 0\}$. Then, $|D|$ is an even number. Furthermore, if $M \in M_2(Z)$ is an expanding matrix such that $\det(M) \in 2\mathbb{Z} \setminus 0$, then there exists infinite families of orthogonal exponentials $E(\Lambda)$ in $L^2(\mu_{M,D})$ with $\Lambda \subseteq \mathbb{Z}^2$.*

Further, he showed the spectrality and non-spectrality of the measure $\mu_{M,D}$ for certain cases of $M$ and two-element digit set $D$.

PROPOSITION 2.5 ([**19**]). *For an expanding integer matrix $M \in M_2(\mathbb{Z})$ and a two-element digit set $D \subset \mathbb{Z}^2$ given by*

$$M = \begin{bmatrix} a & b \\ c & d \end{bmatrix} \quad \text{and} \quad D = \left\{ \begin{pmatrix} 0 \\ 0 \end{pmatrix}, \begin{pmatrix} \alpha \\ \beta \end{pmatrix} \right\}, \tag{2.1}$$

*define $\gamma$ as*

$$\gamma := c\alpha^2 - b\beta^2 + (d - a)\alpha\beta. \tag{2.2}$$

*The case $\gamma = 0$ is equivalent to the fact that $(\alpha, \beta)^t$ is an eigen vector of $M$ with certain eigenvalue $l \in \mathbb{Z} \setminus \{0, \pm 1\}$. If $l$ is odd, then $\mu_{M,D}$ is a non-spectral measure, and there are at most two mutually orthogonal exponential functions in $L^2(\mu_{M,D})$; If $l$ is even, then $\det(M) = ad - bc \in 2\mathbb{Z} \setminus \{0\}$, and $\mu_{M,D}$ is a spectral measure.*

*If $\gamma \neq 0, \det(M) \notin 2\mathbb{Z}$, and $a + d = 0$, then $\mu_{M,D}$ is a non-spectral measure. Furthermore, there are at most four mutually orthogonal exponential functions in $L^2(\mu_{M,D})$; and the number 4 is the best.*

For the planar Sierpinski gasket corresponds to the expanding integer matrix $M \in M_2(\mathbb{Z})$ and the three-element digit set D given by

$$M = \begin{bmatrix} p & 0 \\ 0 & p \end{bmatrix} \ (p \in \mathbb{Z} \setminus \{0, 1\}) \text{ and } D = \left\{ \begin{pmatrix} 0 \\ 0 \end{pmatrix}, \begin{pmatrix} 1 \\ 0 \end{pmatrix} \begin{pmatrix} 0 \\ 1 \end{pmatrix} \right\},$$

Dutkay and Jorgensen [**29**] showed that $\mu_{M,D}$ is a spectral measure if and only if $p \in 3\mathbb{Z} \setminus \{0\}$. Furthermore, for the pair (M, D) given by

$$M = \begin{bmatrix} p_1 & 0 \\ 0 & p_2 \end{bmatrix} (p_1, p_2 \in \mathbb{Z} \setminus \{0, 1\}) \text{ and } D = \left\{ \begin{pmatrix} 0 \\ 0 \end{pmatrix}, \begin{pmatrix} 1 \\ 0 \end{pmatrix} \begin{pmatrix} 0 \\ 1 \end{pmatrix} \right\},$$

Jian-Lin Li [**14**] proved that $\mu_{M,D}$ is a spectral measure if and only if $p_1, p_2 \in 3\mathbb{Z} \setminus \{0\}$. The author [**13**] also showed for an expanding integer matrix $M \in M_2(\mathbb{Z})$ and the three elements digit set D given by

$$M = \begin{bmatrix} a & b \\ c & d \end{bmatrix} \quad \text{and} \quad D = \left\{ \begin{pmatrix} 0 \\ 0 \end{pmatrix}, \begin{pmatrix} 1 \\ 0 \end{pmatrix} \begin{pmatrix} 0 \\ 1 \end{pmatrix} \right\},$$

if $ad - bc \notin 3\mathbb{Z}$, then there exist at most 3 mutually orthogonal exponentials in $L^2(\mu_{M,D})$ and the number 3 is the best. When $ad - bc \in 3\mathbb{Z}$, the spectrality and non spectrality is clear, see [**18**] Theorem 1.1 and [**16**].

When there are four elements in the digit set, the measure $\mu_{M,D}$ is shown to be spectral measure in [**30**] for diagonal matrices i.e

$$M = \begin{bmatrix} 2k & 0 \\ 0 & 2k \end{bmatrix} (k \in \mathbb{Z} \setminus \{0\}) \text{ and } D = \left\{ \begin{pmatrix} 0 \\ 0 \end{pmatrix}, \begin{pmatrix} \alpha \\ \beta \end{pmatrix}, \begin{pmatrix} \gamma \\ \delta \end{pmatrix}, \begin{pmatrix} \alpha + \gamma \\ \beta + \delta \end{pmatrix} \right\}$$

with $a, b, c, d \in \mathbb{R}$ and $\alpha\delta - \beta\gamma \neq 0$, the corresponding measure is spectral measure.

## 3. Main result

The Hadamard matrices that are related to the scaling and translation decisions determine the features of Fourier orthogonality. In this section, we will present new conclusions for a class of two-dimensional matrices.

Consider an expansive matrix $M$

$$M = \begin{bmatrix} a & b \\ c & d \end{bmatrix} \text{ and } D = \left\{ \begin{pmatrix} 0 \\ 0 \end{pmatrix}, \begin{pmatrix} 1 \\ 2 \end{pmatrix}, \begin{pmatrix} 2 \\ 1 \end{pmatrix}, \begin{pmatrix} 3 \\ 3 \end{pmatrix} \right\} \tag{3.1}$$

Note that for $M = \begin{bmatrix} 4 & 0 \\ 0 & 4 \end{bmatrix}$ and above $D$, the attractor of equation (1.1) is cantor dust. The collection of all measures $\mu_{M,D}$ for the digit set $D$ given in (3.1) is called the family of Cantor dust-type measures. In this study, we will show the spectrality of some Cantor-type measures in two dimensions. We shall extract different conditions on the matrix $M$, under which the family of some Cantor dust-type measures $\mu_{M,D}$ is spectral. Additionally, we will identify a generic four-element digit set $D$ for a unique $M$, for which the measure $\mu_{M,D}$ is spectral.

LEMMA 3.1. *If* $\det(M) = ad - bc \notin 2\mathbb{Z}$ *and* $a + d = 0$, *then* $\mu_{M,D}$ *is a non-spectral measure, and there are at most a finite number of orthogonal exponential functions in* $L^2(\mu_{M,D})$.

PROOF. Define $D_1 = \left\{ \begin{pmatrix} 0 \\ 0 \end{pmatrix}, \begin{pmatrix} 1 \\ 2 \end{pmatrix} \right\}$; and $D_2 = \left\{ \begin{pmatrix} 0 \\ 0 \end{pmatrix}, \begin{pmatrix} 2 \\ 1 \end{pmatrix} \right\}$. Then $D = D_1 + D_2 = D_1 \oplus D_2$ which shows that, $\hat{\mu}_{M,D}(\xi) = \hat{\mu}_{M,D_1}(\xi)\, \hat{\mu}_{M,D_2}(\xi)$. Hence, the zeros of Fourier transform $\hat{\mu}_{M,D}$ satisfy $Z(\hat{\mu}_{M,D}) = Z(\hat{\mu}_{M,D_1}) \cup Z(\hat{\mu}_{M,D_2})$. Now, we have the following results:

(i) The space $L^2(\mu_{M,D})$ has infinite families of orthogonal exponentials if and only if one of the two spaces $L^2(\mu_{M,D_1})$, $L^2(\mu_{M,D_2})$ has infinite families of orthogonal exponentials.

(ii) The space $L^2(\mu_{M,D})$ has at most a finite number of orthogonal exponentials if and only if both the spaces $L^2(\mu_{M,D_1})$ and $L^2(\mu_{M,D_2})$ have at most a finite number of orthogonal exponentials.

Since $\det(M) = ad-bc \notin 2\mathbb{Z}$, both the eigenvalues are odd. By Proposition 2.5, $L^2(\mu_{M,D_1})$ and $L^2(\mu_{M,D_2})$ have at most a finite number of orthogonal exponentials which in turn gives $L^2(\mu_{M,D})$ have at most a finite number of orthogonal exponentials. Hence, $\mu_{M,D}$ is a non-spectral measure. □

The solution to the mask polynomial of the digit set $D$ given in Eq. (3.1) is

$$Z(m_D) = \left\{ \begin{pmatrix} \alpha \\ \frac{2n+1}{4} - \frac{\alpha}{2} \end{pmatrix}; \alpha \in \mathbb{R}, n \in \mathbb{Z} \right\} \cup \left\{ \begin{pmatrix} \frac{n+3m+2}{6} \\ \frac{n-3m-1}{6} \end{pmatrix}; m, n \in \mathbb{Z} \right\}. \tag{3.2}$$

Clearly, $(1/2,0), (0,1/2), (1/2,1/2) \in \{x \in \mathbb{R}^2 : m_D(x) = 0\}$. So, by Proposition 2.5, if $\det(M) \in 2\mathbb{Z} \setminus \{0\}$, there exists infinite families of orthogonal exponentials.

Using the residue system of modulo 2, $M$ can be written as:

$$M = \begin{bmatrix} a & b \\ c & d \end{bmatrix} = 2 \begin{bmatrix} l_1 & l_2 \\ l_3 & l_4 \end{bmatrix} + M_i, \tag{3.3}$$

where the entries of the matrix $M_i$ are from the set $\{0,1\}$. Clearly, there are 16 cases for $M_i$'s, among them the following ten cases give $\det(M) \in 2\mathbb{Z}$.

$$M_1 = \begin{bmatrix} 0 & 0 \\ 0 & 0 \end{bmatrix}, M_2 = \begin{bmatrix} 1 & 0 \\ 0 & 0 \end{bmatrix}, M_3 = \begin{bmatrix} 0 & 1 \\ 0 & 0 \end{bmatrix}, M_4 = \begin{bmatrix} 0 & 0 \\ 1 & 0 \end{bmatrix},$$
$$M_5 = \begin{bmatrix} 0 & 0 \\ 0 & 1 \end{bmatrix}, M_6 = \begin{bmatrix} 0 & 1 \\ 0 & 1 \end{bmatrix}, M_7 = \begin{bmatrix} 0 & 0 \\ 1 & 1 \end{bmatrix}, M_8 = \begin{bmatrix} 1 & 0 \\ 1 & 0 \end{bmatrix},$$
$$M_9 = \begin{bmatrix} 1 & 1 \\ 0 & 0 \end{bmatrix}, M_{10} = \begin{bmatrix} 1 & 1 \\ 1 & 1 \end{bmatrix}$$

By Proposition 2.5, it is easily seen that infinite orthogonal exponentials exist for all the above ten cases.

Let $P = P^{-1} = \begin{bmatrix} 0 & 1 \\ 1 & 0 \end{bmatrix}$. Then

$$P^{-1}D = \begin{bmatrix} 0 & 1 \\ 1 & 0 \end{bmatrix} \left\{ \begin{pmatrix} 0 \\ 0 \end{pmatrix}, \begin{pmatrix} 1 \\ 2 \end{pmatrix}, \begin{pmatrix} 2 \\ 1 \end{pmatrix}, \begin{pmatrix} 3 \\ 3 \end{pmatrix} \right\} = \left\{ \begin{pmatrix} 0 \\ 0 \end{pmatrix}, \begin{pmatrix} 1 \\ 2 \end{pmatrix}, \begin{pmatrix} 2 \\ 1 \end{pmatrix}, \begin{pmatrix} 3 \\ 3 \end{pmatrix} \right\} = D.$$

By Proposition 2.3, the ten cases are reduced to the following six cases:

$$M_1 = \begin{bmatrix} 0 & 0 \\ 0 & 0 \end{bmatrix}, M_2 = \begin{bmatrix} 1 & 0 \\ 0 & 0 \end{bmatrix}, M_3 = \begin{bmatrix} 0 & 1 \\ 0 & 0 \end{bmatrix}, M_6 = \begin{bmatrix} 0 & 1 \\ 0 & 1 \end{bmatrix}, M_9 = \begin{bmatrix} 1 & 1 \\ 0 & 0 \end{bmatrix},$$
$$M_{10} = \begin{bmatrix} 1 & 1 \\ 1 & 1 \end{bmatrix}.$$

We will now establish criteria on the matrix $M$ so that the family of certain Cantor dust-type measures $\mu_{M,D}$ are spectral. This is achieved in the following theorem, which is our main result.

THEOREM 3.2. *Let $D$ be as in* (3.1).

(1) *If $l_3$ and $l_4$ in* (3.3) *are even integers then the measure $\mu_{M,D}$ corresponding to each of the four cases $M_i (i = 1,2,3,9)$ is spectral.*

(2) *If $l_1$ and $l_2$ in (3.3) are even integers then the measure $\mu_{M,D}$ corresponding to each of the three cases $M_i (i = 1, 3, 6)$ is spectral.*

(3) *If $M = \begin{bmatrix} 6l_1 & 6l_2 \\ 6l_3 & 6l_4 \end{bmatrix}$ then the measure $\mu_{M,D}$ is spectral.*

(4) *If $M = \begin{bmatrix} 3l_1 & 3l_2 \\ 12l_3 & 12l_4 \end{bmatrix}$ then the measure $\mu_{M,D}$ is spectral.*

PROOF. It is observed that the compatible pair $(M^{-1}D, S)$ is same as compatible pair of $(D, M^{*-1}S)$. Let $L = M^{*-1}S = \left\{ \begin{pmatrix} 0 \\ 0 \end{pmatrix}, \begin{pmatrix} x_1 \\ y_1 \end{pmatrix}, \begin{pmatrix} x_2 \\ y_2 \end{pmatrix}, \begin{pmatrix} x_3 \\ y_3 \end{pmatrix} \right\}$. Then, $(D, L)$ is compatible pair if and only if

$$1 + e^{2\pi i(x_1+2y_1)} + e^{2\pi i(x_2+2y_2)} + e^{2\pi i(x_3+2y_3)} = 0$$
$$1 + e^{2\pi i(2x_1+y_1)} + e^{2\pi i(2x_2+y_2)} + e^{2\pi i(2x_3+y_3)} = 0$$
$$1 + e^{2\pi i(x_1-y_1)} + e^{2\pi i(x_2-y_2)} + e^{2\pi i(x_3-y_3)} = 0$$
$$1 + e^{2\pi i(3x_1+3y_1)} + e^{2\pi i(3x_2+3y_2)} + e^{2\pi i(3x_3+3y_3)} = 0$$

The following are some of the solutions to the above system of equations:

$$L_1 = \left\{ \begin{pmatrix} 0 \\ 0 \end{pmatrix} \begin{pmatrix} \frac{4m_1+1}{4} \\ n_1 \end{pmatrix} \begin{pmatrix} \frac{4m_2-1}{4} \\ n_2 \end{pmatrix} \begin{pmatrix} \frac{2m_3+1}{2} \\ n_3 \end{pmatrix} \right\}$$
$$L_2 = \left\{ \begin{pmatrix} 0 \\ 0 \end{pmatrix} \begin{pmatrix} m_1 \\ \frac{4n_1+1}{4} \end{pmatrix} \begin{pmatrix} m_2 \\ \frac{4n_2-1}{4} \end{pmatrix} \begin{pmatrix} m_3 \\ \frac{2n_3+1}{2} \end{pmatrix} \right\}$$
$$L_3 = \left\{ \begin{pmatrix} 0 \\ 0 \end{pmatrix} \begin{pmatrix} \frac{m}{3} \\ \frac{2m-6n-3}{6} \end{pmatrix} \begin{pmatrix} \alpha \\ \frac{2j+1-2\alpha}{4} \end{pmatrix} \begin{pmatrix} \alpha + k \\ \frac{2j+1-2\alpha}{4} + \frac{2p+1}{2} \end{pmatrix} \right\}$$
$$L_4 = \left\{ \begin{pmatrix} 0 \\ 0 \end{pmatrix} \begin{pmatrix} \frac{2m+6n+3}{6} \\ \frac{m}{3} \end{pmatrix} \begin{pmatrix} \alpha \\ \frac{2j+1-4\alpha}{2} \end{pmatrix} \begin{pmatrix} \alpha + \frac{2k+1}{2} \\ \frac{2j+1-4\alpha}{2} + p \end{pmatrix} \right\}$$
$$L_5 = \left\{ \begin{pmatrix} 0 \\ 0 \end{pmatrix} \begin{pmatrix} \frac{4n_2-2n_1+1}{6} \\ \frac{4n_1-2n_2+1}{6} \end{pmatrix} \begin{pmatrix} \frac{m}{3} \\ \frac{2m-6n-3}{6} \end{pmatrix} \begin{pmatrix} \frac{2k+6\alpha+3}{6} \\ \frac{k}{3} \end{pmatrix} \right\}$$

CASE 1. Since $l_3$ and $l_4$ in Eq.(3.3) are even integers, the four cases of $M_i (i = 1, 2, 3, 9)$ are such that the corresponding $S = M^* L_2$ is a subset of $\mathbb{Z}^2$ and $(M, D, S)$ forms a Hadamard triple. Hence, by Theorem 2.2, the corresponding measure $\mu_{M,D}$ for the four cases

$$M_1 = \begin{bmatrix} 0 & 0 \\ 0 & 0 \end{bmatrix}, M_2 = \begin{bmatrix} 1 & 0 \\ 0 & 0 \end{bmatrix}, M_3 = \begin{bmatrix} 0 & 1 \\ 0 & 0 \end{bmatrix}, M_9 = \begin{bmatrix} 1 & 1 \\ 0 & 0 \end{bmatrix}, \tag{3.4}$$

is spectral.

CASE 2. When $l_1$ and $l_2$ in Eq. (3.3) are even integers, the three cases of $M_i (i = 1, 3, 6)$ are such that the corresponding $S = M^* L_1$ is a subset of $\mathbb{Z}^2$ and $(M, D, S)$ forms a Hadamard triple. Similar to case 1, by Theorem 2.2, the corresponding measure $\mu_{M,D}$ for the three cases

$$M_1 = \begin{bmatrix} 0 & 0 \\ 0 & 0 \end{bmatrix}, M_3 = \begin{bmatrix} 0 & 1 \\ 0 & 0 \end{bmatrix}, M_6 = \begin{bmatrix} 0 & 1 \\ 0 & 1 \end{bmatrix},$$

is spectral.

CASE 3. For $M = \begin{bmatrix} 6l_1 & 6l_2 \\ 6l_3 & 6l_4 \end{bmatrix}$, there is a set $S = M^* L_4$ which is subset of $\mathbb{Z}^2$ and so, $\mu_{M,D}$ is a spectral measure.

CASE 4. For $M = \begin{bmatrix} 3l_1 & 3l_2 \\ 12l_3 & 12l_4 \end{bmatrix}$, the set $S = M^* L_3$ is subset of $\mathbb{Z}^2$ and so $\mu_{M,D}$ is spectral. □

REMARK 3.3. The collection of $M$ in case 1 and case 3 are not subsets of each other. For example, let $A = \begin{bmatrix} 6 & 6 \\ 6 & 12 \end{bmatrix}$ and $B = \begin{bmatrix} 6 & 6 \\ 4 & 8 \end{bmatrix}$. The matrix $A$ belongs only to the matrices of the form given by case 3 and does not belong to the matrices of the form given by case 1. Similarly, the matrix $B$ belongs only to the matrices of the form given by case 1 and does not belong to the matrices of the form given by case 3. However, there are some matrices which belong to both the cases.

COROLLARY 3.4. *The measure $\mu_{M,D}$ is spectral for $M = \begin{bmatrix} 4l_1 & 4l_2 \\ 4l_2 & 4l_1 \end{bmatrix}$ and $D = \left\{ \begin{pmatrix} 0 \\ 0 \end{pmatrix}, \begin{pmatrix} \alpha \\ \beta \end{pmatrix}, \begin{pmatrix} \beta \\ \alpha \end{pmatrix}, \begin{pmatrix} \alpha+\beta \\ \alpha+\beta \end{pmatrix} \right\}$ with $\alpha^2 - \beta^2 \neq 0$.*

PROOF. Choosing $P^{-1} = \begin{bmatrix} \frac{2\beta-\alpha}{3} & \frac{2\alpha-\beta}{3} \\ \frac{2\alpha-\beta}{3} & \frac{2\beta-\alpha}{3} \end{bmatrix}$, $l_1 = 2l_4$, $l_2 = 2l_3$ if $M = \begin{bmatrix} 2l_1 & 2l_2 \\ 4l_3 & 4l_4 \end{bmatrix}$ or $l_4 = 2l_1$, $l_3 = 2l_2$ if $M = \begin{bmatrix} 4l_1 & 4l_2 \\ 2l_3 & 2l_4 \end{bmatrix}$ and applying Proposition 2.3 and Theorem 3.2, the result is obtained. □

COROLLARY 3.5. *The measure $\mu_{M,D}$ is spectral for $M = \begin{bmatrix} 4l_1 & 4l_2 \\ -4l_2 & 4l_1 \end{bmatrix}$ and $D = \left\{ \begin{pmatrix} 0 \\ 0 \end{pmatrix}, \begin{pmatrix} \alpha+2\beta \\ -\beta+2\alpha \end{pmatrix}, \begin{pmatrix} 2\alpha+\beta \\ -2\beta+\alpha \end{pmatrix}, \begin{pmatrix} 3\alpha+3\beta \\ 3\alpha-3\beta \end{pmatrix} \right\}$ with $\alpha \neq 0$ and $\beta \neq 0$.*

PROOF. Choosing $P^{-1} = \begin{bmatrix} \alpha & \beta \\ -\beta & \alpha \end{bmatrix}$, $l_1 = 2l_4$, $l_2 = -2l_3$ if $M = \begin{bmatrix} 2l_1 & 2l_2 \\ 4l_3 & 4l_4 \end{bmatrix}$ or $l_4 = 2l_1$, $l_3 = -2l_2$ if $M = \begin{bmatrix} 4l_1 & 4l_2 \\ 2l_3 & 2l_4 \end{bmatrix}$ and applying Proposition 2.3 and Theorem 3.2, the result is obtained. □

## 4. Conclusion

In this paper, we investigated the spectrality of some Cantor-type measures in two dimensions. When $D = \left\{ \begin{pmatrix} 0 \\ 0 \end{pmatrix}, \begin{pmatrix} 1 \\ 2 \end{pmatrix}, \begin{pmatrix} 2 \\ 1 \end{pmatrix}, \begin{pmatrix} 3 \\ 3 \end{pmatrix} \right\}$, the measure $\mu_{M,D}$ is shown to be a spectral measure for the matrices

(1) $M = \begin{bmatrix} 2l_1 & 2l_2 \\ 4l_3 & 4l_4 \end{bmatrix} + \begin{bmatrix} 0 & 0 \\ 0 & 0 \end{bmatrix}$,

(2) $M = \begin{bmatrix} 2l_1 & 2l_2 \\ 4l_3 & 4l_4 \end{bmatrix} + \begin{bmatrix} 1 & 0 \\ 0 & 0 \end{bmatrix}$,

(3) $M = \begin{bmatrix} 2l_1 & 2l_2 \\ 4l_3 & 4l_4 \end{bmatrix} + \begin{bmatrix} 0 & 1 \\ 0 & 0 \end{bmatrix}$,

(4) $M = \begin{bmatrix} 2l_1 & 2l_2 \\ 4l_3 & 4l_4 \end{bmatrix} + \begin{bmatrix} 1 & 1 \\ 0 & 0 \end{bmatrix}$,

(5) $M = \begin{bmatrix} 4l_1 & 4l_2 \\ 2l_3 & 2l_4 \end{bmatrix} + \begin{bmatrix} 0 & 0 \\ 0 & 0 \end{bmatrix}$,

(6) $M = \begin{bmatrix} 4l_1 & 4l_2 \\ 2l_3 & 2l_4 \end{bmatrix} + \begin{bmatrix} 0 & 1 \\ 0 & 0 \end{bmatrix}$,

(7) $M = \begin{bmatrix} 4l_1 & 4l_2 \\ 2l_3 & 2l_4 \end{bmatrix} + \begin{bmatrix} 0 & 1 \\ 0 & 1 \end{bmatrix}$,

and by Proposition 2.3

$$(8)\ M = \begin{bmatrix} 2l_1 & 2l_2 \\ 4l_3 & 4l_4 \end{bmatrix} + \begin{bmatrix} 0 & 0 \\ 0 & 1 \end{bmatrix}, \qquad (11)\ M = \begin{bmatrix} 4l_1 & 4l_2 \\ 2l_3 & 2l_4 \end{bmatrix} + \begin{bmatrix} 0 & 0 \\ 1 & 0 \end{bmatrix},$$

$$(9)\ M = \begin{bmatrix} 2l_1 & 2l_2 \\ 4l_3 & 4l_4 \end{bmatrix} + \begin{bmatrix} 0 & 0 \\ 1 & 0 \end{bmatrix}, \qquad (12)\ M = \begin{bmatrix} 4l_1 & 4l_2 \\ 2l_3 & 2l_4 \end{bmatrix} + \begin{bmatrix} 1 & 0 \\ 1 & 0 \end{bmatrix}.$$

$$(10)\ M = \begin{bmatrix} 2l_1 & 2l_2 \\ 4l_3 & 4l_4 \end{bmatrix} + \begin{bmatrix} 0 & 0 \\ 1 & 1 \end{bmatrix},$$

Further, the measure $\mu_{M,D}$ is shown to be a spectral measure for the matrices $M = \begin{bmatrix} 4l_1 & 4l_2 \\ 4l_2 & 4l_1 \end{bmatrix}$ and $M = \begin{bmatrix} 4l_1 & 4l_2 \\ -4l_2 & 4l_1 \end{bmatrix}$ with $D = \left\{ \begin{pmatrix} 0 \\ 0 \end{pmatrix}, \begin{pmatrix} \alpha \\ \beta \end{pmatrix}, \begin{pmatrix} \beta \\ \alpha \end{pmatrix}, \begin{pmatrix} \alpha+\beta \\ \alpha+\beta \end{pmatrix} \right\}$ with $\alpha^2 - \beta^2 \neq 0$. and $D = \left\{ \begin{pmatrix} 0 \\ 0 \end{pmatrix}, \begin{pmatrix} \alpha+2\beta \\ -\beta+2\alpha \end{pmatrix}, \begin{pmatrix} 2\alpha+\beta \\ -2\beta+\alpha \end{pmatrix}, \begin{pmatrix} 3\alpha+3\beta \\ 3\alpha-3\beta \end{pmatrix} \right\}$ with $\alpha \neq 0$ and $\beta \neq 0$ respectively.

## Acknowledgments

The authors thank the referees for their constructive evaluation of the paper and valuable suggestions.

## References

[1] B. Fuglede, *Commuting self-adjoint partial differential operators and a group theoretic problem*, J. Functional Analysis **16** (1974), 101–121, DOI 10.1016/0022-1236(74)90072-x. MR470754

[2] J. E. Hutchinson, *Fractals and self-similarity*, Indiana Univ. Math. J. **30** (1981), no. 5, 713–747, DOI 10.1512/iumj.1981.30.30055. MR625600

[3] P. E. T. Jorgensen and S. Pedersen, *Dense analytic subspaces in fractal $L^2$-spaces*, J. Anal. Math. **75** (1998), 185–228, DOI 10.1007/BF02788699. MR1655831

[4] I. Łaba and Y. Wang, *On spectral Cantor measures*, J. Funct. Anal. **193** (2002), no. 2, 409–420, DOI 10.1006/jfan.2001.3941. MR1929508

[5] R. S. Strichartz, *Mock Fourier series and transforms associated with certain Cantor measures*, J. Anal. Math. **81** (2000), 209–238, DOI 10.1007/BF02788990. MR1785282

[6] X.-R. Dai, *Spectra of Cantor measures*, Math. Ann. **366** (2016), no. 3-4, 1621–1647, DOI 10.1007/s00208-016-1374-5. MR3563247

[7] D. E. Dutkay, D. Han, and Q. Sun, *On the spectra of a Cantor measure*, Adv. Math. **221** (2009), no. 1, 251–276, DOI 10.1016/j.aim.2008.12.007. MR2509326

[8] T.-Y. Hu and K.-S. Lau, *Spectral property of the Bernoulli convolutions*, Adv. Math. **219** (2008), no. 2, 554–567, DOI 10.1016/j.aim.2008.05.004. MR2435649

[9] X.-R. Dai, *When does a Bernoulli convolution admit a spectrum?*, Adv. Math. **231** (2012), no. 3-4, 1681–1693, DOI 10.1016/j.aim.2012.06.026. MR2964620

[10] R. S. Strichartz, *Convergence of mock Fourier series*, J. Anal. Math. **99** (2006), 333–353, DOI 10.1007/BF02789451. MR2279556

[11] J.-L. Li, *$\mu_{M,D}$-orthogonality and compatible pair*, J. Funct. Anal. **244** (2007), no. 2, 628–638, DOI 10.1016/j.jfa.2006.10.003. MR2297038

[12] J.-L. Li, *Non-spectral problem for a class of planar self-affine measures*, J. Funct. Anal. **255** (2008), no. 11, 3125–3148, DOI 10.1016/j.jfa.2008.04.001. MR2464572

[13] J.-L. Li, *Non-spectrality of planar self-affine measures with three-elements digit set*, J. Funct. Anal. **257** (2009), no. 2, 537–552, DOI 10.1016/j.jfa.2008.12.012. MR2527028

[14] J.-L. Li, *On the $\mu_{M,D}$-orthogonal exponentials*, Nonlinear Anal. **73** (2010), no. 4, 940–951, DOI 10.1016/j.na.2010.04.017. MR2653762

[15] J.-C. Liu, X.-H. Dong, and J.-L. Li, *Non-spectral problem for the planar self-affine measures*, J. Funct. Anal. **273** (2017), no. 2, 705–720, DOI 10.1016/j.jfa.2017.04.003. MR3648865

[16] L.-X. An, X.-G. He, and L. Tao, *Spectrality of the planar Sierpinski family*, J. Math. Anal. Appl. **432** (2015), no. 2, 725–732, DOI 10.1016/j.jmaa.2015.06.064. MR3378387

[17] J.-L. Li, *Orthogonal exponentials on the generalized plane Sierpinski gasket*, J. Approx. Theory **153** (2008), no. 2, 161–169, DOI 10.1016/j.jat.2008.01.005. MR2450067

[18] J. Li, *Spectral self-affine measures on the planar Sierpinski family*, Sci. China Math. **56** (2013), no. 8, 1619–1628, DOI 10.1007/s11425-012-4490-z. MR3079819

[19] J. Li and Z. Wen, *Spectrality of planar self-affine measures with two-element digit set*, Sci. China Math. **55** (2012), no. 3, 593–605, DOI 10.1007/s11425-011-4278-6. MR2891321

[20] J. Li, *Spectrality of a class of self-affine measures with decomposable digit sets*, Sci. China Math. **55** (2012), no. 6, 1229–1242, DOI 10.1007/s11425-012-4390-2. MR2925589

[21] J.-L. Li, *Analysis of $\mu_M, D$-orthogonal exponentials for the planar four-element digit sets*, Math. Nachr. **287** (2014), no. 2-3, 297–312, DOI 10.1002/mana.201300009. MR3163581

[22] M.-S. Yang and J.-L. Li, *A class of spectral self-affine measures with four-element digit sets*, J. Math. Anal. Appl. **423** (2015), no. 1, 326–335, DOI 10.1016/j.jmaa.2014.10.004. MR3273183

[23] M.-W. Tang and F.-L. Yin, *Spectrality of Moran measures with four-element digit sets*, J. Math. Anal. Appl. **461** (2018), no. 1, 354–363, DOI 10.1016/j.jmaa.2018.01.018. MR3759546

[24] J. Su, Y. Liu, and J.-C. Liu, *Non-spectrality of the planar self-affine measures with four-element digit sets*, Fractals **27** (2019), no. 7, 1950115, 7 pp., DOI 10.1142/S0218348X19501159. MR4040879

[25] Z.-M. Wang, X.-H. Dong, and Z.-Y. Wang, *Orthogonal exponential functions of self-affine measures in $\mathbb{R}^n$*, Fractals **27** (2019), no. 3, 1950029, 6 pp., DOI 10.1142/S0218348X19500294. MR3957184

[26] X.-R. Dai, X.-G. He, and C.-K. Lai, *Spectral property of Cantor measures with consecutive digits*, Adv. Math. **242** (2013), 187–208, DOI 10.1016/j.aim.2013.04.016. MR3055992

[27] D. E. Dutkay, J. Haussermann, and C.-K. Lai, *Hadamard triples generate self-affine spectral measures*, Trans. Amer. Math. Soc. **371** (2019), no. 2, 1439–1481, DOI 10.1090/tran/7325. MR3885185

[28] J.-L. Li, *Spectral self-affine measures in $\mathbb{R}^N$*, Proc. Edinb. Math. Soc. (2) **50** (2007), no. 1, 197–215, DOI 10.1017/S0013091503000324. MR2294013

[29] D. E. Dutkay and P. E. T. Jorgensen, *Analysis of orthogonality and of orbits in affine iterated function systems*, Math. Z. **256** (2007), no. 4, 801–823, DOI 10.1007/s00209-007-0104-9. MR2308892

[30] J.-L. Li, *Spectra of a class of self-affine measures*, J. Funct. Anal. **260** (2011), no. 4, 1086–1095, DOI 10.1016/j.jfa.2010.12.001. MR2747015

Department of Mathematics and Statistics, Indian Institute of Technology, Tirupati

*Email address*: `megala8niya@gmail.com`

Department of Mathematics and Statistics, Indian Institute of Technology, Tirupati

*Email address*: `srijanani@iittp.ac.in`

Contemporary Mathematics
Volume **825**, 2025
https://doi.org/10.1090/conm/825/16515

# General Hewitt-Stromberg Measures: Properties and Their Role in Multifractal Formalism

Tingting Wang, Bilel Selmi, and Zhiming Li

ABSTRACT. In this research, a new general multifractal analysis is introduced, addressing the constraints associated with conventional multifractal models by incorporating specific functions, namely $\varphi$ and $\psi$. This innovative approach not only surpasses the limitations of traditional multifractal forms but also extends the scope of conventional fractal measures. The study focuses on the exploration and analysis of a comprehensive Hewitt-Stromberg measure, introducing a generalized multifractal structure based on $\varphi$ and $\psi$. The research includes the estimation of upper and lower bounds for the general Hewitt-Stromberg dimension, providing insights into the divergence of multifractal functions at the $q$ point. This allows for obtaining results even in scenarios where the general upper and lower multifractal Hewitt-Stromberg dimension functions exhibit divergence.

## 1. Introduction

This paper primarily builds upon the novel definition introduced in [**2**], along with insights from various publications such as Ben Nasr & Peyrière [**8**], Huang et al. [**26**], Samti [**34**], Shen [**44**], Wu & Xiao [**47**], and Yuan [**49**]. Over the past three decades, our research has extensively focused on multifractal measures. These measures manifest as distributions with significant intensity variations, and we draw theoretical analogies between fractal sets and natural measures. Given the pivotal roles played by multifractal analysis and measurement forms in fractal geometry and dynamic systems, our keen interest has driven our research and development efforts. The term 'multifractal' was initially introduced by Frish and Parisi in 1985 [**18**] as part of the statistical description of energy dissipation in turbulence. Kolmogorov and Mandelbrot [**28, 29**] subsequently initiated and developed this concept. The energy dissipation, represented by the measure $\mu$, occurs in the level set $X_\mu(\alpha)$ with a given Hölder index $\alpha$, prompting a natural consideration of the

2020 *Mathematics Subject Classification.* Primary 28A20, 28A75, 28A78, 28A80, 49Q15.
*Key words and phrases.* General multifractal formalism, general Hausdorff measure, general packing measure, Hewitt-Stromberg measures, fractal dimensions.
**Funding:** This work is supported by NSFC (No.12271432) and by Analysis, Probability & Fractals Laboratory (No. LR18ES17).

**Declaration of competing interest:** We confirm that this manuscript has not been published elsewhere and is not under consideration by another journal. I approved the manuscript and agree with submission to the journal and we have no conflicts of interest to declare.

**Data availability:** No data was used for the research described in the article.

size of these sets. The Hausdorff dimension is employed to rank the sizes of these level sets. The significance of multifractal analysis lies in the fact that only when $\mu$ is derived from this set, $X_\mu(\alpha)$, is the dimension of $\alpha$ obtained. However, this result is not generally achievable. The order $\tau_\mu$ serves as the characteristic function of $\mu$, with dim representing the Hausdorff dimension. The empty set is assigned a Hausdorff dimension of $-\infty$. At this juncture, we assert that multifractal formalism proves effective because, for the equation inferred by physicists (Halsey et al. (1986) [**24**]), verifying the equation is only possible under the condition that the measure $\mu$ satisfies

$$\dim\left(X_\mu(\alpha)\right) = \tau_\mu^*(\alpha) := \inf_{q\in\mathbb{R}} \left(\alpha q + \tau_\mu(q)\right).$$

In 1992, Brown et al. [**11**] demonstrated the effectiveness of grid-related multifractal forms for a substantial class of metrics. In 1995, Olsen [**31**] introduced a grid-free framework and established an intrinsic definition for $\mu$ by considering $h_\mu(q,t) = \sum_i \mu\left(B\left(x_i, r_i\right)\right)^q \left(2r_i\right)^t$. Olsen supported this multifractal form by incorporating the theory of Gibbs measures and other measures with large deviation requirements. While such assumptions are natural when the measure stems from a specific dynamical system, they tend to decrease within the rigorous framework of geometric measure theory. In 2002, Ben Nasr et al. [**9**] endeavored to replace these assumptions with more geometric conditions, resulting in more general and natural outcomes that do not necessitate the existence of Gibbs measures. Subsequently, Ben Nasr et al. [**9**], delved deeper into the analysis, striving to establish an almost necessary and sufficient condition for the formalistic validity proposed by Olsen. In 2003, Peyrière [**33**] advocated the use of functions instead of measures, replacing diameters with more general functions (defined on the sphere of the metric space). This approach, covered in studies by Menceur & Ben Mabrouk [**30**], encompassed and unified the aforementioned research endeavors (see also [**10**]). The measure is defined through this innovative perspective

$$h_{\mu,\varphi}(q,t) = \sum_i \mu\left(B\left(x_i, r_i\right)\right)^q e^{t\varphi(r_i)}.$$

In the following examples: Ben Nasr & Peyrière [**8**]; Douzi et al. [**17**]; Huang et al. [**26**]; Li & Selmi [**27**]; Samti [**34**]; Selmi [**38, 42**]; Shen [**44**]; Wu [**46**]; Wu & Xiao [**47**]; Xiao & Wu [**48**] and Yuan [**49**], it is easy to conclude that these measures are not applicable to multifractal formalism. Regarding the following two questions:

(1) A necessary and sufficient condition for determining the validity of multifractal formalism?
(2) What can be said about the failure of traditional multifractal formalism?

We employ two widely recognized fractal measures in our study—namely, the Hausdorff measure and the packing measure. Felix Hausdorff initially introduced the Hausdorff measure in 1919, while the packing measure, pioneered by Tricot [**45**], gained prominence only in the early 1980s. The primary focus of our investigation is the Hewitt-Stromberg measure, first introduced by Hewitt and Stromberg in 1965 as part of an exercise in a textbook [**25**]. Subsequent detailed research on the Hewitt-Stromberg measure was conducted in the mid-1980s by Christoph Bandt and Herman Haase, along with their students, as documented in [**19, 20, 21, 22, 23**]. A significant hiatus in research on the Hewitt-Stromberg measure persisted from the late 1980s to the mid-2010s until Olsen [**32**] initiated the exploration of various multifractal generalizations of the measure. Since then, numerous authors, including Attia & Selmi [**42**], Douzi & Selmi [**14**, **15**], Selmi [**35**, **37**, **39**, **40**, **41**], and Zindulka [**50**], have delved into the study of these measures. In defining these fractal measures, we establish the general Hewitt-Stromberg measure by employing the packing of a sphere with a specific diameter. Likewise, we define the general Hausdorff measure and packing measure by utilizing the covering and packing of a set family with a diameter smaller than a given positive number. A more comprehensive definition of the packing measure and Hausdorff measure for functions $\varphi$ and $\psi$ was recently proposed in [**2**]. Furthermore, the related properties and applications of the general fractal measures are discussed in [**1**, **3**, **4**, **5**, **12**, **13**, **36**, **43**]. Therefore, building upon the Hewitt-Stromberg measure and drawing from the collective insights of Attia & Selmi [**6**, **7**], Douzi & Selmi [**14**], Selmi [**41**], and the advancements in [**2**], this article aims to formulate and examine a naturally occurring multifractal analog of the Hewitt-Stromberg measure for functions $\varphi$ and $\psi$.

This research introduces a novel extension of Hewitt-Stromberg measures and dimensions, drawing inspiration from the definitions of Hausdorff measure and packing measure and dimension outlined in [**2**], specifically tailored to functions $\varphi$ and $\psi$. Within this context, the measures and dimensions are explicitly associated with the $\varphi, \psi$ function. Leveraging the general fractal measures and dimensions outlined in [**16**], we explore various properties and outcomes of multifractal formalism grounded in conventional Hewitt-Stromberg measures. Simultaneously, we compute both upper and lower bounds for the Hewitt-Stromberg dimension. Even in instances where the Hewitt-Stromberg dimension functions for upper and lower multifractals diverge at certain points, we employ a key theorem to derive consequential results.

The following is the main work of this study. In Section 2, we focused on examining the recently formulated definitions of various fractal and multifractal dimensions and measures applicable to functions $\varphi, \psi$, as introduced in [**2**]. Additionally, we extended these definitions to encompass fractal/multifractal variations of Hewitt-Stromberg measures and dimensions, with a specific emphasis on the gauge control functions $\varphi, \psi$. The section commenced by revisiting the novel definitions of Hausdorff and packing measures tailored to functions $\varphi, \psi$, along with the corresponding Hausdorff and packing dimensions. The provided multifractal formulas for Hausdorff and packing dimensions of functions $\varphi, \psi$ further enriched our exploration. Building upon the groundwork laid in the preceding portion, Sections 2.1 and 2.2 delve into the formulations and extensions of Hewitt-Stromberg measures and dimensions for functions $\varphi, \psi$. Within this framework, we analyzed the interplay between these measures and explored the relationships established by these definitions. Moreover, we conducted a comparative analysis between the Hausdorff and packing measures under this specific definition and their counterparts in the less familiar domain of Hewitt-Stromberg measures. Drawing on the foundations established by the Hewitt-Stromberg measures, the pertinent definitions and key theorems are expounded in Section 3. The substantiation of the key theorems unfolds in Section 4 and Section 5. Additionally, Section 5 not only furnishes the proof but also offers estimations for upper and lower bounds concerning the Hewitt-Stromberg dimension limit. To address scenarios where the Hewitt-Stromberg dimension function of a general multifractal displays inconsistencies at a certain point $q$, the outcomes from Section 5 find application in Section 6.

## 2. Preliminaries

Fix $0 < a \leq \infty$, for functions $\varphi$ and $\psi$ defined on the interval $(0, a)$, we consider the following four properties:

(*P1*) The continuous functions $\varphi$ and $\psi$ are strictly monotonically decreasing and positive on the interval $(0, a)$.

(*P2*) $\lim_{x\to 0} \varphi(x) = \lim_{x\to 0} \psi(x) = \infty$ and $\lim_{x\to a} \varphi(x) = 0$.

(*P3*) $\lim_{x\to\infty} \dfrac{\varphi^{-1}(kx + c)}{\varphi^{-1}(x)} = 0$, for any $k > 1$ and $c \geq 0$.

(*P4*) $\limsup_{x\to 0}(\psi(x) - \psi(kx)) < \infty$, for $k > 1$.

Let $P$ be the set of a pair of functions that satisfy the above four properties, i.e.,

$$P = \Big\{(\varphi, \psi) \;\Big|\; \varphi \text{ and } \psi \text{ satisfy the properties } P1, P2, P3 \text{ and } P4\Big\}.$$

Conditions $(P1)$ and $(P2)$ guarantee that $\varphi^{-1}(x)$ is defined on the interval $(0, \infty)$. The following examples of functions are all in set $P$:

(a) $\Big(\varphi(x) = -\log x, \psi(x) = -\log x\Big)$.

(b) $\Big(\varphi(x) = \log_2 \log_2 1/x, \psi(x) = -\log_2(x)\Big)$.

(c) $\Big(\varphi(x) = \log_2 \log_2 1/x, \psi(x) = \log_2 \log_2 1/x\Big)$.

(d) $\Big(\varphi(x) = \log_2\left(\frac{1}{K}\log_2\left(\frac{1}{x}\right)\right), \psi(x) = -\log_2(x)\Big)$, for all $K > 0$.

Throughout this paper, we let

$\mathscr{P}(\mathbb{R}^n)$ : The set of Borel probability measures defined on $\mathbb{R}^n$.
$\mathscr{P}^c(\mathbb{R}^n)$ : The set of Borel probability measures with compact support on $\mathbb{R}^n$.
$B(x, r)$ : A closed ball with the center $x$ and the radius $r > 0$.
$\#A$ : The number of elements in set $A$.
$\mathbb{N}^*$ : The collection of all positive integers.

Definition 2.1. Assuming that $A$ is a subset of $\mathbb{R}^n$ with $A \neq \emptyset$. For $\delta > 0$, then

(1) A collection $\{B(x_i, r_i)\}_i$ is designated as a centered $\delta$-covering of set $A$, if it consists of countable or finite closed balls that cover $A$, the diameter of each ball does not exceed $\delta$. Additionally, the centers of these closed balls are situated within the boundaries of $A$, i.e.,

$$A \subseteq \bigcup_{i=1}^{\infty} B(x_i, r_i) \qquad 0 < 2r_i \leq \delta, \text{ and } \quad x_i \in A \quad \text{for any} \quad i.$$

Furthermore, if $r_i = r > 0$, for all $i$, $\{B(x_i, r)\}_i$ is called a centered covering of $A$.

(2) A collection $\{B(x_i, r_i)\}_i$ is considered a centered $\delta$-packing of set $A$, if it comprises non-overlapping closed balls, the diameter of each ball does not exceed $\delta$. Furthermore, the centers of these closed balls are located within the confines of $A$, i.e.,

$$B(x_i, r_i) \cap B(x_j, r_j) = \emptyset \quad \text{for} \quad i \neq j; \quad 0 < 2r_i \leq \delta \quad \text{and} \quad x_i \in A \quad \text{for any} \quad i.$$

Furthermore, if $r_i = r > 0$, for all $i$, $\{B(x_i, r)\}_i$ is called a centered packing of $A$.

For $q,t \geq 0$, $\mu \in \mathscr{P}(\mathbb{R}^n)$, $(\varphi,\psi) \in P$, $x \in A \subseteq \mathbb{R}^n$ and $r > 0$, put

$$\Phi^{q,t}_{\varphi,\psi,\mu}(x,r) = \varphi^{-1}\left(q\varphi(\mu(B(x,r))) + t\psi(r)\right).$$

According to [**2**], the definitions of the general multifractal Hausdorff measures and packing measures are given as follows:

$$\overline{\mathscr{H}}_\delta^{\Phi^{q,t}_{\varphi,\psi,\mu}}(A) = \inf\left\{\sum_i \Phi^{q,t}_{\varphi,\psi,\mu}(x_i,r_i) \;\middle|\; \{B(x_i,r_i)\}_i \text{ is a centered } \delta\text{-covering of } A\right\},$$

$$\overline{\mathscr{H}}_\delta^{\Phi^{q,t}_{\varphi,\psi,\mu}}(\emptyset) = 0,$$

$$\overline{\mathscr{H}}^{\Phi^{q,t}_{\varphi,\psi,\mu}}(A) = \sup_{\delta>0} \overline{\mathscr{H}}_\delta^{\Phi^{q,t}_{\varphi,\psi,\mu}}(A),$$

$$\mathscr{H}^{\Phi^{q,t}_{\varphi,\psi,\mu}}(A) = \sup_{F\subseteq A} \overline{\mathscr{H}}_\delta^{\Phi^{q,t}_{\varphi,\psi,\mu}}(F),$$

$$\overline{\mathscr{P}}_\delta^{\Phi^{q,t}_{\varphi,\psi,\mu}}(A) = \sup\left\{\sum_i \Phi^{q,t}_{\varphi,\psi,\mu}(x_i,r_i) \;\middle|\; \{B(x_i,r_i)\}_i \text{ is a centered } \delta\text{-packing of } A\right\},$$

$$\overline{\mathscr{P}}_\delta^{\Phi^{q,t}_{\varphi,\psi,\mu}}(\emptyset) = 0,$$

$$\overline{\mathscr{P}}^{\Phi^{q,t}_{\varphi,\psi,\mu}}(A) = \inf_{\delta>0} \overline{\mathscr{P}}_\delta^{\Phi^{q,t}_{\varphi,\psi,\mu}}(A),$$

$$\mathscr{P}^{\Phi^{q,t}_{\varphi,\psi,\mu}}(A) = \inf_{A\subset\bigcup_i A_i} \sum_i \overline{\mathscr{P}}_\delta^{\Phi^{q,t}_{\varphi,\psi,\mu}}(A_i).$$

The general multifractal dimension functions are defined by:

$$b^q_{\varphi,\psi,\mu}(A) = \dim_H^{\varphi,\psi,q}(A) = \sup\{t \mid \mathscr{H}^{\Phi^{q,t}_{\varphi,\psi,\mu}}(A) = \infty\} = \inf\{t \mid \mathscr{H}^{\Phi^{q,t}_{\varphi,\psi,\mu}}(A) = 0\},$$

$$B^q_{\varphi,\psi,\mu}(A) = \dim_P^{\varphi,\psi,q}(A) = \sup\{t \mid \mathscr{P}^{\Phi^{q,t}_{\varphi,\psi,\mu}}(A) = \infty\} = \inf\{t \mid \mathscr{P}^{\Phi^{q,t}_{\varphi,\psi,\mu}}(A) = 0\}$$

and

$$\Delta^q_{\varphi,\psi,\mu}(A) = \sup\{t \mid \overline{\mathscr{P}}^{\Phi^{q,t}_{\varphi,\psi,\mu}}(A) = \infty\} = \inf\{t \mid \overline{\mathscr{P}}^{\Phi^{q,t}_{\varphi,\psi,\mu}}(A) = 0\}.$$

We define the multifractal functions $b_{\varphi,\psi,\mu}(q) = \dim_H^{\varphi,\psi,q}(\mathbf{supp}\,\mu)$, $B_{\varphi,\psi,\mu}(q) = \dim_P^{\varphi,\psi,q}(\mathbf{supp}\,\mu)$ and $\Delta_{\varphi,\psi,\mu}(q) = \Delta^q_{\varphi,\psi}(\mathbf{supp}\,\mu)$. It is obvious that $b_{\varphi,\psi,\mu}(q)$ and $B_{\varphi,\psi,\mu}(q)$ are nonincreasing and so we have

$$b_{\varphi,\psi,\mu}(1) = 0, \quad b_{\varphi,\psi,\mu}(0) = \dim_H^{\varphi,\psi}(\mathbf{supp}\,\mu)$$

and

$$B_{\varphi,\psi,\mu}(1) = 0, \quad B_{\varphi,\psi,\mu}(0) = \dim_P^{\varphi,\psi}(\mathbf{supp}\,\mu).$$

### 2.1. The $(\varphi,\psi)$-Hewitt-Stromberg measures and dimensions.

DEFINITION 2.2. Assuming that $A$ is a totally bounded subset of $\mathbb{R}^n$, $r > 0$ and $(\varphi,\psi) \in P$ with $\varphi^{-1}(0) = a > 0$, then for all $t \geq 0$, let

$$\mathbf{I}^t_{\varphi,\psi}(A) = \liminf_{r\to 0} O_r(A)\varphi^{-1}(t\psi(r))$$

and

$$\overline{\mathbf{P}}^t_{\varphi,\psi}(A) = \limsup_{r\to 0} D_r(A)\varphi^{-1}(t\psi(r)),$$

where

$$O_r(A) = \inf\left\{\#I \;\Big|\; \{B(x_i,r)\}_{i\in I} \text{ is a centered covering of } A\right\},$$

and

$$D_r(A) = \sup\left\{\#I \;\Big|\; \{B(x_i,r)\}_{i\in I} \text{ is a centered packing of } A\right\}.$$

We define now the function $\overline{\mathbf{H}}^t_{\varphi,\psi}$ as

$$\overline{\mathbf{H}}^t_{\varphi,\psi}(A) = \sup_{F\subseteq A} \mathbf{I}^t_{\varphi,\psi}(F).$$

$\overline{\mathbf{H}}^t_{\varphi,\psi}$ and $\overline{\mathbf{P}}^t_{\varphi,\psi}$ are said to be the lower and upper $(\varphi,\psi)$-Hewitt-Stromberg pre-measures.

Next, we take

$$\mathbf{H}^t_{\varphi,\psi}(A) = \inf_{A\subseteq \bigcup_i A_i} \sum_i \overline{\mathbf{H}}^t_{\varphi,\psi}(A_i)$$

and

$$\mathbf{P}^t_{\varphi,\psi}(A) = \inf_{A\subseteq \bigcup_i A_i} \sum_i \overline{\mathbf{P}}^t_{\varphi,\psi}(A_i).$$

Similar to the classical case. For totally bounded set $A \subseteq \mathbb{R}^n$, $r>0$, $t\geq 0$, and $(\varphi,\psi)\in P$ with $\varphi^{-1}(0) = a > 0$, we have some basic inequalities for the $(\varphi,\psi)$-Hewitt-Stromberg measures, general Hausdorff measure and general packing measure:

$$\mathbf{H}^t_{\varphi,\psi}(A) \leq \overline{\mathbf{H}}^t_{\varphi,\psi}(A),$$
$$\mathbf{P}^t_{\varphi,\psi}(A) \leq \overline{\mathbf{P}}^t_{\varphi,\psi}(A),$$
$$\overline{\mathbf{H}}^t_{\varphi,\psi}(A) \leq \xi\overline{\mathbf{P}}^t_{\varphi,\psi}(A) \leq \xi\overline{\mathscr{P}}^{\Phi^t_{\varphi,\psi}}(A)$$

and

$$\mathscr{H}^{\Phi^t_{\varphi,\psi}}(A) \leq \mathbf{H}^t_{\varphi,\psi}(A) \leq \xi\mathbf{P}^t_{\varphi,\psi}(A) \leq \xi\mathscr{P}^{\Phi^t_{\varphi,\psi}}(A),$$

where $\xi$ is a constant determined by Besicovitch's covering theorem.

Now, we define the lower and upper $(\varphi,\psi)$-Hewitt-Stromberg dimensions $\underline{\dim}^{MB}_{\varphi,\psi}(A)$ and $\overline{\dim}^{MB}_{\varphi,\psi}(A)$ as:

$$\underline{\dim}^{MB}_{\varphi,\psi}(A) = \inf\left\{t\geq 0 \mid \mathbf{H}^t_{\varphi,\psi}(A) = 0\right\} = \sup\left\{t\geq 0 \mid \mathbf{H}^t_{\varphi,\psi}(A) = +\infty\right\}$$

and

$$\overline{\dim}^{MB}_{\varphi,\psi}(A) = \inf\left\{t\geq 0 \mid \mathbf{P}^t_{\varphi,\psi}(A) = 0\right\} = \sup\left\{t\geq 0 \mid \mathbf{P}^t_{\varphi,\psi}(A) = +\infty\right\}.$$

The general upper and lower box dimensions of $A$ are defined by:

$$\overline{\dim}^{B}_{\varphi,\psi}(A) = \limsup_{r\to 0} \frac{\varphi(1/O_r(A))}{\psi(r)} = \limsup_{r\to 0} \frac{\varphi(1/D_r(A))}{\psi(r)}$$

and

$$\underline{\dim}^{B}_{\varphi,\psi}(A) = \liminf_{r\to 0} \frac{\varphi(1/O_r(A))}{\psi(r)} = \liminf_{r\to 0} \frac{\varphi(1/D_r(A))}{\psi(r)}.$$

When $\overline{\dim}^{B}_{\varphi,\psi}(A)$ equals $\underline{\dim}^{B}_{\varphi,\psi}(A)$, this value is called the $(\varphi,\psi)$-box dimension and is represented by $\dim^{B}_{\varphi,\psi}(A)$.

For $t \geq 0$, $A \subseteq \mathbb{R}^n$ and $(\varphi, \psi) \in P$ with $\varphi^{-1}(0) = a > 0$, the following inequalities hold:

$$\dim_H^{\varphi,\psi,0}(A) \leq \underline{\dim}_{\varphi,\psi}^{MB}(A) \leq \overline{\dim}_{\varphi,\psi}^{MB}(A) \leq \dim_P^{\varphi,\psi,0}(A),$$

$$\dim_H^{\varphi,\psi,0}(A) \leq \dim_P^{\varphi,\psi,0}(A) \leq \overline{\dim}_{\varphi,\psi}^{B}(A)$$

and

$$\dim_H^{\varphi,\psi,0}(A) \leq \underline{\dim}_{\varphi,\psi}^{B}(A) \leq \overline{\dim}_{\varphi,\psi}^{B}(A).$$

For $\underline{\dim}_{\varphi,\psi}^{MB}$ and $\overline{\dim}_{\varphi,\psi}^{MB}$, their equivalent definitions are represented as:

$$\underline{\dim}_{\varphi,\psi}^{MB}(A) = \inf\left\{ \sup_i \mathscr{S}_{\varphi,\psi}^{B}(A_i) \;\middle|\; A \subseteq \bigcup_i A_i \quad A_i \text{ is bounded in } \mathbb{R}^n \right\},$$

where $\mathscr{S}_{\varphi,\psi}^{B}(A) = \sup_{F \subseteq A} \underline{\dim}_{\varphi,\psi}^{B}(F)$, and

$$\overline{\dim}_{\varphi,\psi}^{MB}(A) = \inf\left\{ \sup_i \overline{\dim}_{\varphi,\psi}^{B}(A_i) \;\middle|\; A \subseteq \bigcup_i A_i \quad A_i \text{ is bounded in } \mathbb{R}^n \right\}.$$

**2.2. The multifractal $(\varphi, \psi)$-Hewitt-Stromberg measures and dimensions.** We provide the definitions of the multifractal Hewitt-Stromberg measures are represented by $\overline{\mathbf{H}}_{\varphi,\psi,\mu}^{q,t}$ and $\overline{\mathbf{P}}_{\varphi,\psi,\mu}^{q,t}$, respectively.

DEFINITION 2.3. Let $\mu \in \mathscr{P}^c(\mathbb{R}^n)$, $(\varphi, \psi) \in P$ with $\varphi^{-1}(0) = a > 0$, $q, t \geq 0$ and $r > 0$. For $A \subseteq \mathbf{supp}\,\mu$, put

$$D_{\varphi,\mu,r}^{q}(A) = \sup\left\{ \sum_i \varphi^{-1}(q\varphi(\mu(B(x_i, r)))) \;\middle|\; \{B(x_i, r)\}_i \text{ is a centered packing of } A \right\}$$

and

$$O_{\varphi,\mu,r}^{q}(A) = \inf\left\{ \sum_i \varphi^{-1}(q\varphi(\mu(B(x_i, r)))) \;\middle|\; \{B(x_i, r)\}_i \text{ is a centered covering of } A \right\}.$$

Then we define

$$\overline{\mathbf{P}}_{\varphi,\psi,\mu}^{q,t}(A) = \limsup_{r \to 0} D_{\varphi,\mu,r}^{q}(A)\varphi^{-1}(t\psi(r))$$

and

$$\mathbf{I}_{\varphi,\psi,\mu}^{q,t}(A) = \liminf_{r \to 0} O_{\varphi,\mu,r}^{q}(A)\varphi^{-1}(t\psi(r)).$$

Obviously, $\overline{\mathbf{P}}_{\varphi,\psi,\mu}^{q,t}$ is increasing and $\overline{\mathbf{P}}_{\varphi,\psi,\mu}^{q,t}(\emptyset) = 0$, it can be seen that $\overline{\mathbf{P}}_{\varphi,\psi,\mu}^{q,t}$ does not satisfy the condition of $\sigma$-additivity. Therefore

$$\mathbf{P}_{\varphi,\psi,\mu}^{q,t}(A) = \inf_{A \subseteq \bigcup_i A_i} \sum_i \overline{\mathbf{P}}_{\varphi,\psi,\mu}^{q,t}(A_i).$$

Since $\mathbf{I}_{\varphi,\psi,\mu}^{q,t}$ is not increasing and does not satisfy $\sigma$-additivity, in order to obtain the outer measure, we set

$$\overline{\mathbf{H}}_{\varphi,\psi,\mu}^{q,t}(A) = \sup_{F \subseteq A} \mathbf{I}_{\varphi,\psi,\mu}^{q,t}(F)$$

and

$$\mathbf{H}_{\varphi,\psi,\mu}^{q,t}(A) = \inf_{A \subseteq \bigcup_i A_i} \sum_i \overline{\mathbf{H}}_{\varphi,\psi,\mu}^{q,t}(A_i).$$

Hence, $\mathbf{H}^{q,t}_{\varphi,\psi,\mu}$ is a multifractal extension of the lower $(\varphi,\psi)$-Hewitt-Stromberg measure $\mathbf{H}^{t}_{\varphi,\psi}$. Similarly, $\mathbf{P}^{q,t}_{\varphi,\psi,\mu}$ is a multifractal generalization of the upper $(\varphi,\psi)$-Hewitt-Stromberg measure $\mathbf{P}^{t}_{\varphi,\psi}$. And for $t \geq 0$, the following equalities hold:

$$\mathbf{H}^{0,t}_{\varphi,\psi,\mu} = a\,\mathbf{H}^{t}_{\varphi,\psi} \quad \text{and} \quad \mathbf{P}^{0,t}_{\varphi,\psi,\mu} = a\,\mathbf{P}^{t}_{\varphi,\psi}.$$

Similar to the classic Hewitt-Stromberg measures and dimensions, there are many properties that hold true for the newly defined multifractal Hewitt-Stromberg measures ($\mathbf{H}^{q,t}_{\varphi,\psi,\mu}$ and $\mathbf{P}^{q,t}_{\varphi,\psi,\mu}$) for $(\varphi,\psi) \in P$ with $\varphi^{-1}(0) = a > 0$. In the following sections, we provide and prove the relevant properties of these new measures and dimensions.

THEOREM 2.4. *Let $\mu \in \mathscr{P}^c(\mathbb{R}^n)$, $(\varphi,\psi) \in P$ with $\varphi^{-1}(0) = a > 0$, $q,t \geq 0$ and $A \subseteq \mathbf{supp}\,\mu$, then*

(1) *The functions $\mathbf{H}^{q,t}_{\varphi,\psi,\mu}$ and $\mathbf{P}^{q,t}_{\varphi,\psi,\mu}$ are outer measures and thus they are measures on the $\sigma$-algebra.*
(2) *The function $\mathbf{H}^{q,t}_{\varphi,\psi,\mu}$ is a metric outer measure and it is a measure on the Borel $\sigma$-algebra.*
(3) *One has*

$$\mathbf{H}^{q,t}_{\varphi,\psi,\mu}(A) \leq \overline{\mathbf{H}}^{q,t}_{\varphi,\psi,\mu}(A) \text{ and } \mathbf{P}^{q,t}_{\varphi,\psi,\mu}(A) \leq \overline{\mathbf{P}}^{q,t}_{\varphi,\psi,\mu}(A).$$

(4) *For the constant $\xi$ determined by Besicovitch's covering theorem, we have*

$$\mathbf{H}^{q,t}_{\varphi,\psi,\mu}(A) \leq \xi \mathbf{P}^{q,t}_{\varphi,\psi,\mu}(A).$$

(5) *If $C_1\varphi^{-1}(x)\varphi^{-1}(y) \leq \varphi^{-1}(x+y) \leq C_2\varphi^{-1}(x)\varphi^{-1}(y)$, for all $x,y \in [0,+\infty)$ and for some $C_2 \geq C_1 > 0$, then there exists $\xi \in \mathbb{N}^*$, such that*

$$\overline{\mathbf{H}}^{q,t}_{\varphi,\psi,\mu}(A) \leq \xi \overline{\mathbf{P}}^{q,t}_{\varphi,\psi,\mu}(A) \leq \frac{\xi}{C_1} \overline{\mathscr{P}}^{\phi^{q,t}_{\varphi,\psi,\mu}}(A)$$

*and*

$$\frac{1}{C_2}\mathscr{H}^{\phi^{q,t}_{\varphi,\psi,\mu}}(A) \leq \mathbf{H}^{q,t}_{\varphi,\psi,\mu}(A) \leq \xi \mathbf{P}^{q,t}_{\varphi,\psi,\mu}(A) \leq \frac{\xi}{C_1}\mathscr{P}^{\phi^{q,t}_{\varphi,\psi,\mu}}(A).$$

PROOF.

(1) It follows immediately from the definitions.
(2) According to the definition of metric outer measure and the last property, we just need to prove

$$\mathbf{H}^{q,t}_{\varphi,\psi,\mu}(E \cup F) \geq \mathbf{H}^{q,t}_{\varphi,\psi,\mu}(E) + \mathbf{H}^{q,t}_{\varphi,\psi,\mu}(F),$$

for any $E, F \subseteq \mathbf{supp}\,\mu$ with $d(E,F) > 0$. So, we can assume that $\mathbf{H}^{q,t}_{\varphi,\psi,\mu}(E \cup F) < \infty$. Then for $\epsilon > 0$, let $\{U_i\}_i$ be a cover of $E \cup F$ such that

$$\sum_i \overline{\mathbf{H}}^{q,t}_{\varphi,\psi,\mu}(U_i) \leq \mathbf{H}^{q,t}_{\varphi,\psi,\mu}(E \cup F) + \epsilon.$$

Next put

$$E_i = E \cap U_i \text{ and } F_i = F \cap U_i.$$

Obviously, $\{E_i\}_i$ is the cover of $E$ and $\{F_i\}_i$ is the cover of $F$, $E_i \cap F_i = \emptyset$, so

$$\begin{aligned}\mathbf{H}^{q,t}_{\varphi,\psi,\mu}(E) + \mathbf{H}^{q,t}_{\varphi,\psi,\mu}(F) &\leq \sum_i \left(\overline{\mathbf{H}}^{q,t}_{\varphi,\psi,\mu}(E_i) + \overline{\mathbf{H}}^{q,t}_{\varphi,\psi,\mu}(F_i)\right)\\ &\leq \sum_i \left(\overline{\mathbf{H}}^{q,t}_{\varphi,\psi,\mu}(U_i)\right)\\ &\leq \mathbf{H}^{q,t}_{\varphi,\psi,\mu}(E \cup F) + \epsilon.\end{aligned}$$

Letting $\epsilon \to 0$, we have

$$\mathbf{H}^{q,t}_{\varphi,\psi,\mu}(E \cup F) \geq \mathbf{H}^{q,t}_{\varphi,\psi,\mu}(E) + \mathbf{H}^{q,t}_{\varphi,\psi,\mu}(F).$$

The proof is complete.

(3) Obviously

$$\mathbf{H}^{q,t}_{\varphi,\psi,\mu}(A) = \inf_{A \subseteq \bigcup_i A_i} \sum_i \overline{\mathbf{H}}^{q,t}_{\varphi,\psi,\mu}(A_i) \leq \overline{\mathbf{H}}^{q,t}_{\varphi,\psi,\mu}(A)$$

and

$$\mathbf{P}^{q,t}_{\varphi,\psi,\mu}(A) = \inf_{A \subseteq \bigcup_i A_i} \sum_i \overline{\mathbf{P}}^{q,t}_{\varphi,\psi,\mu}(A_i) \leq \overline{\mathbf{P}}^{q,t}_{\varphi,\psi,\mu}(A).$$

(4) Let $F \subseteq A$ and take a centered covering $\{B(x_i, r)\}_i$ of $F$. According to the Besicovitch's covering theorem, there exists $\xi$-finite or countable subfamily $\{(B(x_{ij}, r))_j,\ \ 1 \leq i \leq \xi\}$ such that

$$F \subseteq \bigcup_{i=1}^{\xi} \bigcup_j B(x_{ij}, r),$$

and for all $i$, $\{B(x_{ij}, r)\}_j$ is a centered packing of $F$. Then

$$\begin{aligned}O^q_{\varphi,\mu,r}(F)\varphi^{-1}(t\psi(r)) &\leq \varphi^{-1}(t\psi(r)) \sum_{i=1}^{\xi} \sum_j \varphi^{-1}\big(q\varphi(\mu(B(x_{ij}, r)))\big)\\ &\leq \sum_{i=1}^{\xi} D^q_{\varphi,\mu,r}(F)\varphi^{-1}(t\psi(r))\\ &= \xi D^q_{\varphi,\mu,r}(F)\varphi^{-1}(t\psi(r)).\end{aligned}$$

So, we obtain that

$$\mathbf{I}^{q,t}_{\varphi,\psi,\mu}(F) \leq \xi \overline{\mathbf{P}}^{q,t}_{\varphi,\psi,\mu}(F) \leq \xi \overline{\mathbf{P}}^{q,t}_{\varphi,\psi,\mu}(A).$$

Thus

$$\overline{\mathbf{H}}^{q,t}_{\varphi,\psi,\mu}(A) \leq \xi \overline{\mathbf{P}}^{q,t}_{\varphi,\psi,\mu}(A),$$

and then

$$\mathbf{H}^{q,t}_{\varphi,\psi,\mu}(A) \leq \xi \mathbf{P}^{q,t}_{\varphi,\psi,\mu}(A).$$

(5) We only need to prove

$$\mathscr{H}^{\Phi^{q,t}_{\varphi,\psi,\mu}}(A) \leq C_2 \mathbf{H}^{q,t}_{\varphi,\psi,\mu}(A)$$

and

$$C_1 \mathbf{P}^{q,t}_{\varphi,\psi,\mu}(A) \leq \mathscr{P}^{\Phi^{q,t}_{\varphi,\psi,\mu}}(A).$$

Let $F \subseteq A$, for any $r_1, r_2 > 0$, take a centered covering $\{B(x_i, r_1)\}_i$ of $F$, and a centered packing $\{B(y_j, r_2)\}_j$ of $A$. Since

$$C_1\varphi^{-1}(x)\varphi^{-1}(y) \leq \varphi^{-1}(x+y) \leq C_2\varphi^{-1}(x)\varphi^{-1}(y),$$

for all $x, y \in [0, +\infty)$ and for some $C_2 \geq C_1 > 0$, then for all $q, t \geq 0$, we have

$$\sum_j \varphi^{-1}\left(q\varphi(\mu(B(y_j, r_2))) + t\psi(r_2)\right) \geq C_1\varphi^{-1}(t\psi(r_2)) \sum_j \varphi^{-1}\left(q\varphi(\mu(B(y_j, r_2)))\right) \tag{2.1}$$

and

$$\sum_i \varphi^{-1}\left(q\varphi(\mu(B(x_i, r_1))) + t\psi(r_1)\right) \leq C_2\varphi^{-1}(t\psi(r_1)) \sum_i \varphi^{-1}\left(q\varphi(\mu(B(x_i, r_1)))\right). \tag{2.2}$$

According to inequality (2.1), it is obvious that

$$\overline{\mathscr{P}}^{\Phi^{q,t}_{\varphi,\psi,\mu}}_{r_2}(A) \geq C_1 D^q_{\varphi,\mu,r_2}(A)\varphi^{-1}(t\psi(r_2)).$$

Letting $r_2 \to 0$,

$$C_1\overline{\mathbf{P}}^{q,t}_{\varphi,\psi,\mu}(A) \leq \overline{\mathscr{P}}^{\Phi^{q,t}_{\varphi,\psi,\mu}}(A).$$

Put $\{A_i\}_i$ be a cover of $A$. We have

$$C_1\mathbf{P}^{q,t}_{\varphi,\psi,\mu}(A) \leq C_1\sum_i \overline{\mathbf{P}}^{q,t}_{\varphi,\psi,\mu}(A_i) \leq \sum_i \overline{\mathscr{P}}^{\Phi^{q,t}_{\varphi,\psi,\mu}}(A_i),$$

which gives that

$$C_1\mathbf{P}^{q,t}_{\varphi,\psi,\mu}(A) \leq \mathscr{P}^{\Phi^{q,t}_{\varphi,\psi,\mu}}(A).$$

According to inequality (2.2), we obtain

$$\overline{\mathscr{H}}^{\Phi^{q,t}_{\varphi,\psi,\mu}}_{r_1}(F) \leq C_2 O^q_{\varphi,\mu,r_1}(F)\varphi^{-1}(t\psi(r_1)).$$

Letting $r_1 \to 0$,

$$\overline{\mathscr{H}}^{\Phi^{q,t}_{\varphi,\psi,\mu}}(F) \leq C_2\mathbf{I}^{q,t}_{\varphi,\psi,\mu}(F) \leq C_2\overline{\mathbf{H}}^{q,t}_{\varphi,\psi,\mu}(F).$$

Next, let $\{F_i\}_i$ be a cover of $F$, so

$$\overline{\mathscr{H}}^{\Phi^{q,t}_{\varphi,\psi,\mu}}(F) \leq C_2\sum_i \overline{\mathbf{H}}^{q,t}_{\varphi,\psi,\mu}(F_i),$$

and

$$\overline{\mathscr{H}}^{\Phi^{q,t}_{\varphi,\psi,\mu}}(F) \leq C_2\mathbf{H}^{q,t}_{\varphi,\psi,\mu}(F) \leq C_2\mathbf{H}^{q,t}_{\varphi,\psi,\mu}(A).$$

Then

$$\mathscr{H}^{\Phi^{q,t}_{\varphi,\psi,\mu}}(A) \leq C_2\mathbf{H}^{q,t}_{\varphi,\psi,\mu}(A).$$

□

For each subset $A$ of $\mathbb{R}^n$, its dimensions regarding multifractal functions $\mathbf{H}^{q,t}_{\varphi,\psi,\mu}$ and $\mathbf{P}^{q,t}_{\varphi,\psi,\mu}$ and the pre-measure $\overline{\mathbf{P}}^{q,t}_{\varphi,\psi,\mu}$ are represented as $\mathbf{b}^q_{\varphi,\psi,\mu}(A)$, $\mathbf{B}^q_{\varphi,\psi,\mu}(A)$ and $\mathbf{\Delta}^q_{\varphi,\psi,\mu}(A)$.

Proposition 2.1. *Let* $\mu \in \mathscr{P}^c(\mathbb{R}^n)$, $(\varphi, \psi) \in P$ *with* $\varphi^{-1}(0) = a > 0$, $q, t \geq 0$ *and* $A \subseteq \mathbf{supp}\,\mu$, *one has*

(1) *There is a unique number* $\mathbf{b}^q_{\varphi,\psi,\mu}(A) \in [0, +\infty]$ *that satisfies*

$$\mathbf{H}^{q,t}_{\varphi,\psi,\mu}(A) = \begin{cases} \infty & \text{if} \quad t \leq \mathbf{b}^q_{\varphi,\psi,\mu}(A) \\ 0 & \text{if} \quad t > \mathbf{b}^q_{\varphi,\psi,\mu}(A). \end{cases}$$

(2) *There is a unique number* $\mathbf{B}^q_{\varphi,\psi,\mu}(A) \in [0, +\infty]$ *that satisfies*

$$\mathbf{P}^{q,t}_{\varphi,\psi,\mu}(A) = \begin{cases} \infty & \text{if} \quad t \leq \mathbf{B}^q_{\varphi,\psi,\mu}(A) \\ 0 & \text{if} \quad t > \mathbf{B}^q_{\varphi,\psi,\mu}(A). \end{cases}$$

(3) *There is a unique number* $\mathbf{\Delta}^q_{\varphi,\psi,\mu}(A) \in [0, +\infty]$ *that satisfies*

$$\overline{\mathbf{P}}^{q,t}_{\varphi,\psi,\mu}(A) = \begin{cases} \infty & \text{if} \quad t \leq \mathbf{\Delta}^q_{\varphi,\psi,\mu}(A) \\ 0 & \text{if} \quad t > \mathbf{\Delta}^q_{\varphi,\psi,\mu}(A). \end{cases}$$

(4) *There is a unique number* $\mathscr{S}^q_{\varphi,\psi,\mu}(A) \in [0, +\infty]$ *that satisfies*

$$\overline{\mathbf{H}}^{q,t}_{\varphi,\psi,\mu}(A) = \begin{cases} \infty & \text{if} \quad t \leq \mathscr{S}^q_{\varphi,\psi,\mu}(A) \\ 0 & \text{if} \quad t > \mathscr{S}^q_{\varphi,\psi,\mu}(A). \end{cases}$$

(5) *There is a unique number* $\kappa^q_{\varphi,\psi,\mu}(A) \in [0, +\infty]$ *that satisfies*

$$\mathbf{I}^{q,t}_{\varphi,\psi,\mu}(A) = \begin{cases} \infty & \text{if} \quad t \leq \kappa^q_{\varphi,\psi,\mu}(A) \\ 0 & \text{if} \quad t > \kappa^q_{\varphi,\psi,\mu}(A). \end{cases}$$

*Furthermore, according to Theorem 2.4, we have*

$$\mathscr{S}^q_{\varphi,\psi,\mu}(A) \leq \mathbf{\Delta}^q_{\varphi,\psi,\mu}(A) \leq \Delta^q_{\varphi,\psi,\mu}(A)$$

*and*

$$b^q_{\varphi,\psi,\mu}(A) \leq \mathbf{b}^q_{\varphi,\psi,\mu}(A) \leq \mathbf{B}^q_{\varphi,\psi,\mu}(A) \leq B^q_{\varphi,\psi,\mu}(A).$$

The number $\mathbf{b}^q_{\varphi,\psi,\mu}(A)$ is an obvious multifractal analogue of the lower $(\varphi, \psi)$-Hewitt-Stromberg dimension $(\underline{\dim}^{MB}_{\varphi,\psi}(A))$ of $A$, and $\mathbf{B}^q_{\varphi,\psi,\mu}(A)$ is an obvious multifractal analogues of the upper $(\varphi, \psi)$-Hewitt-Stromberg dimension $(\overline{\dim}^{MB}_{\varphi,\psi}(A))$ of $A$. So

$$\mathbf{b}^0_{\varphi,\psi,\mu}(A) = \underline{\dim}^{MB}_{\varphi,\psi}(A) \text{ and } \mathbf{B}^0_{\varphi,\psi,\mu}(A) = \overline{\dim}^{MB}_{\varphi,\psi}(A).$$

The number $\mathscr{S}^q_{\varphi,\psi,\mu}(A)$ is an obvious multifractal analogue of the lower $(\varphi, \psi)$-box dimension $(\underline{\dim}^{B}_{\varphi,\psi}(A))$ of $A$ and $\mathbf{\Delta}^q_{\varphi,\psi,\mu}(A)$ is an obvious multifractal analogues of the upper $(\varphi, \psi)$-box dimension $(\overline{\dim}^{B}_{\varphi,\psi}(A))$ of $A$. Obviously, it follows immediately from the above definitions that

$$\mathscr{S}^0_{\varphi,\psi,\mu}(A) = \underline{\dim}^{B}_{\varphi,\psi}(A) \text{ and } \mathbf{\Delta}^0_{\varphi,\psi,\mu}(A) = \overline{\dim}^{B}_{\varphi,\psi}(A).$$

Next, put

$$\mathbf{\Delta}_{\varphi,\psi,\mu}(q) = \mathbf{\Delta}^q_{\varphi,\psi,\mu}(\mathbf{supp}\,\mu),$$
$$\mathbf{B}_{\varphi,\psi,\mu}(q) = \mathbf{B}^q_{\varphi,\psi,\mu}(\mathbf{supp}\,\mu)$$

and

$$\mathbf{b}_{\varphi,\psi,\mu}(q) = \mathbf{b}^q_{\varphi,\psi,\mu}(\mathbf{supp}\,\mu).$$

For a real-value function $f : \mathbb{R} \to \mathbb{R}$, the Legendre transform $f^* : \mathbb{R} \to [-\infty, +\infty]$ is defined by

$$f^* = \inf_y (xy + f(y)).$$

We understand that the Legendre transform $\tau^*$ of the function $\tau$ is equivalent to the multifractal spectrum, which exhibits characteristics such as decreasing, smooth, and convex. The definition of multifractal dimension functions suggests a contrast with the traditional $\tau$-function in multifractal forms. Consequently, we demonstrate that these functions indeed possess some of these distinctive properties.

PROPOSITION 2.2. *Let $\mu \in \mathscr{P}^c(\mathbb{R}^n)$, $(\varphi, \psi) \in P$ with $\varphi^{-1}(0) = a > 0$, $q, t \geq 0$ and $A \subseteq \mathbf{supp}\,\mu$. Then*

(1) *The functions $q \mapsto \mathbf{H}^{q,t}_{\varphi,\psi,\mu}(A)$, $\mathbf{P}^{q,t}_{\varphi,\psi,\mu}(A)$ and $\overline{\mathbf{P}}^{q,t}_{\varphi,\psi,\mu}(A)$ are decreasing.*
(2) *The functions $t \mapsto \mathbf{H}^{q,t}_{\varphi,\psi,\mu}(A)$, $\mathbf{P}^{q,t}_{\varphi,\psi,\mu}(A)$ and $\overline{\mathbf{P}}^{q,t}_{\varphi,\psi,\mu}(A)$ are decreasing.*
(3) *The functions $q \mapsto \mathbf{b}^q_{\varphi,\psi,\mu}(A)$, $\mathbf{B}^q_{\varphi,\psi,\mu}(A)$ and $\mathbf{\Delta}^q_{\varphi,\psi,\mu}(A)$ are decreasing.*
(4) *If $\varphi^{-1}(x+y) \leq \varphi^{-1}(x)\varphi^{-1}(y)$ for all $x, y \in [0, +\infty)$ and $\varphi^{-1}(\beta x) \leq (\varphi^{-1}(x))^\beta$ for all $0 < \beta < 1$, then the functions $q \mapsto \mathbf{B}^q_{\varphi,\psi,\mu}(A)$ and $\mathbf{\Delta}^q_{\varphi,\psi,\mu}(A)$ are convex.*
(5) *One has*

$$\mathbf{b}_{\varphi,\psi,\mu}(q) \leq \mathbf{B}_{\varphi,\psi,\mu}(q) \leq \mathbf{\Delta}_{\varphi,\psi,\mu}(q).$$

PROOF.

(1) Take $0 \leq q \leq p$, for fixed $r > 0$, let $\{B(x_i, r)\}_i$ be a centered packing of $A$. We have

$$\sum_i \varphi^{-1}\left(p\varphi(\mu(B(x_i, r)))\right) \leq \sum_i \varphi^{-1}\left(q\varphi(\mu(B(x_i, r)))\right) \leq D^q_{\varphi,\mu,r}(A).$$

Thus

$$D^p_{\varphi,\mu,r}(A) \leq D^q_{\varphi,\mu,r}(A)$$

and

$$D^p_{\varphi,\mu,r}(A)\varphi^{-1}(t\psi(r)) \leq D^q_{\varphi,\mu,r}(A)\varphi^{-1}(t\psi(r)).$$

Hence

$$\overline{\mathbf{P}}^{p,t}_{\varphi,\psi,\mu}(A) \leq \overline{\mathbf{P}}^{q,t}_{\varphi,\psi,\mu}(A).$$

Let $\{A_i\}_i$ be a cover of $A$, we obtain

$$\mathbf{P}^{p,t}_{\varphi,\psi,\mu}(A) \leq \sum_i \overline{\mathbf{P}}^{p,t}_{\varphi,\psi,\mu}(A_i) \leq \sum_i \overline{\mathbf{P}}^{q,t}_{\varphi,\psi,\mu}(A_i).$$

This implies that

$$\mathbf{P}^{p,t}_{\varphi,\psi,\mu}(A) \leq \mathbf{P}^{q,t}_{\varphi,\psi,\mu}(A).$$

By similar arguments, we can prove that the function $q \mapsto \mathbf{H}^{q,t}_{\varphi,\psi,\mu}(A)$ is decreasing.

(2) Take $0 \leq t_1 \leq t_2$ and $r > 0$.
   (a) Let F $\subseteq A$ and $\{B(x_i, r)\}_i$ be a centered covering of $F$. Then

$$O^q_{\varphi,\mu,r}(F)\varphi^{-1}(t_2\psi(r)) \leq O^q_{\varphi,\mu,r}(F)\varphi^{-1}(t_1\psi(r)),$$

   and so

$$\mathbf{I}^{q,t_2}_{\varphi,\psi,\mu}(F) \leq \mathbf{I}^{q,t_1}_{\varphi,\psi,\mu}(F) \leq \overline{\mathbf{H}}^{q,t_1}_{\varphi,\psi,\mu}(A).$$

It follows that

$$\overline{\mathbf{H}}^{q,t_2}_{\varphi,\psi,\mu}(A) \le \overline{\mathbf{H}}^{q,t_1}_{\varphi,\psi,\mu}(A),$$

and let $\{A_i\}_i$ be a cover of $A$, then we obtain

$$\mathbf{H}^{q,t_2}_{\varphi,\psi,\mu}(A) \le \sum_i \overline{\mathbf{H}}^{q,t_2}_{\varphi,\psi,\mu}(A_i) \le \sum_i \overline{\mathbf{H}}^{q,t_1}_{\varphi,\psi,\mu}(A_i).$$

So the function $t \mapsto \mathbf{H}^{q,t}_{\varphi,\psi,\mu}(A)$ is decreasing.

(b) Let $\{B(x_i,r)\}_i$ be a centered packing of $A$, we have

$$D^q_{\varphi,\mu,r}(A)\varphi^{-1}(t_1\psi(r)) \ge D^q_{\varphi,\mu,r}(A)\varphi^{-1}(t_2\psi(r)).$$

Hence

$$\overline{\mathbf{P}}^{q,t_1}_{\varphi,\psi,\mu}(A) \ge \overline{\mathbf{P}}^{q,t_2}_{\varphi,\psi,\mu}(A).$$

Assuming that $\{A_i\}_i$ is a cover of $A$, then

$$\mathbf{P}^{q,t_2}_{\varphi,\psi,\mu}(A) \le \sum_i \overline{\mathbf{P}}^{q,t_2}_{\varphi,\psi,\mu}(A_i) \le \sum_i \overline{\mathbf{P}}^{q,t_1}_{\varphi,\psi,\mu}(A_i),$$

which gives that

$$\mathbf{P}^{q,t_2}_{\varphi,\psi,\mu}(A) \ge \mathbf{P}^{q,t_1}_{\varphi,\psi,\mu}(A).$$

So functions $t \mapsto \overline{P}^{\Phi^{q,t}_{\varphi,\psi,\mu}}(A)$ and $t \mapsto P^{\Phi^{q,t}_{\varphi,\psi,\mu}}(A)$ are decreasing.

(3) According to the definition and the results above, this is obvious.

(4) Let $\alpha \in [0,1]$ and $p,q \ge 0$.

(a) Put $t > \mathbf{\Delta}^q_{\varphi,\psi,\mu}(A)$ and $s > \mathbf{\Delta}^p_{\varphi,\psi,\mu}(A)$. So

$$\overline{\mathbf{P}}^{q,t}_{\varphi,\psi,\mu}(A) = 0$$

and

$$\overline{\mathbf{P}}^{p,s}_{\varphi,\psi,\mu}(A) = 0.$$

Next, let $r > 0$ and $\{B(x_i,r)\}_i$ be a centered packing of $A$. According to

$$\varphi^{-1}(x+y) \le \varphi^{-1}(x)\varphi^{-1}(y), \text{ for all } x,y \in [0,+\infty),$$
$$\varphi^{-1}(\beta x) \le (\varphi^{-1}(x))^\beta, \text{ for all } 0 < \beta < 1$$

and the Hölder inequality, we have inequality:

$$\begin{aligned}
&\sum_i \varphi^{-1}\big((\alpha q + (1-\alpha)p)\varphi(\mu(B(x_i,r)))\big)\\
&\le \sum_i \varphi^{-1}\big(\alpha q\varphi(\mu(B(x_i,r)))\big)\varphi^{-1}\big((1-\alpha)p\varphi(\mu(B(x_i,r)))\big)\\
&\le \left(\sum_i \varphi^{-1}(q\varphi(\mu(B(x_i,r))))\right)^\alpha \left(\sum_i \varphi^{-1}(p\varphi(\mu(B(x_i,r))))\right)^{1-\alpha}\\
&\le \big(D^q_{\varphi,\mu,r}(A)\big)^\alpha \big(D^p_{\varphi,\mu,r}(A)\big)^{1-\alpha}.
\end{aligned}$$

This shows that

$$\begin{aligned}
&D^{\alpha q+(1-\alpha)p}_{\varphi,\mu,r}(A)\,\varphi^{-1}\big((\alpha t + (1-\alpha)s)\psi(r)\big)\\
&\qquad \le \big(D^q_{\varphi,\mu,r}(A)\,\varphi^{-1}(t\psi(r))\big)^\alpha \big(D^p_{\varphi,\mu,r}(A)\,\varphi^{-1}(s\psi(r))\big)^{1-\alpha}.
\end{aligned}$$

Letting $r$ tend to 0, one has

$$\overline{\mathbf{P}}^{\alpha q+(1-\alpha)p,\alpha t+(1-\alpha)s}_{\varphi,\psi,\mu}(A) \leq \left(\overline{\mathbf{P}}^{q,t}_{\varphi,\psi,\mu}(A)\right)^{\alpha}\left(\overline{\mathbf{P}}^{p,s}_{\varphi,\psi,\mu}(A)\right)^{1-\alpha} = 0. \tag{2.3}$$

Consequently, for all $\alpha \in [0,1]$, $t > \mathbf{\Delta}^{q}_{\varphi,\psi,\mu}(A)$ and $s > \mathbf{\Delta}^{p}_{\varphi,\psi,\mu}(A)$, we have

$$\mathbf{\Delta}^{\alpha q+(1-\alpha)p}_{\varphi,\psi,\mu}(A) \leq \alpha t + (1-\alpha)s.$$

We conclude that

$$\mathbf{\Delta}^{\alpha q+(1-\alpha)p}_{\varphi,\psi,\mu}(A) \leq \alpha\mathbf{\Delta}^{q}_{\varphi,\psi,\mu}(A) + (1-\alpha)\mathbf{\Delta}^{p}_{\varphi,\psi,\mu}(A).$$

(b) Let $\epsilon > 0$ and $t = \mathbf{B}^{q}_{\varphi,\psi,\mu}(A)$ and $s = \mathbf{B}^{p}_{\varphi,\psi,\mu}(A)$. Hence

$$\mathbf{P}^{q,t+\epsilon}_{\varphi,\psi,\mu}(A) = \mathbf{P}^{p,s+\epsilon}_{\varphi,\psi,\mu}(A) = 0.$$

We can assume that $\{H_i\}_i$ and $\{G_i\}_i$ are two covers of $A$, such that

$$\sum_i \overline{\mathbf{P}}^{q,t+\epsilon}_{\varphi,\psi,\mu}(H_i) \leq 1 \quad \text{and} \quad \sum_i \overline{\mathbf{P}}^{p,s+\epsilon}_{\varphi,\psi,\mu}(G_i) \leq 1.$$

Next, for $n \in \mathbb{N}^*$, put $A_n = \bigcup_{i,j=1}^{n} (H_i \cap G_j)$, according to the inequality (2.3) and the Hölder inequality, we have

$$\begin{aligned}
\mathbf{P}^{\alpha q+(1-\alpha)p,\alpha t+(1-\alpha)s+\epsilon}_{\varphi,\psi,\mu}(A_n) &\leq \sum_{i,j=1}^{n} \mathbf{P}^{\alpha q+(1-\alpha)p,\alpha t+(1-\alpha)s+\epsilon}_{\varphi,\psi,\mu}(H_i \cap G_j) \\
&\leq \sum_{i,j=1}^{n} \overline{\mathbf{P}}^{\alpha q+(1-\alpha)p,\alpha t+(1-\alpha)s+\epsilon}_{\varphi,\psi,\mu}(H_i \cap G_j) \\
&\leq \sum_{i,j=1}^{n} \left(\overline{\mathbf{P}}^{q,t+\epsilon}_{\varphi,\psi,\mu}(H_i \cap G_j)\right)^{\alpha}\left(\overline{\mathbf{P}}^{p,s+\epsilon}_{\varphi,\psi,\mu}(H_i \cap G_j)\right)^{1-\alpha} \\
&\leq \left(\sum_{i,j=1}^{n} \overline{\mathbf{P}}^{q,t+\epsilon}_{\varphi,\psi,\mu}(H_i \cap G_j)\right)^{\alpha}\left(\sum_{i,j=1}^{n} \overline{\mathbf{P}}^{p,s+\epsilon}_{\varphi,\psi,\mu}(H_i \cap G_j)\right)^{1-\alpha} \\
&\leq \left(\sum_{i,j=1}^{n} \overline{\mathbf{P}}^{q,t+\epsilon}_{\varphi,\psi,\mu}(H_i)\right)^{\alpha}\left(\sum_{i,j=1}^{n} \overline{\mathbf{P}}^{p,s+\epsilon}_{\varphi,\psi,\mu}(G_j)\right)^{1-\alpha} \\
&\leq \left(n\sum_{i=1}^{n} \overline{\mathbf{P}}^{q,t+\epsilon}_{\varphi,\psi,\mu}(H_i)\right)^{\alpha}\left(n\sum_{j=1}^{n} \overline{\mathbf{P}}^{p,s+\epsilon}_{\varphi,\psi,\mu}(G_j)\right)^{1-\alpha} \\
&\leq n^{\alpha}\, n^{1-\alpha} \\
&= n < \infty,
\end{aligned}$$

which yields that

$$\mathbf{B}^{\alpha q+(1-\alpha)p}_{\varphi,\psi,\mu}(A_n) \leq \alpha t + (1-\alpha)s + \epsilon.$$

Since $A \subseteq \bigcup_n A_n$, we can conclude that

$$\begin{aligned}\mathbf{B}^{\alpha q+(1-\alpha)p}_{\varphi,\psi,\mu}(A) &\le \mathbf{B}^{\alpha q+(1-\alpha)p}_{\varphi,\psi,\mu}\Big(\bigcup_n A_n\Big)\\ &\le \sup_n \mathbf{B}^{\alpha q+(1-\alpha)p}_{\varphi,\psi,\mu}(A_n)\\ &\le \alpha\mathbf{B}^{q}_{\varphi,\psi,\mu}(A) + (1-\alpha)\mathbf{B}^{p}_{\varphi,\psi,\mu}(A) + \epsilon.\end{aligned}$$

Letting $\epsilon$ tend to 0, so

$$\mathbf{B}^{\alpha q+(1-\alpha)p}_{\varphi,\psi,\mu}(A) \le \alpha\mathbf{B}^{q}_{\varphi,\psi,\mu}(A) + (1-\alpha)\mathbf{B}^{p}_{\varphi,\psi,\mu}(A).$$

(5) According to the definition and above assertions, this result is obvious.

□

PROPOSITION 2.3. *Let $\mu \in \mathscr{P}^c(\mathbb{R}^n)$, $(\varphi,\psi) \in P$ with $\varphi^{-1}(0) = a > 0$, $r > 0$, $q \ge 0$ and $A \subseteq$ **supp** $\mu$, we have*

$$\kappa^q_{\varphi,\psi,\mu}(A) = \liminf_{r\to 0} \frac{\varphi\left(\frac{1}{O^q_{\varphi,\mu,r}(A)}\right)}{\psi(r)} \qquad \textit{and} \qquad \mathbf{\Delta}^q_{\varphi,\psi,\mu}(A) = \limsup_{r\to 0} \frac{\varphi\left(\frac{1}{D^q_{\varphi,\mu,r}(A)}\right)}{\psi(r)}.$$

*In addition,*

$$\mathscr{S}^q_{\varphi,\psi,\mu}(A) = \sup_{F\subseteq A} \kappa^q_{\varphi,\psi,\mu}(F).$$

PROOF. Assuming that

$$\liminf_{r\to 0} \frac{\varphi\left(\frac{1}{O^q_{\varphi,\mu,r}(A)}\right)}{\psi(r)} > \kappa^q_{\varphi,\psi,\mu}(A) + \epsilon, \quad \text{for some } \epsilon > 0.$$

Then we can find $r_0 > 0$ such that

$$1 < O^q_{\varphi,\mu,r}(A)\,\varphi^{-1}\big((\kappa^q_{\varphi,\psi,\mu}(A) + \epsilon)\psi(r)\big),$$

for $0 < r < r_0$, so

$$1 \le \mathbf{I}^{q,\kappa^q_{\varphi,\psi,\mu}(A)+\epsilon}_{\varphi,\psi,\mu} = 0,$$

which is a contradiction. Hence

$$\liminf_{r\to 0} \frac{\varphi\left(\frac{1}{O^q_{\varphi,\mu,r}(A)}\right)}{\psi(r)} \le \kappa^q_{\varphi,\psi,\mu}(A) + \epsilon, \quad \text{for any } \epsilon > 0,$$

and then

$$\kappa^q_{\varphi,\psi,\mu}(A) = \liminf_{r\to 0} \frac{\varphi\left(\frac{1}{O^q_{\varphi,\mu,r}(A)}\right)}{\psi(r)}.$$

Similarly, we have

$$\mathbf{\Delta}^q_{\varphi,\psi,\mu}(A) = \limsup_{r\to 0} \frac{\varphi\left(\frac{1}{D^q_{\varphi,\mu,r}(A)}\right)}{\psi(r)}.$$

In addition,

$$\mathscr{S}^q_{\varphi,\psi,\mu}(A) = \sup_{F\subseteq A} \kappa^q_{\varphi,\psi,\mu}(F),$$

which is obvious by definition. □

PROPOSITION 2.4. *Let $\mu \in \mathscr{P}^c(\mathbb{R}^n)$, $(\varphi, \psi) \in P$ with $\varphi^{-1}(0) = a > 0$, $q \geq 0$ and $A \subseteq \mathbf{supp}\,\mu$, we have*

$$\mathbf{b}^q_{\varphi,\psi,\mu}(A) = \inf\left\{ \sup_i \mathscr{S}^q_{\varphi,\psi,\mu}(A_i) \;\middle|\; A \subseteq \bigcup_i A_i \right\},$$

*and*

$$\mathbf{B}^q_{\varphi,\psi,\mu}(A) = \inf\left\{ \sup_i \mathbf{\Delta}^q_{\varphi,\psi,\mu}(A_i) \;\middle|\; A \subseteq \bigcup_i A_i \right\}.$$

PROOF. Put

$$\gamma := \inf\left\{ \sup_i \mathscr{S}^q_{\varphi,\psi,\mu}(A_i) \;\middle|\; A \subseteq \bigcup_i A_i \right\}.$$

If $\gamma < \mathbf{b}^q_{\varphi,\psi,\mu}(A)$, we can take $\alpha \in (\gamma, \mathbf{b}^q_{\varphi,\psi,\mu}(A))$. Then there is a cover $\{A_i\}_i$ of $A$ that satisfies

$$\sup_i \mathscr{S}^q_{\varphi,\psi,\mu}(A_i) < \alpha.$$

This implies that

$$\overline{\mathbf{H}}^{q,\alpha}_{\varphi,\psi,\mu}(A_i) = 0 \quad \text{for all } i \Rightarrow \mathbf{H}^{q,\alpha}_{\varphi,\psi,\mu}(A) = 0,$$

which is a contradiction.

If $\mathbf{b}^q_{\varphi,\psi,\mu}(A) < \gamma$, take $\alpha \in (\mathbf{b}^q_{\varphi,\psi,\mu}(A), \gamma)$. One has

$$\mathbf{H}^{q,\alpha}_{\varphi,\psi,\mu}(A) = 0.$$

Then there exists $\{A_i\}_i$ with $A \subseteq \bigcup_i A_i$, such that $\sup_i \overline{\mathbf{H}}^{q,\alpha}_{\varphi,\psi,\mu}(A_i) < \infty$. Hence,

$$\sup_i \mathscr{S}^q_{\varphi,\psi,\mu}(A_i) \leq \alpha,$$

which is also a contradiction. In summary, $\beta = \mathbf{b}^q_{\varphi,\psi,\mu}(A)$.

Similarly, we have

$$\mathbf{B}^q_{\varphi,\psi,\mu}(A) = \inf\left\{ \sup_i \mathbf{\Delta}^q_{\varphi,\psi,\mu}(A_i) \;\middle|\; A \subseteq \bigcup_i A_i \right\}.$$

□

## 3. A general multifractal formalism for $(\varphi, \psi)$Hewitt-Stromberg measures

For $\mu \in \mathscr{P}^c(\mathbb{R}^n)$, $\delta > 0$ and $(\varphi, \psi) \in P$ with $\varphi^{-1}(0) = a > 0$, consider also generalized local dimensions of the measure $\mu$ at $x$ which defined by

$$\underline{\alpha}^{\varphi,\psi}_\mu(x) = \liminf_{\delta \to 0} \frac{\varphi(\mu(B(x,\delta)))}{\psi(\delta)}, \quad \overline{\alpha}^{\varphi,\psi}_\mu(x) = \limsup_{\delta \to 0} \frac{\varphi(\mu(B(x,\delta)))}{\psi(\delta)}.$$

For $\alpha, \beta \geq 0$, put

$$\underline{X}_\alpha = \{x \in \mathbf{supp}\,\mu \mid \underline{\alpha}^{\varphi,\psi}_\mu(x) \geq \alpha\}, \qquad \overline{X}^\beta = \{x \in \mathbf{supp}\,\mu \mid \overline{\alpha}^{\varphi,\psi}_\mu(x) \leq \beta\},$$
$$\underline{X}^\alpha = \{x \in \mathbf{supp}\,\mu \mid \underline{\alpha}^{\varphi,\psi}_\mu(x) \leq \alpha\}, \qquad \overline{X}_\beta = \{x \in \mathbf{supp}\,\mu \mid \overline{\alpha}^{\varphi,\psi}_\mu(x) \geq \beta\},$$

and

$$X(\alpha, \beta) = \underline{X}_\alpha \cap \overline{X}^\beta, \quad X(\alpha) = \underline{X}_\alpha \cap \overline{X}^\alpha.$$

This section will demonstrate the multifractal formalism based on the measures $\mathbf{H}^{q,t}_{\varphi,\psi,\mu}$, $\mathbf{P}^{q,t}_{\varphi,\psi,\mu}$ and their dimension functions. The multifractal spectrum functions:

$$\alpha \mapsto \mathsf{f}^{\varphi,\psi}_{\mu}(\alpha) := \underline{\dim}^{MB}_{\varphi,\psi}(X(\alpha))$$

and

$$\alpha \mapsto \mathsf{F}^{\varphi,\psi}_{\mu}(\alpha) := \overline{\dim}^{MB}_{\varphi,\psi}(X(\alpha)).$$

The following results will give some close relationship between the spectral functions ($\mathsf{f}^{\varphi,\psi}_{\mu}$ and $\mathsf{F}^{\varphi,\psi}_{\mu}$) and the dimensional functions ($\mathbf{b}_{\varphi,\psi,\mu}$ and $\mathbf{B}_{\varphi,\psi,\mu}$), and the dimension function $\mathbf{\Delta}_{\varphi,\psi,\mu}$ is related to the upper box spectrum. Therefore, if

$$\underline{\dim}^{MB}_{\varphi,\psi}(X(\alpha)) = \overline{\dim}^{MB}_{\varphi,\psi}(X(\alpha)) = \mathbf{b}^*_{\varphi,\psi,\mu}(\alpha) = \mathbf{B}^*_{\varphi,\psi,\mu}(\alpha) \quad \text{for some } \alpha \geq 0,$$

we say that the Hewitt-Stromberg-based generic multifractal formalism is valid and holds true.

Let

$$\alpha_{min} = \sup_{q>0} -\frac{\mathbf{b}_{\varphi,\psi,\mu}(q)}{q} \quad \text{and} \quad \beta_{min} = \sup_{q>0} -\frac{\mathbf{B}_{\varphi,\psi,\mu}(q)}{q}.$$

THEOREM 3.1. *Let $\alpha \geq 0$, $\mu \in \mathscr{P}^c(\mathbb{R}^n)$, $A \subseteq \mathbf{supp}\,\mu$ and $(\varphi,\psi) \in P$ with $\varphi^{-1}(0) = a > 0$, if $\varphi^{-1}(x+y) \leq C\,\varphi^{-1}(x)\varphi^{-1}(y)$, for all $x, y \in [0,+\infty)$ and for some positive constant $C$, then we have*

(1)

$$\alpha_{min} \leq \inf_{x\in A} \overline{\alpha}^{\varphi,\psi}_{\mu}(x) \leq \sup_{x\in A} \overline{\alpha}^{\varphi,\psi}_{\mu}(x)$$

*and*

$$\beta_{min} \leq \inf_{x\in A} \underline{\alpha}^{\varphi,\psi}_{\mu}(x) \leq \sup_{x\in A} \underline{\alpha}^{\varphi,\psi}_{\mu}(x).$$

(2)

$$\underline{\dim}^{MB}_{\varphi,\psi}(X(\alpha)) \begin{cases} = 0 & \text{if } \alpha < \alpha_{min} \\ \leq \mathbf{b}^*_{\varphi,\psi,\mu}(\alpha) & \text{if } \alpha \geq \alpha_{min}. \end{cases}$$

(3)

$$\overline{\dim}^{MB}_{\varphi,\psi}(X(\alpha)) \begin{cases} = 0 & \text{if } \alpha < \beta_{min} \\ \leq \mathbf{B}^*_{\varphi,\psi,\mu}(\alpha) & \text{if } \alpha \geq \beta_{min}. \end{cases}$$

The dimensions of the sets mentioned in Theorem 3.1 are more challenging to minorate. The major goal of the studies that follow is to suggest a sufficient condition that yields the lower bound. For $n \in \mathbb{N}^*$, $\delta, q > 0$ and $\alpha \geq 0$, put

$$X_n(\alpha) = \left\{ x \in \overline{X}^{\alpha} \;\middle|\; \frac{\varphi((\mu(B(x,r)))}{\psi(r)} \leq \alpha + \frac{\delta}{q} \text{ for } 0 < r < \frac{1}{n} \right\},$$

$$Y_n(\alpha) = \left\{ x \in \underline{X}_{\alpha} \;\middle|\; \frac{\varphi((\mu(B(x,r)))}{\psi(r)} \geq \alpha - \frac{\delta}{q} \text{ for } 0 < r < \frac{1}{n} \right\}$$

and

$$\mathbf{X}_n(\alpha) = X_n(\alpha) \cap Y_n(\alpha).$$

It is clear that

$$\overline{X}(\alpha) = \bigcup_n X_n(\alpha),$$

$$\underline{X}(\alpha) = \bigcup_n Y_n(\alpha)$$

and

$$X(\alpha) = \bigcup_n \mathbf{X}_n(\alpha).$$

THEOREM 3.2. *Let $\mu \in \mathscr{P}^c(\mathbb{R}^n)$, $(\varphi, \psi) \in P$ with $\varphi^{-1}(0) = a > 0$, $q, t \geq 0$ and $\alpha > 0$ such that $\alpha q + t \geq 0$. For Borel set $A$, if $\varphi^{-1}(x+y) \geq C' \varphi^{-1}(x)\varphi^{-1}(y)$, for all $x, y \in [0, +\infty)$ and some positive constant $C'$, then*

(1) *If $A \subseteq X(\alpha)$ with $\mathbf{H}^{q,t}_{\varphi,\psi,\mu}(A) > 0$, then*

$$\underline{\dim}^{MB}_{\varphi,\psi}(X(\alpha)) \geq \alpha q + t.$$

*In addition, if $\mathbf{b}_{\varphi,\psi,\mu}$ is differentiable at $q$, we suppose that $\mathbf{b}^*_{\varphi,\psi,\mu}(-\mathbf{b}'_{\varphi,\psi,\mu}(q)) \geq 0$ and $\mathbf{H}^{q,\mathbf{b}_{\varphi,\psi,\mu}(q)}_{\varphi,\psi,\mu}(X(-\mathbf{b}'_{\varphi,\psi,\mu}(q))) > 0$, so*

$$\underline{\dim}^{MB}_{\varphi,\psi}\left(X\left(-\mathbf{b}'_{\varphi,\psi,\mu}(q)\right)\right) \geq \mathbf{b}^*_{\varphi,\psi,\mu}\left(-\mathbf{b}'_{\varphi,\psi,\mu}(q)\right).$$

(2) *If there exists $n \in \mathbb{N}^*$ such that $A \subseteq \mathbf{X}_n(\alpha)$ and $\mathbf{P}^{q,t}_{\varphi,\psi,\mu}(A) > 0$, then*

$$\overline{\dim}^{MB}_{\varphi,\psi}(X(\alpha)) \geq \alpha q + t.$$

*In particular, if $\mathbf{B}_{\varphi,\psi,\mu}$ is differentiable at $q$, then, suppose*

$$\mathbf{B}^*_{\varphi,\psi,\mu}\left(-\mathbf{B}'_{\varphi,\psi,\mu}(q)\right) \geq 0$$

*and*

$$\mathbf{P}^{q,\mathbf{B}_{\varphi,\psi,\mu}(q)}_{\varphi,\psi,\mu}(\mathbf{X}_n(-\mathbf{B}'_{\varphi,\psi,\mu}(q))) > 0,$$

*for some $n \in \mathbb{N}^*$. We have*

$$\overline{\dim}^{MB}_{\varphi,\psi}\left(X\left(-\mathbf{B}'_{\varphi,\psi,\mu}(q)\right)\right) \geq \mathbf{B}^*_{\mu,h,hg}\left(-\mathbf{B}'_{\varphi,\psi,\mu}(q)\right).$$

THEOREM 3.3. *Let $\mu \in \mathscr{P}^c(\mathbb{R}^n)$, $(\varphi, \psi) \in P$ with $\varphi^{-1}(0) = a > 0$ and $q, t \geq 0$, if $\mathbf{H}^{q,\mathbf{B}_{\varphi,\psi,\mu}(q)}_{\varphi,\psi,\mu}(\mathbf{supp}\,\mu) > 0$ and $\varphi^{-1}(x+y) \leq C\varphi^{-1}(x)\varphi^{-1}(y)$, for all $x, y \in [0, +\infty)$ and some positive constant $C$, then*

$$\underline{\dim}^{MB}_{\varphi,\psi}(X(-\mathbf{B}'_{\varphi,\psi,\mu_+}(q), -\mathbf{B}'_{\varphi,\psi,\mu_-}(q))) \geq -q\mathbf{B}'_{\varphi,\psi,\mu_+}(q) + \mathbf{B}_{\varphi,\psi,\mu}(q).$$

*In particular, if $\mathbf{B}_{\varphi,\psi,\mu}$ is differentiable at $q$, put $\alpha = -\mathbf{B}'_{\varphi,\psi,\mu}(q)$ one has*

$$\underline{\dim}^{MB}_{\varphi,\psi}(X(\alpha)) = \overline{\dim}^{MB}_{\varphi,\psi}(X(\alpha)) = \mathbf{B}^*_{\varphi,\psi,\mu}(\alpha) = \mathbf{b}^*_{\varphi,\psi,\mu}(\alpha).$$

*Conversely, if*

$$\underline{\dim}^{MB}_{\varphi,\psi}(X(\alpha)) \geq \mathbf{B}^*_{\varphi,\psi,\mu}(\alpha),$$

*then*

$$\mathbf{b}_{\varphi,\psi,\mu}(q) = \mathbf{B}_{\varphi,\psi,\mu}(q).$$

This theorem gives a new sufficient condition for multifractal formalism. In addition, calling $\mathbf{b}_{\varphi,\psi,\mu}(q) = \mathbf{B}_{\varphi,\psi,\mu}(q)$ is an analogy of Taylor's regularity requirement, and at the same time, when $\mathbf{B}'_{\varphi,\psi,\mu}(q)$ exists, it is also a prerequisite for the validity of multifractal formalism. However, we are not sure whether the weaker condition holds true for the first half of Theorem 3.3.

## 4. Proof of Theorems 3.1 and 3.2

The following lemmas are important for proving Theorems 3.1 and 3.2.

LEMMA 4.1. *Let $\mu \in \mathscr{P}^c(\mathbb{R}^n)$, $(\varphi, \psi) \in P$ with $\varphi^{-1}(0) = a > 0$ and $\alpha > 0$ if $\varphi^{-1}(x+y) \le C\,\varphi^{-1}(x)\varphi^{-1}(y)$, for all $x, y \in [0, +\infty)$ and some positive constant $C$, then*

$$\underline{X}^\alpha = \emptyset \quad \textit{for} \quad \alpha < \beta_{\min} \quad \textit{and} \quad \overline{X}^\alpha = \emptyset \quad \textit{for} \quad \alpha < \alpha_{\min}.$$

PROOF. Fix $x \in \underline{X}^\alpha$ and $\alpha < \beta_{\min}$. Then, there exist $\epsilon > 0$ and $q > 0$, such that

$$-q(\alpha + \epsilon) > \mathbf{B}_{\varphi,\psi,\mu}(q).$$

We can choose a sequence $(r_k)$ such that $r_k \searrow 0$, $0 < r_k < \frac{1}{k}$ and

$$\frac{\varphi(\mu(B(x, r_k))}{\psi(r_k)} < \alpha + \epsilon.$$

Let $t = -q(\alpha + \epsilon)$. Since there is a real number $a > 0$ that satisfies $\varphi^{-1}(0) = a$, and $\varphi^{-1}(x+y) \le C\,\varphi^{-1}(x)\varphi^{-1}(y)$, then

$$\begin{aligned}
\overline{\mathbf{P}}^{q,t}_{\varphi,\psi,\mu}(\{x\}) &\ge \varphi^{-1}\left(q\varphi(\mu(B(x, r_k)))\right)\varphi^{-1}\left(t\psi(r_k)\right)\\
&\ge \varphi^{-1}\left(q(\alpha+\epsilon)\psi(r_k)\right)\varphi^{-1}\left(t\psi(r_k)\right)\\
&\ge \frac{1}{C}\varphi^{-1}\left((q(\alpha+\epsilon)+t)\psi(r_k)\right)\\
&= \frac{1}{C}\varphi^{-1}(0)\\
&= \frac{a}{C} > 0,
\end{aligned}$$

which implies that

$$\mathbf{P}^{q,t}_{\varphi,\psi,\mu}(\{x\}) \ge \frac{a}{C} > 0 \Longrightarrow t \le \mathbf{B}^q_{\varphi,\psi,\mu}(\{x\}) \le \mathbf{B}_{\varphi,\psi,\mu}(q),$$

which contradicts the fact that $-q(\alpha+\epsilon) > \mathbf{B}_{\varphi,\psi,\mu}(q)$.

Similarly, for $\alpha < \alpha_{\min}$, one has

$$\overline{X}^\alpha = \emptyset.$$

□

LEMMA 4.2. *Let $\mu \in \mathscr{P}^c(\mathbb{R}^n)$, $(\varphi, \psi) \in P$ with $\varphi^{-1}(0) = a > 0$, $\alpha, q, t \ge 0$ and $\delta > 0$ such that $\delta \le \alpha q + t$. If $\varphi^{-1}(x+y) \le C\,\varphi^{-1}(x)\varphi^{-1}(y)$ for all $x, y \in [0, +\infty)$ and some positive constant $C$, then the following hold up:*

(1) (a) $\mathbf{H}^{\alpha q+t+\delta}_{\varphi,\psi}(\overline{X}^\alpha) \le C\,\mathbf{H}^{q,t}_{\varphi,\psi,\mu}(\overline{X}^\alpha)$
 (b) *If* $0 \le \alpha q + \mathbf{b}_{\varphi,\psi,\mu}(q)$, *then*

$$\underline{\dim}^{MB}_{\varphi,\psi}(\overline{X}^\alpha) \le \inf_q \alpha q + \mathbf{b}_{\varphi,\psi,\mu}(q).$$

*In particular, we have* $\underline{\dim}^{MB}_{\varphi,\psi}(\overline{X}^\alpha) \le \alpha$.

(2) *Let $n \in \mathbb{N}^*$, we have*
 (a)

$$\mathbf{P}^{\alpha q+t+\delta}_{\varphi,\psi}(\overline{X}^\alpha) \le C\,\mathbf{P}^{q,t}_{\varphi,\psi,\mu}(\overline{X}^\alpha) \qquad \textit{for } q = 0.$$

$$\mathbf{P}^{\alpha q+t+\delta}_{\varphi,\psi}(X_n(\alpha)) \le C\,\mathbf{P}^{q,t}_{\varphi,\psi,\mu}(\overline{X}^\alpha) \qquad \textit{for } 0 < q.$$

(b) $\mathbf{P}_{\varphi,\psi}^{\alpha q+t+\delta}(\underline{X}_\alpha) \le C\,\mathbf{P}_{\varphi,\psi,\mu}^{q,t}(\underline{X}_\alpha)$ *for* $q=0$.

(c) *If* $0 \le \alpha q + \mathbf{B}_{\varphi,\psi,\mu}(q)$, *then*

$$\overline{\dim}_{\varphi,\psi}^{MB}(\overline{X}^\alpha) \le \inf_q \alpha q + \mathbf{B}_{\varphi,\psi,\mu}(q).$$

*In particular, we have* $\overline{\dim}_{\varphi,\psi}^{MB}(\overline{X}^\alpha) \le \alpha$.

PROOF.

(1) (a) The statement holds true when $q=0$, so we can assume $q>0$. For any $n \in \mathbb{N}^*$, express as follows

$$X_n(\alpha) = \left\{ x \in \overline{X}^\alpha \;\middle|\; \frac{\varphi(\mu(B(x,r)))}{\psi(r)} \le \alpha + \frac{\delta}{q} \text{ for } 0<r<\frac{1}{n} \right\}.$$

Fix $n \in \mathbb{N}^*$ and $0 < r < \frac{1}{n}$. Let $\{B(x_i,r)\}_i$ be a centered covering of $A \subseteq X_n(\alpha)$. Since $\varphi^{-1}(x+y) \le C\,\varphi^{-1}(x)\varphi^{-1}(y)$, so we have

$$\begin{aligned}
\frac{\varphi(\mu(B(x_i,r)))}{\psi(r)} \le \alpha + \frac{\delta}{q} &\Longrightarrow q\varphi(\mu(B(x,r)) \le (\alpha q+\delta)\psi(r) \\
&\Longrightarrow \sum_i \varphi^{-1}\left(q\varphi(\mu(B(x,r))\right) \ge O_r(A)\,\varphi^{-1}\left((\alpha q+\delta)\psi(r)\right) \\
&\Longrightarrow O_{\varphi,\mu,r}^q(A) \ge O_r(A)\,\varphi^{-1}\left((\alpha q+\delta)\psi(r)\right) \\
&\Longrightarrow O_{\varphi,\mu,r}^q(A)\varphi^{-1}(t\psi(r)) \ge \frac{1}{C} O_r(A)\,\varphi^{-1}\left((\alpha q+\delta+t)\psi(r)\right) \\
&\Longrightarrow C\,\mathbf{I}_{\varphi,\psi,\mu}^{q,t}(A) \ge \mathbf{I}_{\varphi,\psi}^{\alpha q+\delta+t}(A).
\end{aligned}$$

It follows that

$$C\,\overline{\mathbf{H}}_{\varphi,\psi,\mu}^{q,t}(A) \ge \overline{\mathbf{H}}_{\varphi,\psi}^{\alpha q+\delta+t}(A).$$

Put $X_n(\alpha) \subseteq \bigcup_i F_i$, so

$$\begin{aligned}
\mathbf{H}_{\varphi,\psi}^{\alpha q+\delta+t}(X_n(\alpha)) &= \mathbf{H}_{\varphi,\psi}^{\alpha q+\delta+t}\Big(\bigcup_i (X_n(\alpha)\cap F_i)\Big) \\
&\le \sum_i \mathbf{H}_{\varphi,\psi}^{\alpha q+\delta+t}\Big(X_n(\alpha)\cap F_i\Big) \\
&\le \sum_i \overline{\mathbf{H}}_{\varphi,\psi}^{\alpha q+\delta+t}\Big(X_n(\alpha)\cap F_i\Big) \\
&\le C\sum_i \overline{\mathbf{H}}_{\varphi,\psi,\mu}^{q,t}\Big(X_n(\alpha)\cap F_i\Big) \\
&\le C\sum_i \overline{\mathbf{H}}_{\varphi,\psi,\mu}^{q,t}\Big(F_i\Big)
\end{aligned}$$

which implies that

$$C\,\mathbf{H}_{\varphi,\psi,\mu}^{q,t}(X_n(\alpha)) \ge \mathbf{H}_{\varphi,\psi}^{q\alpha+\delta+t}(X_n(\alpha)).$$

Finally, since $\overline{X}^\alpha = \bigcup_n X_n(\alpha)$, we get

$$C\,\mathbf{H}_{\varphi,\psi,\mu}^{q,t}(\overline{X}^\alpha) \ge \mathbf{H}_{\varphi,\psi}^{q\alpha+\delta+t}(\overline{X}^\alpha).$$

(b) It is easy to derive from (a), so this proof is omitted.

(2) (a) The result is true for $q=0$. Hence for $q>0$ and $n\in\mathbb{N}^*$, consider

$$X_n(\alpha)=\left\{x\in\overline{X}^{\alpha}\ \Big|\ \frac{\varphi(\mu(B(x,r)))}{\psi(r)}\le\alpha+\frac{\delta}{q}\ \text{for } 0<r<\frac{1}{n}\right\}.$$

Next fix $n\in\mathbb{N}^*$ and $0<r<\dfrac{1}{n}$. Let $\{B(x_i,r)\}_i$ be a centered packing of $F\subseteq X_n(\alpha)$. Since $\varphi^{-1}(x+y)\le C\,\varphi^{-1}(x)\varphi^{-1}(y)$, then one has

$$\begin{aligned}\frac{\varphi(\mu(B(x_i,r)))}{\psi(r)}\le\alpha+\frac{\delta}{q}&\Longrightarrow q\varphi(\mu(B(x_i,r))\le(\alpha q+\delta)\psi(r)\\&\Longrightarrow\sum_i\varphi^{-1}\left(q\varphi(\mu(B(x_i,r)))\ge D_r(F)\,\varphi^{-1}\left((\alpha q+\delta)\psi(r)\right)\right.\\&\Longrightarrow D^q_{\varphi,\mu,r}(F)\,\varphi^{-1}\left(t\psi(r)\right)\ge\frac{1}{C}D_r(F)\,\varphi^{-1}\left((\alpha q+\delta+t)\psi(r)\right)\\&\Longrightarrow C\,\overline{\mathbf{P}}^{q,t}_{\varphi,\psi,\mu}(F)\ge\overline{\mathbf{P}}^{\alpha q+\delta+t}_{\varphi,\psi}(F).\end{aligned}$$

Consequently,

$$C\,\overline{\mathbf{P}}^{q,t}_{\varphi,\psi,\mu}(F)\ge\overline{\mathbf{P}}^{\alpha q+\delta+t}_{\varphi,\psi}(F),\quad\text{for all } F\subseteq X_n(\alpha).$$

Next for any cover $\{F_i\}_i$ of $X_n(\alpha)$, one gets

$$\begin{aligned}\mathbf{P}^{q\alpha+\delta+t}_{\varphi,\psi}(X_n(\alpha))&\le\sum_i\mathbf{P}^{q\alpha+\delta+t}_{\varphi,\psi}(X_n(\alpha)\cap F_i)\\&\le\sum_i\overline{\mathbf{P}}^{q\alpha+\delta+t}_{\varphi,\psi}\left(X_n(\alpha)\cap F_i\right)\\&\le C\sum_i\overline{\mathbf{P}}^{q,t}_{\varphi,\psi,\mu}\left(X_n(\alpha)\cap F_i\right)\\&\le C\sum_i\overline{\mathbf{P}}^{q,t}_{\varphi,\psi,\mu}\left(F_i\right).\end{aligned}$$

It results that

$$\mathbf{P}^{\alpha q+t+\delta}_{\varphi,\psi}(X_n(\alpha))\le C\,\mathbf{P}^{q,t}_{\varphi,\psi,\mu}(X_n(\alpha)).$$

However, since $\overline{X}^{\alpha}=\bigcup_n X_n(\alpha)$, we have

$$C\,\mathbf{P}^{q,t}_{\varphi,\psi,\mu}(\overline{X}^{\alpha})\ge\mathbf{P}^{q\alpha+\delta+t}_{\varphi,\psi}(X_n(\alpha).$$

(b) For $q=0$, it is obvious.
(c) It is easy to derive from (a).

□

LEMMA 4.3. *Let $\mu\in\mathscr{P}^c(\mathbb{R}^n)$, $(\varphi,\psi)\in P$ with $\varphi^{-1}(0)=a>0$, $\alpha,q,t\ge0$ and $\delta>0$ such that $\delta\le\alpha q+t$. If $\varphi^{-1}(x+y)\ge C'\,\varphi^{-1}(x)\varphi^{-1}(y)$, for all $x,y\in[0,+\infty)$ and some positive constant $C'$, then*

(1) *For $n\in\mathbb{N}^*$, we have*
 (a) *If $A\subseteq\overline{X}^{\alpha}$, then $\mathbf{P}^{\alpha q+t-\delta}_{\varphi,\psi}(A)\ge C'\,\mathbf{P}^{q,t}_{\varphi,\psi,\mu}(A)$ for $q=0$.*
 (b) *If $A\subseteq\underline{X}_{\alpha}$, then $\mathbf{P}^{\alpha q+t-\delta}_{\varphi,\psi}(A)\ge C'\,\mathbf{P}^{q,t}_{\varphi,\psi,\mu}(A)$ for $q=0$.*
 *If $A\subseteq Y_n(\alpha)$, then $\mathbf{P}^{\alpha q+t-\delta}_{\varphi,\psi}(A)\ge C'\,\mathbf{P}^{q,t}_{\varphi,\psi,\mu}(A)$ for $0<q$.*
 *In particular, if $A$ is Borel with $\mu(A)>0$ then $\overline{\dim}^{MB}_{\varphi,\psi}(A)\ge\alpha$.*

(2) *If* $A \subseteq \underline{X}_\alpha$ *is Borel then* $\mathbf{H}^{\alpha q+t-\delta}_{\varphi,\psi}(A) \geq C'\,\mathbf{H}^{q,t}_{\varphi,\psi,\mu}(A)$ *for* $0 \leq q$.
*In particular, if* $\mu(A) > 0$ *then* $\underline{\dim}^{MB}_{\varphi,\psi}(A) \geq \alpha$.

PROOF.

(1) (a) We observe that the outcome holds when $q = 0$.
(b) Put $q, r > 0$, $F \subseteq A$ and $\{B(x_i, r)\}_i$ be a centered packing of $F$, Fix $n \in \mathbb{N}^*$ and $0 < r < \dfrac{1}{n}$. one has

$$\begin{aligned}\frac{\varphi(\mu(B(x_i,r)))}{\psi(r)} \geq \alpha - \frac{\delta}{q} &\Longrightarrow q\varphi(\mu(B(x_i,r))) \geq (\alpha q - \delta)\psi(r)\\ &\Longrightarrow \sum_i \varphi^{-1}\big(q\varphi(\mu(B(x_i,r)))\big) \leq D_r(F)\,\varphi^{-1}\big((\alpha q - \delta)\psi(r)\big)\\ &\Longrightarrow D^q_{\varphi,\mu,r}(F)\varphi^{-1}\big(t\psi(r)\big) \leq \frac{1}{C'}\,D_r(F)\,\varphi^{-1}\big((\alpha q + t - \delta)\psi(r)\big)\\ &\qquad \leq \frac{1}{C'}\overline{\mathbf{P}}^{\alpha q+t-\delta}_{\varphi,\psi}(F)\\ &\Longrightarrow \overline{\mathbf{P}}^{q,t}_{\varphi,\psi,\mu}(F) \leq \frac{1}{C'}\,\overline{\mathbf{P}}^{\alpha q+t-\delta}_{\varphi,\psi}(F),\end{aligned}$$

which means that

$$C'\,\overline{\mathbf{P}}^{q,t}_{\varphi,\psi,\mu}(F) \leq \overline{\mathbf{P}}^{\alpha q+t-\delta}_{\varphi,\psi}(F).$$

Now let $A \subseteq \bigcup_i A_i$, we have

$$\begin{aligned}C'\,\mathbf{P}^{q,t}_{\varphi,\psi,\mu}(A) &\leq C'\sum_i \mathbf{P}^{q,t}_{\varphi,\psi,\mu}(A \cap A_i)\\ &\leq C'\sum_i \overline{\mathbf{P}}^{q,t}_{\varphi,\psi,\mu}(A \cap A_i\big)\\ &\leq \sum_i \overline{\mathbf{P}}^{\alpha q-\delta+t}_{\varphi,\psi}\big(A \cap A_i\big)\\ &\leq \sum_i \overline{\mathbf{P}}^{\alpha q-\delta+t}_{\varphi,\psi}\big(A_i\big).\end{aligned}$$

We can deduce now that

$$C'\,\mathbf{P}^{q,t}_{\varphi,\psi,\mu}(A) \leq \mathbf{P}^{\alpha q+t-\delta}_{\varphi,\psi}(A).$$

(2) (a) The validity of the assertion for $q = 0$ is widely acknowledged. For any $n \in \mathbb{N}^*$, we contemplate

$$A_n = \left\{x \in A \;\middle|\; \frac{\varphi(\mu(B(x,r)))}{\psi(r)} \geq \alpha - \frac{\delta}{q} \text{ for } 0 < r < \frac{1}{n}\right\}.$$

Fix $n \in \mathbb{N}^*$, $F \subseteq A_n$ and $0 < r < \dfrac{1}{n}$. Let $\{B(x_i, r)\}_i$ be a centered covering of $F$, then we have

$$\begin{aligned}
\frac{\varphi(\mu(B(x_i,r)))}{\psi(r)} \geq \alpha - \frac{\delta}{q} &\Longrightarrow q\varphi(\mu(B(x_i,r))) \geq (\alpha q - \delta)\psi(r) \\
&\Longrightarrow \sum_i \varphi^{-1}\big(q\varphi(\mu(B(x_i,r)))\big) \leq O_r(F)\,\varphi^{-1}\big((\alpha q - \delta)\psi(r)\big) \\
&\Longrightarrow C'\,O^q_{\varphi,\mu,r}(F)\varphi^{-1}\big(t\psi(r)\big) \leq O_r(F)\,\varphi^{-1}\big((\alpha q + t - \delta)\psi(r)\big) \\
&\Longrightarrow C'\,\mathbf{I}^{q,t}_{\varphi,\psi,\mu}(F) \leq \mathbf{I}^{\alpha q + t - \delta}_{\varphi,\psi}(F).
\end{aligned}$$

Hence

$$C'\,\overline{\mathbf{H}}^{q,t}_{\varphi,\psi,\mu}(F) \leq \overline{\mathbf{H}}^{\alpha q + t - \delta}_{\varphi,\psi}(F).$$

Let $\{F_i\}_i$ be a cover of $A_n$. One gets

$$\begin{aligned}
C'\,\mathbf{H}^{q,t}_{\varphi,\psi,\mu}(A_n) &\leq C' \sum_i \mathbf{H}^{q,t}_{\varphi,\psi,\mu}(A_n \cap F_i) \\
&\leq C' \sum_i \overline{\mathbf{H}}^{q,t}_{\varphi,\psi,\mu}\big(A_n \cap F_i\big) \\
&\leq \sum_i \overline{\mathbf{H}}^{\alpha q - \delta + t}_{\varphi,\psi}\Big(A_n \cap F_i\Big) \\
&\leq \sum_i \overline{\mathbf{H}}^{\alpha q - \delta + t}_{\varphi,\psi}\Big(F_i\Big)
\end{aligned}$$

and which leads to

$$C'\,\mathbf{H}^{q,t}_{\varphi,\psi,\mu}(A_n) \leq \mathbf{H}^{\alpha q + t - \delta}_{\varphi,\psi}(A_n).$$

Finally, we derive $\mathbf{H}^{\alpha q + t - \delta}_{\varphi,\psi}(A) \geq C'\,\mathbf{H}^{q,t}_{\varphi,\psi,\mu}(A)$ from $A = \bigcup_n A_n$.

□

## 5. Proof of Theorem 3.3

**5.1. Bounding the dimensions related to the $(\varphi, \psi)$-Hewitt-Stromberg measures.** For $\nu \in \mathscr{P}^c(\mathbb{R}^n)$, $(\varphi, \psi) \in P$ with $\varphi^{-1}(0) = a > 0$ and $A \subseteq \mathbf{supp}\,\nu$, put

$$\Xi_{\varphi,\psi}(A) = \mathbf{H}^{1,0}_{\nu,\varphi,\psi}(A).$$

THEOREM 5.1. *Let $\nu \in \mathscr{P}^c(\mathbb{R}^n)$, $(\varphi, \psi) \in P$ with $\varphi^{-1}(0) = a > 0$ and bounded subset $A \subseteq \mathbb{R}^n$, if $\mathbf{B}^1_{\nu,\varphi,\psi}(A) \leq 0$ and $\varphi^{-1}(x+y) \leq C\,\varphi^{-1}(x)\varphi^{-1}(y)$ for all $x, y \in [0, +\infty)$ and some positive constant $C$, then*

$$\underline{\dim}^{MB}_{\varphi,\psi}(A) \leq \sup_{x \in A} \underline{\alpha}^{\varphi,\psi}_\nu(x) \quad \textit{and} \quad \overline{\dim}^{MB}_{\varphi,\psi}(A) \leq \sup_{x \in A} \overline{\alpha}^{\varphi,\psi}_\nu(x).$$

PROOF. Take $\alpha > \sup_{x \in A} \underline{\alpha}^{\varphi,\psi}_\nu(x) \geq 0$ and $\varepsilon > 0$. Since $\mathbf{B}^1_{\nu,\varphi,\psi}(A) \leq 0$ we have $\mathbf{P}^{1,\varepsilon}_{\nu,\varphi,\psi}(A) = 0$, then there exists a sequence $(A_k)_k$ such that $A = \bigcup_k A_k$ and

$$\sum_k \overline{\mathbf{P}}^{1,\frac{\varepsilon}{2}}_{\nu,\varphi,\psi}(A_k) < 1, \quad \sum_k \overline{\mathbf{P}}^{1,\varepsilon}_{\nu,\varphi,\psi}(A_k) = 0.$$

Fix $k \in \mathbb{N}$, put $\delta > 0$ and $F \subseteq A_k$. For all $x \in F$, there exists $c_x \geq 2$ and $\frac{\delta}{c_x} < r_x < \delta$ such that

$$\nu(B(x, r_x)) > \varphi^{-1}(\alpha\psi(r_x)).$$

Thus, the collection $\{B(x, r_x)\}_{x\in F}$ forms a centered $2\delta$-covering of the closure $\overline{F}$. So, we can choose a finite subfamily $\{B(x_i, r_{x_i})\}_i$ of $\{B(x, r_x)\}_{x\in F}$, and it is a centered $2\delta$-covering of $F$.

In addition, if we take $r = \max_i\{r_{x_i}\}$, so $\{B(x_i, r)\}_i$ is a centered covering of $F$. According to the Besicovitch's covering theorem, there exists $\xi_n$-finite or countable subfamily $\{(B(x_{ij}, r))_j,\ 1 \leq i \leq \xi_n\}$ such that

$$F \subseteq \bigcup_{i=1}^{\xi_n}\bigcup_j B(x_{ij}, r)$$

and for all $i$, $\{B(x_{ij}, r)\}_j$ is a centered packing of $F$, with $\nu\big(B_{ij}\big) > \varphi^{-1}(\alpha\psi(r))$. We have

$$\begin{aligned} O_r(F)\varphi^{-1}((\alpha+\varepsilon)\psi(r)) &\leq \sum_{i,j}\varphi^{-1}((\alpha+\varepsilon)\psi(r)) \\ &\leq \sum_{i,j}\nu\big(B_{ij}\big)\varphi^{-1}(\varepsilon\psi(r)) \\ &\leq \xi_n D^1_{\nu,\delta}(F)\varphi^{-1}(\varepsilon\psi(r)). \end{aligned}$$

This implies that

$$\mathbf{I}^{\alpha+\varepsilon}_{\varphi,\psi}(F) \leq \xi_n\overline{\mathbf{P}}^{1,\varepsilon}_{\nu,\varphi,\psi}(F) \leq \xi_n\overline{\mathbf{P}}^{1,\varepsilon}_{\nu,\varphi,\psi}(A_k).$$

It follows that

$$\overline{\mathbf{H}}^{\alpha+\varepsilon}_{\varphi,\psi}(A_k) \leq \xi_n\overline{\mathbf{P}}^{1,\varepsilon}_{\nu,\varphi,\psi}(A_k)$$

and

$$\mathbf{H}^{\alpha+\varepsilon}_{\varphi,\psi}(A) \leq \sum_k\overline{\mathbf{H}}^{\alpha+\varepsilon}_{\varphi,\psi}(A_k) \leq \xi_n\sum_k\overline{\mathbf{P}}^{1,\varepsilon}_{\nu,\varphi,\psi}(A_k) = 0.$$

We deduce that

$$\underline{\dim}^{MB}_{\varphi,\psi}(A) \leq \alpha+\varepsilon \quad \text{for all} \quad \varepsilon > 0.$$

Hence

$$\underline{\dim}^{MB}_{\varphi,\psi}(A) \leq \sup_{x\in A}\underline{\alpha}^{\varphi,\psi}_{\nu}(x).$$

Put $\varepsilon > 0$ and $\alpha > \sup_{x\in A}\overline{\alpha}^{\varphi,\psi}_{\nu}(x) \geq 0$. According to $\mathbf{B}^1_{\nu,\varphi,\psi}(A) \leq 0$, one has $\mathbf{P}^{1,\varepsilon}_{\nu,h,,g}(A) = 0$. So we can choose $\{H\}_j$ such that $A = \bigcup_j H_j$,

$$\sum_j\overline{\mathbf{P}}^{1,\frac{\varepsilon}{2}}_{\nu,\varphi,\psi}(H_j) < 1 \quad \text{and} \quad \sum_j\overline{\mathbf{P}}^{1,\varepsilon}_{\nu,\varphi,\psi}(H_j) = 0.$$

Let $x \in A$, we can find $\delta > 0$, and so for all $0 < r < \delta$, we have

$$\nu\big(B(x, r)\big) > \varphi^{-1}(\alpha\psi(r)).$$

Now, we consider

$$A_m = \left\{x \in A\ \middle|\ \text{for all } r < \frac{1}{m},\ ;\nu\big(B(x, r)\big) > \varphi^{-1}(\alpha\psi(r))\right\}.$$

Fix $m \in \mathbb{N}^*$, for $0 < r < \min(\delta, \frac{1}{m})$, let $\{B_i = B(x_i, r)\}_i$ be a centered packing of $H_j \cap A_m$. Then

$$D_r(H_j \cap A_m)\varphi^{-1}((\alpha+\varepsilon)\psi(r)) \le \sum_i \nu(B_i)\varphi^{-1}(\varepsilon\psi(r)) \le D^1_{\nu,r}(H_j)\varphi^{-1}(\varepsilon\psi(r)),$$

which implies that

$$\overline{\mathbf{P}}^{\alpha+\varepsilon}_{\varphi,\psi}(H_j \cap A_m) \le \overline{\mathbf{P}}^{1,\varepsilon}_{\nu,\varphi,\psi}(H_j) = 0.$$

Hence

$$\mathbf{P}^{\alpha+\varepsilon}_{\varphi,\psi}(A_m) \le \sum_j \overline{\mathbf{P}}^{\alpha+\varepsilon}_{\varphi,\psi}(H_j \cap A_m) \le \sum_j \overline{\mathbf{P}}^{1,\varepsilon}_{\nu,\varphi,\psi}(H_j) = 0.$$

Since $A = \bigcup_m A_m$, we deduce that

$$\overline{\dim}^{MB}_{\varphi,\psi}(A) \le \alpha + \varepsilon \quad \text{for all} \quad \varepsilon > 0,$$

and so

$$\overline{\dim}^{MB}_{\varphi,\psi}(A) \le \sup_{x\in A} \overline{\alpha}^{\varphi,\psi}_\nu(x).$$

□

THEOREM 5.2. *Let $\nu \in \mathscr{P}^c(\mathbb{R}^n)$, $(\varphi,\psi) \in P$ with $\varphi^{-1}(0) = a > 0$, $A \subseteq$ **supp** $\nu$ and $\Xi_{\varphi,\psi}(A) > 0$, then*

$$\underline{\dim}^{MB}_{\varphi,\psi}(A) \ge \operatorname*{ess\,sup}_{x\in A} \underline{\alpha}^{\varphi,\psi}_\nu(x) \quad \textit{and} \quad \overline{\dim}^{MB}_{\varphi,\psi}(A) \ge \operatorname*{ess\,sup}_{x\in A} \overline{\alpha}^{\varphi,\psi}_\nu(x).$$

*Obviously, the right side of the inequalities is related to the measure $\Xi_{\varphi,\psi}$.*

PROOF. Fix $\alpha < \operatorname*{ess\,sup}_{x\in A} \underline{\alpha}^{\varphi}_\nu(x)$ and let

$$F = \left\{ x \in A \;\middle|\; \liminf_{r\to 0} \frac{\varphi\big(\nu(B(x,r))\big)}{\psi(r)} > \alpha \right\}.$$

Hence $\Xi_{\varphi,\psi}(F) > 0$. If $x \in A$, we can find $\delta > 0$ such that

$$\nu\big(B(x,r)\big) \le \varphi^{-1}(\alpha\psi(r)),$$

for all $0 < r < \delta$. Let $F = \bigcup_j F_j$ and

$$F_{j_p} = \left\{ x \in F_j \;\middle|\; \nu\big(B(x,r)\big) \le \varphi^{-1}(\alpha\psi(r)) \text{ for } 0 < r < \frac{1}{p} \right\}.$$

Fix $p \in \mathbb{N}^*$, let $G \subseteq F_{j_p}$, $0 < r < \min(\delta, \frac{1}{p})$ and $\{B(x_i, r)\}_i\}$ be a centered covering of $G$. Since $\varphi^{-1}(0) = a > 0$, we have

$$\frac{1}{a} O^1_{\nu,r}(G)\varphi^{-1}(0\psi(r)) \le \sum_i \nu(B(x_i, r)) \le O_r(G)\varphi^{-1}(\alpha\psi(r)).$$

This shows that

$$\frac{1}{a}\mathbf{I}^{1,0}_{\nu,\varphi,\psi}(G) \le \mathbf{I}^{\alpha}_{\varphi,\psi}(G) \le \overline{\mathbf{H}}^{\alpha}_{\varphi,\psi}(G) \le \overline{\mathbf{H}}^{\alpha}_{\varphi,\psi}(F_{j_p})$$

and then

$$\frac{1}{a}\overline{\mathbf{H}}^{1,0}_{\nu,\varphi,\psi}(F_{j_p}) \le \overline{\mathbf{H}}^{\alpha}_{\varphi,\psi}(F_{j_p}).$$

Since

$$F_j = \bigcup_p F_{j_p},$$

for all $j$ and $\Xi_{\varphi,\psi}(F) > 0$, one has

$$0 < \frac{1}{a}\mathbf{H}^{1,0}_{\nu,\varphi,\psi}(F) \leq \frac{1}{a}\sum_{j,p}\overline{\mathbf{H}}^{1,0}_{\nu,\varphi,\psi}(F_{j_p}) \leq \sum_{j,p}\overline{\mathbf{H}}^{\alpha}_{\varphi,\psi}(F_{j_p}),$$

which implies that

$$0 < \mathbf{H}^{\alpha}_{\varphi,\psi}(F) \leq \mathbf{H}^{\alpha}_{\varphi,\psi}(A).$$

Hence

$$\underline{\dim}^{MB}_{\varphi,\psi}(A) \geq \alpha,$$

and then

$$\underline{\dim}^{MB}_{\varphi,\psi}(A) \geq \operatorname*{ess\,sup}_{x\in A}\underline{\alpha}^{\varphi,\psi}_{\nu}(x).$$

Let $\alpha < \operatorname*{ess\,sup}_{x\in E}\overline{\alpha}^{\varphi,\psi}_{\nu}(x)$ and put

$$F = \left\{ x \in A \;\middle|\; \limsup_{r\to 0}\frac{\varphi(\nu(B(x,r)))}{\psi(r)} > \alpha \right\},$$

so $\Xi_{\varphi,\psi}(F) > 0$. Let $G \subseteq F$ and $\delta > 0$, for all $x \in G$, take $0 < r_x < \delta$, such that

$$\nu(B(x,r_x)) \leq \varphi^{-1}(\alpha\psi(r_x)).$$

Hence, the collection $\{B(x,r_x)\}_{x\in G}$ forms a centered $2\delta$-covering of the closure $\overline{G}$. Consequently, we can select a finite subfamily $\{B(x_i,r_{x_i})\}_{i\in I}$ from $\{B(x,r_x)\}_{x\in G}$ that still constitutes a centered $2\delta$-covering of $G$. In addition, if we take $r = \max_i\{r_{x_i}\}$, so $\{B(x_i,r)\}_{i\in I}$ is a centered covering of $G$. According to the Besicovitch's covering theorem, there exists $\gamma_n$-finite or countable subfamily $\{(B(x_{ij},r))_j,\ 1 \leq i \leq \gamma_n\}$ such that

$$G \subseteq \bigcup_{i=1}^{\gamma_n}\bigcup_{j} B(x_{ij},r)$$

and for all $i$, $\{B(x_{ij},r)\}_j$ is a centered packing of $G$, with $\nu(B(x_{ij},r)) \leq \varphi^{-1}(\alpha\psi(r))$. Since $\varphi^{-1}(0) = a > 0$, it follows that

$$\frac{1}{a}O^{1}_{\nu,r}(G)\varphi^{-1}(0\psi(r)) \leq \sum_{i,j}\nu(B(x_{ij},r)) \leq \gamma_n D_r(G)\varphi^{-1}(\alpha\psi(r)).$$

Then

$$\frac{1}{a}\mathbf{I}^{1,0}_{\nu,\varphi,\psi}(G) \leq \gamma_n\overline{\mathbf{P}}^{\alpha}_{\varphi,\psi}(G) \leq \gamma_n\overline{\mathbf{P}}^{\alpha}_{\varphi,\psi}(F),$$

which gives that

$$\frac{1}{a}\overline{\mathbf{H}}^{1,0}_{\nu,\varphi,\psi}(F) \leq \gamma_n\overline{\mathbf{P}}^{\alpha}_{\varphi,\psi}(F).$$

Next, if $F = \bigcup_i F_i$, we have

$$0 < \frac{1}{a}\mathbf{H}^{1,0}_{\nu,\varphi,\psi}(F) \leq \frac{1}{a}\sum_{i}\overline{\mathbf{H}}^{1,0}_{\nu,\varphi,\psi}(F_i) \leq \gamma_n\sum_{i}\overline{\mathbf{P}}^{\alpha}_{\varphi,\psi}(F_i),$$

and then

$$\mathbf{P}^{\alpha}_{\varphi,\psi}(A) \geq \mathbf{P}^{\alpha}_{\varphi,\psi}(F) > 0,$$

which implies that

$$\overline{\dim}^{\varphi,\psi}_{MB}(A) \geq \alpha,$$

for all $\alpha < \operatorname{ess\,sup}_{x\in E} \overline{\alpha}_{\nu}^{\varphi,\psi}(x)$. Hence

$$\overline{\dim}_{\varphi,\psi}^{MB}(A) \geq \operatorname{ess\,sup}_{x\in A} \overline{\alpha}_{\nu}^{\varphi,\psi}(x).$$

□

DEFINITION 5.1. Consider $\mu \in \mathscr{P}^c(\mathbb{R}^n)$ and a finite Borel measure $\nu$ on $\mathbb{R}^n$ with their common support denoted as $K := \mathbf{supp}\,\mu = \mathbf{supp}\,\nu$, $(\varphi,\psi) \in P$ with $\varphi^{-1}(0) = a > 0$, for $\zeta = (\mu,\nu)$, $A \subseteq K$, $r > 0$ and $q, t \geq 0$, put

$$D^q_{\zeta,r}(A) = \sup\left\{\sum_i \varphi^{-1}\left(q\varphi(\mu(B(x_i,r))) + \varphi(\nu(B(x_i,r)))\right) \;\middle|\; \{B(x_i,r)\}_i \text{ is a centered packing of } A\right\}$$

and

$$O^q_{\zeta,r}(A) = \inf\left\{\sum_i \varphi^{-1}\left(q\varphi(\mu(B(x_i,r))) + \varphi(\nu(B(x_i,r)))\right) \;\middle|\; \{B(x_i,r)\}_i \text{ is a centered covering of } A\right\}.$$

Then, let

$$\overline{\mathbf{P}}^{q,t}_{\zeta,\varphi,\psi}(A) = \limsup_{r\to 0} D^q_{\zeta,r}(A)\varphi^{-1}(t\psi(r))$$

and

$$\mathbf{I}^{q,t}_{\zeta,\varphi,\psi}(A) = \liminf_{r\to 0} O^q_{\zeta,r}(A)\varphi^{-1}(t\psi(r)),$$

$$\overline{\mathbf{H}}^{q,t}_{\zeta,\varphi,\psi}(A) = \sup_{F\subseteq A} \mathbf{I}^{q,t}_{\zeta,\varphi,\psi}(F).$$

$\overline{\mathbf{P}}^{q,t}_{\zeta,\varphi,\psi}$ and $\overline{\mathbf{H}}^{q,t}_{\zeta,\varphi,\psi}$ are said to be $(\varphi,\psi)$-Hewitt-Stromberg pre-measures about $\zeta = (\mu,\nu)$.

REMARK 5.2. It is clear that $\mathbf{I}^{q,t}_{\zeta,\varphi,\psi}$ is not monotonically increasing and does not satisfy $\sigma$-additivity.

Next, put

$$\mathbf{P}^{q,t}_{\zeta,\varphi,\psi}(A) = \inf\left\{\sum_i \overline{\mathbf{P}}^{q,t}_{\zeta,\varphi,\psi}(A_i) \;\middle|\; A \subseteq \bigcup_i A_i\right\}$$

and

$$\mathbf{H}^{q,t}_{\zeta,\varphi,\psi}(A) = \inf\left\{\sum_i \overline{\mathbf{H}}^{q,t}_{\zeta,\varphi,\psi}(A_i) \;\middle|\; A \subseteq \bigcup_i A_i\right\}.$$

Obviously, $\mathbf{H}^{q,t}_{\zeta,\varphi,\psi}$ and $\mathbf{P}^{q,t}_{\zeta,\varphi,\psi}$ are metric outer measures. In addition, for $A \subseteq \mathbb{R}^n$, their dimensions are respectively denoted by $\mathbf{b}^q_{\zeta,\varphi,\psi}(A)$, $\mathbf{B}^q_{\zeta,\varphi,\psi}(A)$ and $\mathbf{\Delta}^q_{\zeta,\varphi,\psi}(A)$:

$$\mathbf{b}^q_{\zeta,\varphi,\psi}(A) = \inf\left\{t \geq 0 \mid \mathbf{H}^{q,t}_{\zeta,\varphi,\psi}(A) = 0\right\},$$

$$\mathbf{B}^q_{\zeta,\varphi,\psi}(A) = \inf\left\{t \geq 0 \mid \mathbf{P}^{q,t}_{\zeta,\varphi,\psi}(A) = 0\right\},$$

$$\mathbf{\Delta}^q_{\zeta,\varphi,\psi}(A) = \inf\left\{t \geq 0 \mid \overline{\mathbf{P}}^{q,t}_{\zeta,\varphi,\psi}(A) = 0\right\}.$$

It follows that

$$\mathbf{b}^q_{\zeta,\varphi,\psi}(A) \le \mathbf{B}^q_{\zeta,\varphi,\psi}(A) \le \mathbf{\Delta}^q_{\zeta,\varphi,\psi}(A).$$

Now, we define the dimension functions by

$$\mathbf{b}_{\zeta,\varphi,\psi} : q \mapsto \mathbf{b}^q_{\zeta,\varphi,\psi}(K),$$
$$\mathbf{B}_{\zeta,\varphi,\psi} : q \mapsto \mathbf{B}^q_{\zeta,\varphi,\psi}(K)$$

and

$$\mathbf{\Delta}_{\zeta,\varphi,\psi} :, q \mapsto \mathbf{\Delta}^q_{\zeta,\varphi,\psi}(K).$$

PROPOSITION 5.1. *For $\mu \in \mathscr{P}^c(\mathbb{R}^n)$ and a finite Borel measure $\nu$ on $\mathbb{R}^n$ with their common support, let $\zeta = (\mu, \nu)$, $(\varphi, \psi) \in P$ with $\varphi^{-1}(0) = a > 0$, $g(t) = \mathbf{B}_{\zeta,\varphi,\psi}(t)$ with $g(0) = 0$ and $\Xi_{\varphi,\psi}(K) > 0$. If $\varphi^{-1}(x + y) \le \varphi^{-1}(x)\varphi^{-1}(y)$, for all $x, y \in [0, +\infty)$, then*

$$\Xi_{\varphi,\psi}(X(-g'_+(0), -g'_-(0))) = 0,$$

*where $f'_-$ and $f'_+$ are the left and right derivatives of the function $g$, respectively.*

PROOF. Just prove

$$\Xi_{\varphi,\psi}\left(\left\{x \in K \;\middle|\; \overline{\alpha}^{\varphi,\psi}_{\mu}(x) > -g'_-(0)\right\}\right) = 0 \tag{5.1}$$

and

$$\Xi_{\varphi,\psi}\left(\left\{x \in K \;\middle|\; \underline{\alpha}^{\varphi,\psi}_{\mu}(x) < -g'_+(0)\right\}\right) = 0. \tag{5.2}$$

Here, we only prove (5.1).

Fix $\alpha > -g'_-(0)$, then we can find $\beta, t > 0$ such that

$$\alpha > \beta > -g'_-(0) \text{ and } g(-t) < \beta t.$$

Hence $\mathbf{P}^{-t,\beta t}_{\zeta,\varphi,\psi}(K) = 0$. Then we can find a countable partition $\{K_j\}_j$ of $K$, with

$$\sum_j \overline{\mathbf{P}}^{-t,\beta t}_{\zeta,\varphi,\psi}(K_j) \le 1 \quad \text{and} \quad \overline{\mathbf{P}}^{-t,\alpha t}_{\zeta,\varphi,\psi}(K_j) = 0, \quad \forall\, j.$$

For

$$F = \left\{x \in K \;\middle|\; \limsup_{r\to 0} \frac{\varphi(\mu(B(x,r)))}{\psi(r)} > \alpha\right\}.$$

Let $x \in F$ and $\delta > 0$, we choose $0 < r_x < \delta$ such that

$$\mu(B(x, r_x)) \le \varphi^{-1}(\alpha\psi(r_x)).$$

Put $F_j = K_j \cap F$ and $G_j \subseteq F_j$. Obviously, $\{B(x, r_x)\}_{x\in G_j}$ is a centered $2\delta$-covering of $\overline{G_j}$. So, we can choose a finite subfamily $\{B(x_i, r_{x_i})\}_{i\in I}$ of $\{B(x, r_x)\}_{x\in G_j}$, and it is a centered $2\delta$-covering of $G_j$.

In addition, if we take $r_j = \max_{i\in I}\{r_{x_i}\}$, so $\{B(x_i, r_j)\}_{i\in I}$ is a centered covering of $G_j$. According to the Besicovitch's covering theorem, there exists $\xi_n$-finite or countable subfamily $\{(B(x_{ij}, r_j))_j,\ 1 \le i \le \xi_n\}$ such that

$$G_j \subseteq \bigcup_{i=1}^{\xi_n} \bigcup_j B(x_{ij}, r_j)$$

and for all $i$, $\{B(x_{ij}, r_j)\}_j$ is a centered packing of $G_j$ and

$$\mu(B(x_{ij}, r_j)) \le \varphi^{-1}(\alpha\psi(r_j)).$$

Then

$$\begin{aligned} O^1_{\nu,r_j}(G_j)\varphi^{-1}(0\,\psi(r)) &\leq \sum_i \varphi^{-1}\big(\varphi(\nu(B(x_{ij},r_j)))\big) \\ &= \sum_i \varphi^{-1}\big(t\varphi(\mu(B(x_{ij},r_j))) - t\varphi(\mu(B(x_{ij},r_j))) + \varphi(\nu(B(x_{ij},r_j)))\big) \\ &\leq \sum_i \varphi^{-1}\big(-t\varphi(\mu(B(x_{ij},r_j))) + \varphi(\nu(B(x_{ij},r_j)))\big)\big)\varphi^{-1}(\alpha t\psi(r_j)). \end{aligned}$$

So

$$\mathbf{I}^{1,0}_{\nu,\varphi,\psi}(G_j) \leq \xi_n \overline{\mathbf{P}}^{-t,\alpha t}_{\zeta,\varphi,\psi}(K_j).$$

This implies that

$$\overline{\mathbf{H}}^{1,0}_{\nu,\varphi,\psi}(F_j) = \sup_{G_j\subseteq F_j} \mathbf{I}^{1,0}_{\nu,\varphi,\psi}(G_j) \leq \xi_n \overline{\mathbf{P}}^{-t,\alpha t}_{\zeta,\varphi,\psi}(K_j) = 0,$$

and then

$$\Xi_{\varphi,\psi}(F) = \mathbf{H}^{1,0}_{\nu,\varphi,\psi}(F) \leq \sum_j \overline{\mathbf{H}}^{1,0}_{\nu,\varphi,\psi}(F_j) = 0.$$

□

PROPOSITION 5.2. *For $\mu \in \mathscr{P}^c(\mathbb{R}^n)$ and a finite Borel measure $\nu$ on $\mathbb{R}^n$ with their common support, let $\zeta = (\mu,\nu)$, $(\varphi,\psi) \in P$ with $\varphi^{-1}(0) = a > 0$, $g(t) = \mathbf{B}_{\zeta,\varphi,\psi}(t)$ with $g(0) = 0$ and $\Xi_{\varphi,\psi}(K) > 0$. If $\varphi^{-1}(x+y) \leq \varphi^{-1}(x)\varphi^{-1}(y)$, for all $x, y \in [0,+\infty)$, one has*

$$\underline{\dim}^{MB}_{\varphi,\psi}\big(X\big(-g'_+(0), -g'_-(0)\big)\big) \geq \inf\left\{ \underline{\alpha}^{\varphi,\psi}_\nu(x) \;\middle|\; x \in X\big(-g'_+(0), -g'_-(0)\big)\right\}$$

*and*

$$\overline{\dim}^{MB}_{\varphi,\psi}\big(X\big(-g'_+(0), -g'_-(0)\big)\big) \geq \inf\left\{ \overline{\alpha}^{\varphi,\psi}_\nu(x) \;\middle|\; x \in X\big(-g'_+(0), -g'_-(0)\big)\right\}.$$

PROOF. The immediate consequence stems from Theorem 5.2 and Proposition 5.1. □

**5.2. Proof of Theorem 3.3.** For $q \geq 0$ and $r > 0$, consider

$$\nu\big(B(x,r)\big) = \varphi^{-1}\big(q\varphi(\mu(B(x,r)))\big)\varphi^{-1}\big(\mathbf{B}_{\varphi,\psi,\mu}(q)\psi(r)\big).$$

We have

$$\mathbf{B}_{\zeta,\varphi,\psi}(t) = \mathbf{B}_{\varphi,\psi,\mu}(q+t) - \mathbf{B}_{\varphi,\psi,\mu}(q).$$

For $x \in \overline{X}^\alpha$, one has

$$\overline{\alpha}^{\varphi,\psi}_\nu(x) = q\overline{\alpha}^{\varphi,\psi}_\mu(x) + \mathbf{B}_{\varphi,\psi,\mu}(q) \leq q\alpha + \mathbf{B}_{\varphi,\psi,\mu}(q).$$

According to Theorem 5.1 that

$$\overline{\dim}^{MB}_{\varphi,\psi}\big(\overline{X}^\alpha\big) \leq \inf_{q\geq 0} q\alpha + \mathbf{B}_{\varphi,\psi,\mu}(q).$$

(1) Next, we suppose that $\mathbf{H}^{q,\mathbf{B}_{\varphi,\psi,\mu}(q)}_{\varphi,\psi,\mu}\big(K\big) > 0$, then $\Xi_{\varphi,\psi}(K) > 0$. Proposition 5.1 shows that

$$\Xi_{\varphi,\psi}\big(X\big(-\mathbf{B}'_{\varphi,\psi,\mu_+}(q), -\mathbf{B}'_{\varphi,\psi,\mu_-}(q)\big)\big) > 0.$$

According to Proposition 5.2, if $q \geq 0$, we have

$$\underline{\dim}^{MB}_{\varphi,\psi}\big(X\big(-\mathbf{B}'_{\varphi,\psi,\mu_+}(q), -\mathbf{B}'_{\varphi,\psi,\mu_-}(q)\big)\big) \geq -q\mathbf{B}'_{\varphi,\psi,\mu_+}(q) + \mathbf{B}_{\varphi,\psi,\mu}(q).$$

Finally, since $\mathbf{B}_{\varphi,\psi,\mu}$ is differentiable at $q$, we obtain

$$\begin{aligned}\underline{\dim}^{MB}_{\varphi,\psi}(X(-\mathbf{B}'_{\varphi,\psi,\mu}(q))) &= \overline{\dim}^{MB}_{\varphi,\psi}(X(-\mathbf{B}'_{\varphi,\psi,\mu}(q)))\\ &= \mathbf{B}^*_{\varphi,\psi,\mu}\big(-\mathbf{B}'_{\varphi,\psi,\mu}(q)\big)\\ &= \mathbf{b}^*_{\varphi,\psi,\mu}\big(-\mathbf{B}'_{\varphi,\psi,\mu}(q)\big).\end{aligned}$$

(2) For $q \geq 0$,

$$X\big(-\mathbf{B}'_{\varphi,\psi,\mu_+}(q), -\mathbf{B}'_{\varphi,\psi,\mu_-}(q)\big) = \underline{X}_{-\mathbf{B}'_{\varphi,\psi,\mu_+}(q)} \cap \overline{X}^{-\mathbf{B}'_{\varphi,\psi,\mu_-}(q)} \subseteq \overline{X}^{-\mathbf{B}'_{\varphi,\psi,\mu_-}(q)}.$$

It follows immediately that

$$\begin{aligned}-\mathbf{B}'_{\varphi,\psi,\mu_-}(q)q + \mathbf{B}_{\varphi,\psi,\mu}(q) &\leq \underline{\dim}^{MB}_{\varphi,\psi}\big(X\big(-\mathbf{B}'_{\varphi,\psi,\mu_+}(q), -\mathbf{B}'_{\varphi,\psi,\mu_-}(q)\big)\big)\\ &\leq \underline{\dim}^{MB}_{\varphi,\psi}\big(\overline{X}^{(-\mathbf{B}'_{\varphi,\psi,\mu_-}(q))}\big).\end{aligned}$$

Suppose that $\alpha = -\mathbf{B}'_{\varphi,\psi,\mu_-}(q))$. We only consider $q \geq 0$, so

$$\underline{\dim}^{MB}_{\varphi,\psi}(\overline{X}^{\alpha}) \geq \alpha q + \mathbf{B}_{\varphi,\psi,\mu}(q).$$

According to the fact $\mathbf{b}_{\varphi,\psi,\mu}(q) \leq \mathbf{B}_{\varphi,\psi,\mu}(q)$, next we prove that

$$\mathbf{b}_{\varphi,\psi,\mu}(q) \geq \mathbf{B}_{\varphi,\psi,\mu}(q).$$

Fix $t < \mathbf{B}_{\varphi,\psi,\mu}(q)$ and $\beta \in \mathbb{R}$, with $\beta > \alpha$ and $\beta q + t < \alpha q + \mathbf{B}_{\varphi,\psi,\mu}(q)$. For $m \in \mathbb{N}^*$, we consider

$$A_m = \left\{x \in \overline{X}^{\alpha} \;\middle|\; \mu(B(x,r)) \geq \varphi^{-1}(\beta\psi(r)),\ 0 < r < \frac{1}{m}\right\}.$$

letting $p \to \infty$, so We have $A_m \nearrow \overline{X}^{\alpha}$. Then there exists $m > 0$ such that

$$\underline{\dim}^{MB}_{\varphi,\psi}(A_m) > \beta q + t,$$

and then

$$\mathbf{H}^{\beta q+t}_{\varphi,\psi}(A_m) > 0.$$

Fix $0 < r < \frac{1}{m}$ and put $\{B(x_i,r)\}_i$ be a centered covering of $A_m$. Hence

$$\begin{aligned}\sum_i \varphi^{-1}(q\varphi(\mu(B(x_i,r)))\,\varphi^{-1}(t\psi(r)) &\geq \sum_i \varphi^{-1}((\beta q+t)\psi(r))\\ &\geq O_r(A_m)\,\varphi^{-1}((\beta q+t)\psi(r)).\end{aligned}$$

Then

$$O^q_{\mu,r}(A_m)\,\varphi^{-1}(t\psi(r)) \geq O_r(A_m)\,\varphi^{-1}((\beta q+t)\psi(r))$$

and

$$\mathbf{I}^{q,t}_{\varphi,\psi,\mu}(A_m) \geq \mathbf{I}^{\beta q+t}_{\varphi,\psi}(A_m),$$

which implies that

$$\mathbf{H}^{q,t}_{\varphi,\psi,\mu}(K) \geq \mathbf{H}^{q,t}_{\varphi,\psi,\mu}(\overline{X}^{\alpha}) \geq \mathbf{H}^{q,t}_{\varphi,\psi,\mu}(A_m) \geq \mathbf{H}^{\beta q+t}_{\varphi,\psi}(A_m) > 0.$$

Hence

$$\mathbf{b}_{\varphi,\psi,\mu}(q) \geq t,$$

for all $t < \mathbf{B}_{\varphi,\psi,\mu}(q)$. Finally, we obtain

$$\mathbf{b}_{\varphi,\psi,\mu}(q) = \mathbf{B}_{\varphi,\psi,\mu}(q).$$

THEOREM 5.3. *The preceding findings remain applicable when substituting the multifractal function* $\mathbf{B}(.,\varphi,\psi)$ *with the function* $\mathbf{\Delta}(.,\varphi,\psi)$.

**5.3. An example.** The usual multifractal formalism corresponds to $\varphi(x) = \psi(x) = -\log x$. We will consider multifractal formalism which corresponds to $\varphi(x) = -\alpha \log x$ and $\psi(x) = \alpha \log(1 - \log x)$ for any $\alpha > 0$. Obviously $(\varphi(x), \psi(x)) \in P$,

$$\begin{aligned}\varphi^{-1}(x+y) &= e^{-\frac{x+y}{\alpha}} \\ &= e^{-\frac{x}{\alpha}} e^{-\frac{y}{\alpha}} \\ &= \varphi^{-1}(x)\varphi^{-1}(y)\end{aligned}$$

and

$$\varphi^{-1}(x) = e^{-\frac{1}{\alpha}x} \Longrightarrow \varphi^{-1}(0) = e^0 = 1 > 0.$$

The example constructed here is basically consistent with [**30**]. Let $b = (b_n^m)_{n,m}$ be a sequence of non-negative real numbers. We start as in the construction of Cantor-type set by interval $I = I_0 = [0,1]$, affected with an initial total measure $\mu_0^0 = 1$. The first iteration intercepts an open interval with a length of $b_1^1$ from the original interval $I = I_0 = [0,1]$, which is far from the two endpoints. Therefore, we obtain two disjoint intervals, denoted as $I_1$ and $I_2$, their measures are respectively determined by $\mu_1^1$ and $\mu_2^1$ represents. Next, we take the interval with a length of $b_1^2$ from $I_1$ and the interval with a length of $b_2^2$ from $I_2$. Thus, four intervals $I_{11}$, $I_{12}$, $I_{21}$ and $I_{22}$ were obtained, with measures $\mu_1^1\mu_1^2$, $\mu_1^1\mu_2^2$, $\mu_2^1\mu_1^2$ and $\mu_2^1\mu_2^2$, respectively. Continuing to do this, we obtain that for $k \in \mathbb{N}$, we have $2^k$ intervals $I_{i_1 i_2 \dots i_k}$, $(i_1 i_2 \dots i_k) \in \{1,2\}^k$ affected with a mass $\mu_{i_1 i_2 \dots i_k} = \mu_{i_1}^1 \mu_{i_2}^2 \cdots \mu_{i_k}^k$, which leads thus to a probability measure $\mu$ supported by the Cantor set. Now, assume for example that all the cells $I_{i_1, i_2, \dots, i_n}$ have the same length

$$d_n = |I_{i_1, i_2, \dots, i_n}| = e^{-e^{n\xi}},$$

for some $\xi \in \mathbb{R}$ fixed. Since functions

$$\varphi(x) = -\alpha \log x, \quad \psi(x) = \alpha \log(1 - \log x)$$

for any $\alpha > 0$, so we have definitions:

$$\underline{\alpha}_\mu(x) = \liminf_{r \to 0} \frac{-\log(\mu(B(x,r)))}{\log(1 - \log \delta)}, \quad \overline{\alpha}_\mu(x) = \limsup_{r \to 0} \frac{-\log(\mu(B(x,r)))}{\log(1 - \log r)}.$$

Suppose that $\mu_i^n$ satisfy

$$\mu_1^n = \beta \pm \varepsilon_n$$

and

$$\mu_2^n = \gamma \mp \varepsilon_n,$$

for some constants $\beta, \gamma \in (0,1)$ with $\beta + \gamma = 1$ and $(\varepsilon_n)_n \subseteq (0,1)$ such that

$$\beta \pm \varepsilon_n \in (0,1) \text{ and } \gamma \mp \varepsilon_n \in (0,1).$$

Similar to the approach in [**30**], it yields that

$$\mathbf{b}_{\varphi,\psi,\mu}(q) = \mathbf{B}_{\varphi,\psi,\mu}(q) = \mathbf{\Delta}_{\varphi,\psi,\mu}(q) = \frac{\log(\beta^q + \gamma^q)}{\xi}.$$

Next, for $q \geq 0$, we have

$$-\mathbf{B}'_{\varphi,\psi,\mu}(q) = -\frac{\beta^q \log \beta + \gamma^q \log \gamma}{\xi(\beta^q + \gamma^q)}.$$

It follows that

$$\begin{aligned}\underline{\dim}^{MB}_{\varphi,\psi}(X(-\mathbf{B}'_{\varphi,\psi,\mu}(q))) &= \overline{\dim}^{MB}_{\varphi,\psi}(X(-\mathbf{B}'_{\varphi,\psi,\mu}(q)))\\ &= \mathbf{B}^*_{\varphi,\psi,\mu}\big(-\mathbf{B}'_{\varphi,\psi,\mu}(q)\big)\\ &= \mathbf{b}^*_{\varphi,\psi,\mu}\big(-\mathbf{B}'_{\varphi,\psi,\mu}(q)\big)\\ &= \inf_q\left(-q\frac{\beta^q\log\beta+\gamma^q\log\gamma}{\xi(\beta^q+\gamma^q)}+\frac{\log(\beta^q+\gamma^q)}{\xi}\right).\end{aligned}$$

## 6. A result for which the $(\varphi,\psi)$-Hewitt-Stromberg dimensions differ

For $\mu\in\mathscr{P}^c(\mathbb{R}^n)$, $q,t\geq 0$ and $(\varphi,\psi)\in P$ with $\varphi^{-1}(0)=a>0$. We will now discuss scenarios in which the upper and lower $(\varphi,\psi)$-Hewitt-Stromberg dimension functions may not always be equal, i.e.

$$\mathbf{B}_{\varphi,\psi,\mu}(q)\neq\mathbf{b}_{\varphi,\psi,\mu}(q),$$

for all $q\neq 1$. Then for all $\alpha,\beta\geq 0$, put

$$X(\alpha,\beta)=\Big\{x\in\mathbf{supp}\,\mu\ \Big|\ \underline{\alpha}^{\varphi,\psi}_{\mu}(x)\leq\alpha\text{ and }\beta\leq\overline{\alpha}^{\varphi,\psi}_{\mu}(x)\Big\}\quad\text{and}\quad X(\alpha)=X(\alpha,\alpha).$$

We set

$$f^-(q)=\limsup_{t\to 0}\frac{f(q-t)-f(q)}{-t}$$

and

$$f^+(q)=\limsup_{t\to 0}\frac{f(q+t)-f(q)}{t},$$

where $f$ is a real function.

THEOREM 6.1. *Let $\mu\in\mathscr{P}^c(\mathbb{R}^n)$, $q\geq 0$, $(\varphi,\psi)\in P$ with $\varphi^{-1}(0)=a>0$ and $\mathbf{H}^{q,\mathbf{b}_{\varphi,\psi,\mu}(q)}_{\varphi,\psi,\mu}(\mathbf{supp}\,\mu)>0$. If $\varphi^{-1}(x+y)\leq\varphi^{-1}(x)\varphi^{-1}(y)$, for all $x,y\in[0,+\infty)$, then we have*

$$\overline{\dim}^{MB}_{\varphi,\psi}\big(X(-\mathbf{b}^-_{\varphi,\psi,\mu}(q),\ -\mathbf{b}^+_{\varphi,\psi,\mu}(q))\big)\geq -q\mathbf{b}^+_{\varphi,\psi,\mu}(q)+\mathbf{b}_{\varphi,\psi,\mu}(q).$$

*In addition, if $\mathbf{b}_{\varphi,\psi,\mu}$ is differentiable at $q$, then*

$$\overline{\dim}^{MB}_{\varphi,\psi}\big(X(-\mathbf{b}'_{\varphi,\psi,\mu}(q))\big)\geq -q\mathbf{b}'_{\varphi,\psi,\mu}(q)+\mathbf{b}_{\varphi,\psi,\mu}(q).$$

The immediate consequence of Theorem 5.2 and the subsequent proposition is the proof of Theorem 6.1.

PROPOSITION 6.1. *For $\mu\in\mathscr{P}^c(\mathbb{R}^n)$ and a finite Borel measure $\nu$ on $\mathbb{R}^n$ with their common support, let $\zeta=(\mu,\nu)$, $(\varphi,\psi)\in P$ with $\varphi^{-1}(0)=a>0$, $f(t)=\mathbf{b}_{\zeta,\varphi,\psi}(t)$ with $f(0)=0$ and $\Xi(\mathbf{supp}\,\mu)>0$. If $\varphi^{-1}(x+y)\leq\varphi^{-1}(x)\varphi^{-1}(y)$, for all $x,y\in[0,+\infty)$, then*

$$\Xi_{\varphi,\psi}\left(\Big\{x\in\mathbf{supp}\,\mu\ \Big|\ \underline{\alpha}^{\varphi,\psi}_{\mu}(x)>-f^-(0)\Big\}\right)=0$$

*and*

$$\Xi_{\varphi,\psi}\left(\Big\{x\in\mathbf{supp}\,\mu\ \Big|\ \overline{\alpha}^{\varphi,\psi}_{\mu}(x)<-f^+(0)\Big\}\right)=0.$$

PROOF. We will establish the validity of the initial statement. Assume $\alpha > -f^-(0) = \liminf_{t\to 0} \frac{f(-t)}{t}$. In this case, there exists $t > 0$ such that $\alpha t > f(-t)$. It is evident that

$$\mathbf{H}^{-t,\alpha t}_{\zeta,\varphi,\psi}(\mathbf{supp}\,\mu) = 0.$$

Now, we define the set

$$A = \left\{ x \in \mathbf{supp}\,\mu \;\middle|\; \liminf_{r\to 0} \frac{\varphi\big(\mu(B(x,r))\big)}{\psi(r)} > \alpha \right\}.$$

If $x$ belongs to $A$, then we can select a positive value for $\delta$ such that, for all $0 < r < \delta$,

$$\mu\big(B(x,r)\big) < \varphi^{-1}(\alpha\psi(r)).$$

Let $\{A_j\}_j$ with $A = \bigcup_j A_j$. We consider

$$A_{j_p} = \left\{ x \in A_j \;\middle|\; \mu\big(B(x,r)\big) < \varphi^{-1}(\alpha\psi(r)), \text{ for all } r < \frac{1}{p} \right\}.$$

Consider $F \subseteq A_{j_p}$ and $0 < r < \min(\delta, \frac{1}{p})$. Let $\{B_i = B(x_i,r)\}_i$ be a centered covering of $F$. We then obtain

$$\begin{aligned} O^1_{\nu,r}(F)\varphi^{-1}(0\psi(r)) \le \sum_i \nu(B_i) &= \sum_i \varphi^{-1}\big(t\varphi(\mu(B_i)) - t\varphi(\mu(B_i)) + \varphi(\nu(B_i))\big) \\ &\le \sum_i \varphi^{-1}\big(-t\varphi(\mu(B_i)) + \varphi(\nu(B_i))\big)\big)\varphi^{-1}(\alpha t\psi(r)). \end{aligned}$$

This implies that

$$O^1_{\nu,r}(F)\varphi^{-1}(0\psi(r)) \le O^{-t}_{\zeta,r}(F)\varphi^{-1}(\alpha t\psi(r)).$$

Letting $r$ tend to 0,

$$\mathbf{I}^{1,0}_{\nu,\varphi,\psi}(F) \le \mathbf{I}^{-t,\alpha t}_{\zeta,\varphi,\psi}(F) \le \overline{\mathbf{H}}^{-t,\alpha t}_{\zeta,\varphi,\psi}(A_{j_p}).$$

We deduce that

$$\overline{\mathbf{H}}^{1,0}_{\nu,\varphi,\psi}(A_{j_p}) \le 2^{-\alpha t}\overline{\mathbf{H}}^{-t,\alpha t}_{\zeta,\varphi,\psi}(A_{j_p})$$

and so

$$\mathbf{H}^{1,0}_{\nu,\varphi,\psi}(A) \le \sum_{j,p} \overline{\mathbf{H}}^{1,0}_{\nu,\varphi,\psi}(A_{j_p}) \le \sum_{j,p} \overline{\mathbf{H}}^{-t,\alpha t}_{\zeta,\varphi,\psi}(A_{j_p}).$$

Finally, we obtain

$$\mathbf{H}^{1,0}_{\nu,\varphi,\psi}(A) \le \mathbf{H}^{-t,\alpha t}_{\zeta,\varphi,\psi}(A) \le \mathbf{H}^{-t,\alpha t}_{\zeta,\varphi,\psi}(\mathbf{supp}\,\mu) = 0.$$

The demonstration of the second statement mirrors that of the corresponding statement in the first part and is consequently omitted. This concludes the proof of Proposition 6.1. □

REMARK 6.1. Our multifractal framework serves as a seamless expansion of the traditional formalism, as delineated in [**7**, **39**] and the approach established in [**16**]. This expansion is realized by adopting the settings $\varphi(x) = \psi(x) = -\log x$ in line with the method presented in [**7**, **39**], and further extending the formalism as per the conditions outlined in [**16**], where $\varphi(x) = -\log x$. Notably, classical fractal geometry is encapsulated as a particular case within this comprehensive multifractal analysis, specifically when $\varphi(x) = \psi(x) = -\log x$ and $q = 0$.

## References


[1] R. Achour, J. Hattab, and B. Selmi, *New fractal dimensions of measures and decompositions of singularly continuous measures*, Fuzzy Sets and Systems **479** (2024), Paper No. 108859, 28, DOI 10.1016/j.fss.2024.108859. MR4688638

[2] R. Achour, Z. Li, B. Selmi, and T. Wang, *A multifractal formalism for new general fractal measures*, Chaos Solitons Fractals **181** (2024), Paper No. 114655, 34 pp., DOI 10.1016/j.chaos.2024.114655. MR4710914

[3] R. Achour, Z. Li, B. Selmi, and T. Wang, *General fractal dimensions of graphs of products and sums of continuous functions and their decompositions*, J. Math. Anal. Appl. **538** (2024), no. 2, Paper No. 128400, 26, DOI 10.1016/j.jmaa.2024.128400. MR4735205

[4] R. Achour and B. Selmi, *Some properties of new general fractal measures*, Monatsh. Math. **204** (2024), no. 4, 659–678, DOI 10.1007/s00605-024-01979-7. MR4772957

[5] R. Achour and B. Selmi, *General fractal dimensions of typical sets and measures*, Fuzzy Sets and Systems **490** (2024), Paper No. 109039, 16, DOI 10.1016/j.fss.2024.109039. MR4759490

[6] N. Attia and B. Selmi, *Regularities of multifractal Hewitt-Stromberg measures*, Commun. Korean Math. Soc. **34** (2019), no. 1, 213–230, DOI 10.4134/CKMS.c180030. MR3912188

[7] N. Attia and B. Selmi, *A multifractal formalism for Hewitt-Stromberg measures*, J. Geom. Anal. **31** (2021), no. 1, 825–862, DOI 10.1007/s12220-019-00302-3. MR4203668

[8] F. Ben Nasr and J. Peyrière, *Revisiting the multifractal analysis of measures*, Rev. Mat. Iberoam. **29** (2013), no. 1, 315–328, DOI 10.4171/RMI/721. MR3010130

[9] F. Ben Nasr, I. Bhouri, and Y. Heurteaux, *The validity of the multifractal formalism: results and examples*, Adv. Math. **165** (2002), no. 2, 264–284, DOI 10.1006/aima.2001.2025. MR1887585

[10] I. Ben Omrane, *Multifractal analysis of anisotropic and directional pointwise regularities for measures*, Chaos Solitons Fractals **183** (2024), Paper No. 114934, 9 pp., DOI 10.1016/j.chaos.2024.114934. MR4744083

[11] G. Brown, G. Michon, and J. Peyrière, *On the multifractal analysis of measures*, J. Statist. Phys. **66** (1992), no. 3-4, 775–790, DOI 10.1007/BF01055700. MR1151978

[12] D. Cheng, Z. Li, and B. Selmi, *On the general fractal dimensions of hyperspace of compact sets*, Fuzzy Sets and Systems **488** (2024), Paper No. 108998, 14 pp., DOI 10.1016/j.fss.2024.108998. MR4754653

[13] S. Doria and B. Selmi, *Conditional aggregation operators defined by the Choquet integral and the Sugeno integral with respect to general fractal measures*, Fuzzy Sets and Systems **477** (2024), Paper No. 108811, 13 pp., DOI 10.1016/j.fss.2023.108811. MR4673581

[14] Z. Douzi and B. Selmi, *On the mutual singularity of Hewitt-Stromberg measures for which the multifractal functions do not necessarily coincide*, Ric. Mat. **72** (2023), no. 1, 1–32, DOI 10.1007/s11587-021-00572-6. MR4583879

[15] ______, *Projection theorems for Hewitt-Stromberg and modified intermediate dimensions.* Results Math., **77** (2022), no.4, Paper No. 158, 14 pp.

[16] ______, *Multifractal analysis of Hewitt-Stromberg measures with respect to gauge control functions.* Topol. Methods Nonlinear Anal., (to appear).

[17] Z. Douzi, B. Selmi & A. Ben Mabrouk, *The refined multifractal formalism of some homogeneous Moran measures.* Eur. Phys. J. Spec. Top., **230** (2021), 3815–3834.

[18] U. Frisch, *Fully developed turbulence: where do we stand?*, Dynamical systems: a renewal of mechanism, World Sci. Publ., Teaneck, NJ, 1986, pp. 13–28. MR1024082

[19] H. Haase, *A contribution to measure and dimension of metric spaces*, Math. Nachr. **124** (1985), 45–55, DOI 10.1002/mana.19851240104. MR827889

[20] H. Haase, *Open-invariant measures and the covering number of sets*, Math. Nachr. **134** (1987), 295–307, DOI 10.1002/mana.19871340121. MR918685

[21] H. Haase, *The dimension of analytic sets*, Acta Univ. Carolin. Math. Phys. **29** (1988), no. 2, 15–18. MR983445

[22] H. Haase, *Dimension functions*, Math. Nachr. **141** (1989), 101–107, DOI 10.1002/mana.19891410112. MR1014419

[23] H. Haase, *Fundamental theorems of calculus for packing measures on the real line*, Math. Nachr. **148** (1990), 293–302, DOI 10.1002/mana.3211480119. MR1127348

[24] T. C. Halsey, M. H. Jensen, L. P. Kadanoff, I. Procaccia, and B. I. Shraiman, *Fractal measures and their singularities: the characterization of strange sets*, Phys. Rev. A (3) **33** (1986), no. 2, 1141–1151, DOI 10.1103/PhysRevA.33.1141. MR823474

[25] E. Hewitt and K. Stromberg, *Real and abstract analysis. A modern treatment of the theory of functions of a real variable*, Springer-Verlag, New York, 1965. MR188387

[26] L. Huang, Q. Liu, and G. Wang, *Multifractal analysis of Bernoulli measures on a class of homogeneous Cantor sets*, J. Math. Anal. Appl. **491** (2020), no. 2, 124362, 15, DOI 10.1016/j.jmaa.2020.124362. MR4123243

[27] Z. Li and B. Selmi, *On the multifractal analysis of measures in a probability space*, Illinois J. Math. **65** (2021), no. 3, 687–718, DOI 10.1215/00192082-9446058. MR4312199

[28] B. Mandelbrot, *Les objets fractals* (French), Nouvelle Bibliothèque Scientifique. [New Scientific Library], Flammarion, Éditeur, Paris, 1975. Forme, hasard et dimension. MR462040

[29] ———, *The fractal geometry of nature*. W. H. Freeman, (1982).

[30] M. Menceur and A. Ben Mabrouk, *A joint multifractal analysis of vector valued non Gibbs measures*, Chaos Solitons Fractals **126** (2019), 203–217, DOI 10.1016/j.chaos.2019.05.010. MR3962802

[31] L. Olsen, *A multifractal formalism*, Adv. Math. **116** (1995), no. 1, 82–196, DOI 10.1006/aima.1995.1066. MR1361481

[32] L. Olsen, *On average Hewitt-Stromberg measures of typical compact metric spaces*, Math. Z. **293** (2019), no. 3-4, 1201–1225, DOI 10.1007/s00209-019-02239-3. MR4024583

[33] J. Peyrière, *A vectorial multifractal formalism*, Fractal geometry and applications: a jubilee of Benoît Mandelbrot, Part 2, Proc. Sympos. Pure Math., vol. 72, Amer. Math. Soc., Providence, RI, 2004, pp. 217–230, DOI 10.1090/pspum/072.2/2112124. MR2112124

[34] A. Samti, *Multifractal formalism of an inhomogeneous multinomial measure with various parameters*, Extracta Math. **35** (2020), no. 2, 229–252, DOI 10.17398/2605-5686.35.2.229. MR4173597

[35] B. Selmi, *On the projections of the multifractal Hewitt-Stromberg dimensions*, Filomat **37** (2023), no. 15, 4869–4880. MR4576621

[36] ———, *Subsets of positive and finite $\Psi_t$-Hausdorff measures and applications.* J. Geom. Anal., **34** (2024), no.3, Paper No. 79, 16 pp.

[37] B. Selmi, *A note on the multifractal Hewitt-Stromberg measures in a probability space*, Korean J. Math. **28** (2020), no. 2, 323–341, DOI 10.11568/kjm.2020.28.2.323. MR4128263

[38] B. Selmi, *The relative multifractal analysis, review and examples*, Acta Sci. Math. (Szeged) **86** (2020), no. 3-4, 635–666, DOI 10.14232/actasm-020-801-8. MR4174797

[39] B. Selmi, *A review on multifractal analysis of Hewitt-Stromberg measures*, J. Geom. Anal. **32** (2022), no. 1, Paper No. 12, 44, DOI 10.1007/s12220-021-00753-7. MR4349456

[40] B. Selmi, *Average Hewitt-Stromberg and box dimensions of typical compact metric spaces*, Quaest. Math. **46** (2023), no. 3, 411–444, DOI 10.2989/16073606.2022.2033338. MR4576823

[41] B. Selmi, *Slices of Hewitt-Stromberg measures and co-dimensions formula*, Analysis (Berlin) **42** (2022), no. 1, 23–39, DOI 10.1515/anly-2021-1005. MR4385597

[42] B. Selmi, *The mutual singularity of multifractal measures for some non-regular Moran fractals*, Bull. Pol. Acad. Sci. Math. **69** (2021), no. 1, 21–35, DOI 10.4064/ba210216-9-10. MR4336308

[43] B. Selmi, *On the projections of the multifractal Hewitt-Stromberg dimensions*, Filomat **37** (2023), no. 15, 4869–4880. MR4576621

[44] S. Shen, *Multifractal analysis of some inhomogeneous multinomial measures with distinct analytic Olsen's b and B functions*, J. Stat. Phys. **159** (2015), no. 5, 1216–1235, DOI 10.1007/s10955-015-1223-z. MR3345417

[45] C. Tricot Jr., *Two definitions of fractional dimension*, Math. Proc. Cambridge Philos. Soc. **91** (1982), no. 1, 57–74, DOI 10.1017/S0305004100059119. MR633256

[46] M. Wu, *The singularity spectrum $f(\alpha)$ of some Moran fractals*, Monatsh. Math. **144** (2005), no. 2, 141–155, DOI 10.1007/s00605-004-0254-3. MR2123961

[47] M. Wu and J. Xiao, *The singularity spectrum of some non-regularity Moran fractals*, Chaos Solitons Fractals **44** (2011), no. 7, 548–557, DOI 10.1016/j.chaos.2011.05.002. MR2813033

[48] J. Xiao and M. Wu, *The multifractal dimension functions of homogeneous Moran measure*, Fractals **16** (2008), no. 2, 175–185, DOI 10.1142/S0218348X08003892. MR2421995

[49] Z. Yuan, *Multifractal spectra of Moran measures without local dimension*, Nonlinearity **32** (2019), no. 12, 5060–5086, DOI 10.1088/1361-6544/ab45d7. MR4030604

[50] O. Zindulka, *Packing measures and dimensions on Cartesian products*, Publ. Mat. **57** (2013), no. 2, 393–420, DOI 10.5565/PUBLMAT_57213_06. MR3114775

School of Mathematics, Northwest University Xi'an, Shannxi, 710127, People's Republic of China

*Email address*: `wtingting@stumail.nwu.edu.cn`

Analysis, Probability and Fractals Laboratory LR18ES17, Department of Mathematics, Faculty of Sciences of Monastir, University of Monastir, 5000-Monastir, Tunisia

*Email address*: `bilel.selmi@fsm.rnu.tn; bilel.selmi@isetgb.rnu.tn`

School of Mathematics, Northwest University Xi'an, Shannxi, 710127, People's Republic of China

*Email address*: `china-lizhiming@163.com`

Contemporary Mathematics
Volume **825**, 2025
https://doi.org/10.1090/conm/825/16516

# Proof of Wojtkowski's Falling Particle Conjecture

Nandor Simanyi

ABSTRACT. In this paper we present an unconditional proof of Wojtkowski's Ergodicity Conjecture for almost every system of $1D$ perfectly elastic balls falling down in a half line under constant gravitational acceleration. Namely, by introducing a new algebraic approach, we prove that almost every such system is (completely hyperbolic and) ergodic.

## 1. Introduction/Prerequisites

In order to introduce the subject of our investigation, the system of $1D$ falling balls subjected to constant gravitation, along with the employed technicalities, we will be closely following the first two sections of [**S2024**]. In order to make this presentation self-contained and easier to read, we quote below two passages of those two sections of [**S2024**], essentially verbatim.

In his paper [**W1990a**] M. Wojtkowski introduced the following Hamiltonian dynamical system with discontinuities: There is a vertical half line $\{q \mid q \geq 0\}$ given and $n$ ($\geq 2$) point particles with masses $m_1 \geq m_2 \geq \cdots \geq m_n > 0$ and positions $0 \leq q_1 \leq q_2 \leq \cdots \leq q_n$ are moving on this half line so that they are subjected to a constant gravitational acceleration $a = -1$ (they fall down), they collide elastically with each other, and the first (lowest) particle also collides elastically with the hard floor $q = 0$. We fix the total energy

$$H = \sum_{i=1}^{n} \left( m_i q_i + \frac{1}{2} m_i v_i^2 \right),$$

by taking $H = 1$. The arising Hamiltonian flow with collisions $(\mathbf{M}, \{\psi^t\}, \mu)$ ($\mu$ is the Liouville measure) is the studied model of this paper.

Before formulating the result of this article, however, it is worth mentioning here three important facts:

(1) Since the phase space $\mathbf{M}$ is compact, the Liouville measure $\mu$ is finite.
(2) The phase points $x \in \mathbf{M}$ for which the trajectory $\{\psi^t(x)|, t \in \mathbb{R}\}$ hits at least one singularity (i. e. a multiple collision) are contained in a countable union of proper, smooth submanifolds of $\mathbf{M}$ and, therefore, such points form a set of $\mu$ measure zero.

2020 *Mathematics Subject Classification.* Primary 37D05.

(3) For $\mu$-almost every phase point $x \in \mathbf{M}$ the collision times of the trajectory $\{\psi^t(x)|, t \in \mathbb{R}\}$ do not have any finite accumulation point, see Proposition A.1 of [**S1996**].

In the paper [**W1990a**] Wojtkowski formulated his main conjecture pertaining to the dynamical system $(\mathbf{M}, \{\psi^t\}, \mu)$:

CONJECTURE 1.1 (Wojtkowski's Conjecture). *If $m_1 \geq m_2 \geq \cdots \geq m_n > 0$ and $m_1 \neq m_n$, then all but one characteristic (Lyapunov) exponents of the flow $(\mathbf{M}, \{\psi^t\}, \mu)$ are nonzero. Futhermore, the system is ergodic.*

REMARK 1.1.

1. The only exceptional exponent zero must correspond to the flow direction.
2. The condition of nonincreasing masses (as above) is essential for establishing the invariance of the symplectic cone field — an important condition for obtaining nonzero characteristic exponents. As Wojtkowski pointed out in Proposition 4 of [**W1990a**], if $n = 2$ and $m_1 < m_2$, then there exists a linearly stable periodic orbit, thus dimming the chances of proving ergodicity.

In the paper [**S2024**] we proved Wojtkowski's Ergodicity Conjecture 1.1 for almost every selection of masses $m_1 > m_2 > \cdots > m_n$, provided that the Transversality Conditions (Claim 3.1 of [**S2024**]) is verified, i. e. singularities of different order are transversal to each other and, analogously, the stable and unstable local invariant manifolds are transversal to all singularities.

Here our main result is to prove the above Transversality Conditions and, as the main corollary, we obtain our main result.

THEOREM 1.2 (Main Theorem). *For almost every selection of masses $m_1 > m_2 > \cdots > m_n$ the falling ball flow $(M, \{\psi^t\}, \mu)$ is (completely hyperbolic and) ergodic.*

We recall that the corresponding billiard map (Poincaré section) $(\partial M, T, \nu)$ is an invertible dynamical system $T$ mapping the boundary

$$\partial M = \{(q, v) \in M \mid q \in \partial Q\}$$

of the phase space $M$ onto itself and preserving the finite measure $\nu$ on $\partial M$ that can be obtained by projecting along the flow the invariant measure $\mu$ of the flow onto $\partial M$. Also, as usual, in $\partial M$ one identifies the pre-collision phase point $(q, v^-) \in \partial M$ with the post-collision phase point $(q, v^+) \in \partial M$.

Finally, denote by

$$\Pi : \partial M \to \partial Q$$

the natural projection of $\partial M$ onto $\partial Q$, i. e. $\Pi(q, v) = q$.

## 2. The Geometry of the Symplectic Flow

We will be working with the symplectic coordinates $(\delta h, \delta v)$ for the tangent vectors of the reduced phase space $\mathbf{M}$ satisfying the usual reduction equations $\sum_{i=1}^n \delta h_i = 0 = \sum_{i=1}^n \delta v_i$.

REMARK 2.1. The coordinates $\delta h_i$ and $\delta v_i$ $(i = 1, 2, \ldots, n)$ serve as suitable symplectic coordinates in the codimension-one subspace $\mathcal{T}_x$ of the full tangent space

$\mathcal{T}_x\mathbf{M}$ of $\mathbf{M}$ at $x$. Recall that the $(2n-2)$-dimensional vector space $\mathcal{T}_x$ is transversal to the flow direction, and the restriction of the canonical symplectic form

$$\omega = \sum_{i=1}^{n} \delta q_i \wedge \delta p_i = \sum_{i=1}^{n} \delta h_i \wedge \delta v_i$$

of $\mathbf{M}$ is non-degenerate on $\mathcal{T}_x$, see [**W1990a**]. We also recall that

$$\delta h_i = m_i \delta q_i + m_i v_i \delta v_i = m_i \delta q_i + v_i \delta p_i.$$

Corresponding to the above choice of symplectic coordinates, the considered monotone Q-form will be

$$Q_1(\delta h, \delta v) = \sum_{i=1}^{n} \delta h_i \delta v_i. \tag{2.1}$$

It is clear that the evolution of $DS^t(\delta h(0), \delta v(0)) = (\delta h(t), \delta v(t))$ between colisions is

$$\frac{d}{dt}(\delta h(t), \delta v(t)) = (0, 0). \tag{2.2}$$

If a collision of type $(i, i+1)$ $(i = 1, 2, \dots, n-1)$ takes place at time $t_k$, then the derivative of the flow at the collision $\delta h^-(t_k) \mapsto \delta h^+(t_k)$, $\delta v^-(t_k) \mapsto \delta v^+(t_k)$ is given by the matrices

$$\begin{aligned} \delta h^+(t_k) &= R_i^* \left[\delta h^-(t_k) + S_i \delta v^-(t_k)\right] \\ \delta v^+(t_k) &= R_i \delta v^-(t_k), \end{aligned} \tag{2.3}$$

where the matrix $R_i$ is the $n \times n$ identity matrix, except that its $2 \times 2$ submatrix at the crossings of the $i$-th and $(i+1)$-st rows and columns is

$$R_i^{(i,i+1)} = \begin{bmatrix} \gamma_i & 1-\gamma_i \\ 1+\gamma_i & -\gamma_i \end{bmatrix}$$

with $\gamma_i = \dfrac{m_i - m_{i+1}}{m_i + m_{i+1}}$. The matrix $S_i$ is, similarly, the $n \times n$ zero matrix, except its $2 \times 2$ submatrix at the crossings of the $i$-th and $(i+1)$-st rows and columns, which takes the form of

$$S_i^{(i,i+1)} = \begin{bmatrix} \alpha_i & -\alpha_i \\ -\alpha_i & \alpha_i \end{bmatrix}$$

with

$$\alpha_i = \frac{2m_i m_{i+1}(m_i - m_{i+1})}{(m_i + m_{i+1})^2}(v_i^- - v_{i+1}^-) > 0. \tag{2.4}$$

These formulas can be found, for example, in Sention 4 of [**W1990a**].
Concerning a floor collision $(0, 1)$ at time $t_k$, the transformations are

$$\begin{aligned} \delta h_1^+(t_k) &= \delta h_1^-(t_k) \\ \delta v_1^+(t_k) &= \delta v_1^-(t_k) + \frac{2\delta h_1^-(t_k)}{m_1 v_1^+(t_k)}, \end{aligned} \tag{2.5}$$

see, for instance, Section 4 of [**W1990a**] or [**W1998**].

## 3. Proof of the Transversality Conditions

If one closely studies the ergodicity proofs based upon the Birkhoff-Sinai Zig-zag Method, like the one in [**L-W1995**], one realizes that, whenever a property is needed to be proved for singular phase points $x_0 = (q_0, v_0) \in \mathcal{S}_0$ (like the transversality of $\mathcal{S}_k$ to $\mathcal{S}_0$ at $x_0$, or the transversality of the local stable manifold $\gamma^s(x_0)$ to $\mathcal{S}_0$), it is always enough to estabslish the required property for almost every point $x_0$ of $\mathcal{S}_0$ with respect to the hypersurface measure of $\mathcal{S}_0$. This is what we do in this section.

We will be focusing on the billiard map $(\partial M, T, \nu)$. Our first result asserts that the system $(\partial M, T, \nu)$ has no focal points almost surely almost everywhere.

PROPOSITION 3.1. *For almost every selection of the masses* $m_1 > m_2 > \cdots > m_n$ *it is true that for almost every phase point* $(q_0, v_0) \in \partial M$, *for every positive integer* $k$, *and for every small enough* $\epsilon > 0$ *the map*

$$\Pi \circ T^k : C_\epsilon(x_0) \to \partial Q \tag{3.2}$$

*is of full rank (i. e. locally onto) at* $x_0$, *where*

$$C_\epsilon(x_0) = \{(q_0, v) \in \partial M \mid \|v - v_0\| < \epsilon\} \tag{3.3}$$

*is the so called "candle manifold", and*

$$\Pi : \partial M \to \partial Q \tag{3.4}$$

*is the natural projection, taking* $\Pi(q, v) = q$ *for* $(q, v) \in \partial M$.

PROOF. It is enough to prove the proposition for a given $k$ and a given symbolic collision sequence $\Sigma = (\sigma_0, \sigma_1, \dots, \sigma_k)$, where $\sigma_l(i_l, i_l + 1)$ tells that the collision $(i_l, i_l + 1)$ takes place at $T^l(x_0)$, where $i_l = 0$ indicates a floor collision.

According to (9) of [**W1990a**], the time evolution (abrupt change) at a ball-to-ball collision $(i, i + 1)$ is given by

$$\begin{aligned} v_i^+ &= \gamma_i v_i^- + (1 - \gamma_i) v_{i+1}^- \\ v_{i+1}^+ &= (1 + \gamma_i) v_i^- - \gamma_i v_{i+1}^-, \end{aligned} \tag{3.5}$$

where

$$\gamma_i = \frac{m_i - m_{i+1}}{m_i + m_{i+1}},$$

and

$$v_1^+ = -v_1^-, \tag{3.6}$$

for any floor collision. It is obvious that the time evolution between collisions is given by

$$\frac{d}{dt} q_t = v_t, \ \frac{d}{dt} v_t = (-1, -1, \dots, -1). \tag{3.7}$$

This means that the entire trajectory segment $\{T^l(x_0) \mid l = 0, 1, \dots, k\}$ is governed by rational functions of the initial data $(q_0, v_0, \vec{m})$, including the moments $t_l$ of the collisions $\sigma_l$.

Furthermore, by taking derivative of the flow (3.5)–(3.7), one obtains that for any tangent vector

$$\tau_0 = (\delta q_0, \delta v_0) \in \mathcal{T}_{x_0} \partial M,$$

the images $D\psi^t[\tau_0] = \tau_t$ are evolving in time as follows:

$$\begin{aligned} \delta v_i^+ &= \gamma_i \delta v_i^- + (1-\gamma_i)\delta v_{i+1}^- \\ \delta v_{i+1}^+ &= (1+\gamma_i)\delta v_i^- - \gamma_i \delta v_{i+1}^-, \end{aligned} \tag{3.8}$$

$$\begin{aligned} \delta q_i^+ &= \gamma_i \delta q_i^- + (1-\gamma_i)\delta q_{i+1}^- \\ \delta q_{i+1}^+ &= (1+\gamma_i)\delta q_i^- - \gamma_i \delta q_{i+1}^-, \end{aligned} \tag{3.9}$$

for $i > 0$, where $\delta v_j^\pm = (\delta v^\pm)_{t_j}$, $\delta q_j^\pm = (\delta q^\pm)_{t_j}$, for $j = i, i+1$, see (5) in [**W1990b**].

At a floor collision $(0,1)$ we clearly have

$$\delta q_1^+ = -\delta q_1^-, \ \delta v_1^+ = -\delta v_1^-. \tag{3.10}$$

It is also clear that the time evolution of the image tangent vector $\tau_t = D\psi^t[\tau_0] = (\delta q_t, \delta v_t)$ between collisions is given by

$$\frac{d}{dt}\delta q_t = \delta v_t, \ \frac{d}{dt}\delta v_t = 0. \tag{3.11}$$

It follows from (3.8)–(3.11) that the time evolution of the tangent vectors $\tau_t = D\psi^t[\tau_0] = (\delta q_t, \delta v_t)$ is also fully governed by finitely many rational functions of the initial data $(q_0, v_0, \vec{m})$.

Finally, the negation of the assertion of the Proposition for a particular phase point $x_0 = (q_0, v_0)$ means that the system of homogeneous linear equations

$$\delta q_{t_k} = 0 \tag{3.12}$$

has a nontrivial solution $\tau_0 = (0, \delta v_0)$. This, in turn, means that cetain minors of this system vanish, i.e.

$$R_j(q_0, v_0, \vec{m}) = 0, \ j = 1, 2, \dots, m \tag{3.13}$$

for certain rational functions $R_1$, $R_2$, ..., $R_m$ of the initial variables.

Consider now the limiting system with

$$m_1 = m_2 = \cdots = m_n > 0. \tag{3.14}$$

This system may not possess a trajectory segment with our prescribed symbolic collision sequence $\Sigma = (\sigma_0, \sigma_1, \dots, \sigma_k)$, yet all the rational functions $R_j$ above are well defined, and they correspond to the time evolution of this limiting system, so that the unwanted collisions are annihilated in such a way that any two particles, about to making an unwanted collision, are let to penetrate through each other without interaction. This limiting system is essentially integrable, after the dynamic change of labels at collisions, as if the particles just penetrate through each other without interaction, see also the paragraph right after Corollary 2.23 in [**S2024**].

In this limiting case, however, we have that $\delta q_t = t\delta v_0$, after the mentioned dynamic change of labels. This means, in turn, that the system of homogeneous linear equations (3.12) only has the trivial solution $\tau_0 = 0$, therefore, not all rational functions $R_j$ are identically zero, even if we assume $m_1 = m_2 = \cdots = m_n$.

This completes the proof of Proposition 3.1. □

Our next result is analogous to the statement of Proposition 3.1, claimed here for almost every singular phase point $x_0 = (q_0, v_0) \in \mathcal{S}_0$.

PROPOSITION 3.15. *For almost every selection of the masses $m_1 > m_2 > \cdots > m_n$ and for every positive integer $k$ it is true that for almost every singular phase point $x_0 = (q_0, v_0) \in \mathcal{S}_0$ and for all small enough $\epsilon > 0$ the map in (3.2) is of maximum rank (i. e. locally onto) at $x_0$.*

PROOF. In Proposition 3.1, $q_0$ possessed $n-1$ independent coordinates. (Keep in mind, that $q_0 \in \partial Q$, so the $q$ coordinates of two colliding particles are equal, or $q_1 = 0$.) When a singular collision takes place at time zero, i. e. $x_0 = (q_0, v_0) \in \mathcal{S}_0$, then one more equation holds true for the configuration coordinates, so one more coordinate should be eliminated to work with independent configuration coordinates. The arising vector with $n-2$ coordinates is denoted by $\tilde{q}_0$. After this, the proof of this proposition is verbatim the same as that of Proposition 3.1, except that $q_0$ needs to be replaced everywhere by $\tilde{q}_0$. □

COROLLARY 3.16 ((Transversality of singularities of different order)). *For almost every selection of the masses $m_1 > m_2 > \cdots > m_n$ and for every positive integer $k$ it is true that for almost every singular phase point $x_0 = (q_0, v_0) \in \mathcal{S}_0$ if $x_k = T^k(x_0)$ happens to belong to $\mathcal{S}_0$, then $T^k(\mathcal{S}_0)$ and $\mathcal{S}_0$ are transversal at $x_k$.*

PROOF. The statement immediately follows from the facts that

($i$) $C_\epsilon(x_0) \subset \mathcal{S}_0$,
($ii$) $\Pi \circ T^k : C_\epsilon(x_0) \to \partial Q$ is locally onto,
($iii$) $\mathcal{S}_0$ is defined purely in terms of the $q$ coordinates. □

COROLLARY 3.17. *For almost every selection of the masses $m_1 > m_2 > \cdots > m_n$ and for almost every singular phase point $x_0 = (q_0, v_0) \in \mathcal{S}_0$ the local stable manifold $\gamma^s(x_0)$ of $x_0$ is transversal to the singularity $\mathcal{S}_0$.*

PROOF. This statement follows from the fact that $\gamma^s(x_0)$ is the $C^1$-uniform limit (locally, near $x_0$) in the following way:

$$\gamma^s(x_0) = \lim_{k\to\infty} T^{-k}\left[C_\epsilon(T^k(x_0))\right],$$

the inverse images $T^{-k}\left[C_\epsilon(T^k(x_0))\right]$ project locally onto $\partial Q$ at $x_0$, and, finally, the maximum rank property is a $C^1$ open property. Hence the projection

$$\Pi : \gamma^s(x_0) \to \partial Q$$

is locally onto, and $\mathcal{S}_0$ is defined purely in terms of the $q$ coordinates in $\partial Q$. □

This completes the proof of our Main Theorem.

## References

[Ch1993] N. I. Chernov, *Local ergodicity of hyperbolic systems with singularities* (Russian), Funktsional. Anal. i Prilozhen. **27** (1993), no. 1, 60–64, DOI 10.1007/BF01768668; English transl., Funct. Anal. Appl. **27** (1993), no. 1, 51–54. MR1225911

[L-W1995] C. Liverani and M. P. Wojtkowski, *Ergodicity in Hamiltonian systems*, Dynamics reported, Dynam. Report. Expositions Dynam. Systems (N.S.), vol. 4, Springer, Berlin, 1995, pp. 130–202. MR1346498

[K-S-Sz1992] A. Krámli, N. Simányi, and D. Szász, *The K-property of four billiard balls*, Comm. Math. Phys. **144** (1992), no. 1, 107–148. MR1151248

[S1996] N. Simányi, *The characteristic exponents of the falling ball model*, Comm. Math. Phys. **182** (1996), no. 2, 457–468. MR1447300

[S2024] N. Simanyi, *Conditional proof of the ergodic conjecture for falling ball systems*, Recent developments in fractal geometry and dynamical systems, Contemp. Math., vol. 797, Amer. Math. Soc., Providence, RI, 2024, pp. 75–83, DOI 10.1090/conm/797/15935. MR4730763

[W1985] M. Wojtkowski, *Invariant families of cones and Lyapunov exponents*, Ergodic Theory Dynam. Systems **5** (1985), no. 1, 145–161, DOI 10.1017/S0143385700002807. MR782793

[W1986] M. Wojtkowski, *Principles for the design of billiards with nonvanishing Lyapunov exponents*, Comm. Math. Phys. **105** (1986), no. 3, 391–414. MR848647

[W1990a] M. P. Wojtkowski, *A system of one-dimensional balls with gravity*, Comm. Math. Phys. **126** (1990), no. 3, 507–533. MR1032871

[W1990b] M. P. Wojtkowski, *The system of one-dimensional balls in an external field. II*, Comm. Math. Phys. **127** (1990), no. 2, 425–432. MR1037112

[W1998] M. P. Wojtkowski, *Hamiltonian systems with linear potential and elastic constraints*, Fund. Math. **157** (1998), no. 2-3, 305–341, DOI 10.4064/fm-157-2-3-305-341. Dedicated to the memory of Wiesław Szlenk. MR1636897

1402 10th Avenue South, Birmingham Alabama 35294-1241
*Email address*: simanyi@uab.edu

Contemporary Mathematics
Volume **825**, 2025
https://doi.org/10.1090/conm/825/16517

# Further analysis of Hausdorff dimension and separation conditions

Saurabh Verma and Amit Priyadarshi

ABSTRACT. The Hausdorff dimension of sets of Euclidean spaces has received a lot of attention in the area of Fractal Geometry. Many researchers have made efforts to relate this dimension to some topological and measure-theoretic properties such as connectedness, smoothness, denseness, length, area, and volume. In this paper, building on some interesting properties of the Hausdorff measure and Hausdorff dimension, we construct some classes of compact sets that are dense in the class of all non-empty compact subsets with respect to the Hausdorff metric. Further, we prove a weaker version of Hochman's theorem for the Hausdorff dimension of self-similar sets and also obtain some remarks and examples for attractors generated by iterated function systems (IFS) consisting of all bi-Lipschitz or affine mappings under various separation conditions. We also collect some interesting dense subsets in the class of Borel probability measures and study separation conditions and invariant measures for the product of two IFSs.

## 1. Introduction

Exact computation or estimation of the Hausdorff dimension of fractal sets has been the central theme of the dimension theory in fractal geometry. In literature, many articles are available on the study of the dimension of the attractor of an Iterated Function System (IFS); see, for instance, [**1, 2, 8, 9, 11, 13, 24**]. To see the fractal dimension of various functions in the different function spaces or generated as attractors of some IFSs with their applications, one may consult [**5, 10, 15, 25, 28–30**].

Let $\mathbb{R}^m$ be equipped with the Euclidean metric $d$, where $m \in \mathbb{N}$. The system $\{\mathbb{R}^m; f_1, f_2, \cdots, f_N\}$ is called an IFS [**13**], where $N \geq 2 \in \mathbb{N}$ and each $f_i$ is a contraction map, that is, for some constant $0 \leq c_i < 1$ (known as contraction ratio of $f_i$), $d(f_i(x), f_i(y)) \leq c_i d(x,y)$ for all $x, y \in \mathbb{R}^m$. Let $(p_i)_{i=1}^N$ be probability vector, that is, $\sum_{i=1}^N p_i = 1$ and $p_i \geq 0$. Hutchinson [**13**] showed the existence of a unique compact set $A$ (widely known as attractor or invariant/fractal set) and a unique Borel probability measure $\mu_p$ (called as invariant/fractal measure) such that $A = \cup_{i=1}^N f_i(A)$ and $\mu_p = \sum_{i=1}^N p_i \mu_p \circ f_i^{-1}$. It is also known that if $p_i > 0$ then the

2020 *Mathematics Subject Classification.* Primary 28A80; Secondary 28A78.
*Key words and phrases.* Separation condition, self-similar sets, invariant measures, Hausdorff measure, Hausdorff dimension, box dimension.
The first author is supported by the SEED grant project of IIIT Allahabad, India.

support of $\mu_p$ will be the attractor $A$. Researchers establish certain results when all mappings $f_i$ of an IFS belong to certain class of functions. Main interests are in the following type of classes. A map $f_i : \mathbb{R}^m \to \mathbb{R}^m$ is called

(1) similarity transformation or similitude if $d(f_i(x), f_i(y)) = c_i d(x, y)$, $\forall\, x$, $y \in \mathbb{R}^m$. Here $c_i$ is called the similarity ratio of $f_i$.
(2) bi-Lipschitz if $c_i d(x, y) \le d(f_i(x), f_i(y)) \le C_i d(x, y)$, $\forall\, x, y \in \mathbb{R}^m$.
(3) an affine transformation if $f_i(x) = A_i x + b_i$, where $A_i$ is a linear transformation on $\mathbb{R}^m$ and $b_i \in \mathbb{R}^m$ is a fixed vector.

Based on all $f_i$, we call the associated attractor/invariant measure a self-similar, self-affine, and self-conformal set/measure, respectively. With the *open set condition* (OSC), Moran [**18**] and Hutchinson [**13**] proved that $\dim_H(A) = \min\{s, m\}$, where $s$ (often called similarity dimension of $A$) is a unique number such that $\sum_{i=1}^{N} c_i^s = 1$, $c_i$ is the similarity ratio of $f_i$, and $\dim_H(.)$ denotes the Hausdorff dimension. In general, without any separation conditions, we have $\dim_H(A) \le \min\{s, m\}$, and in case of exact/total overlap of two cylinders, we have a dimension drop, that is, $\dim_H(A) < \min\{s, m\}$. In 1996, Simon [**26**] posed that the dimension drop can only occur due to exact overlappings. This became a conjecture in $\mathbb{R}$, widely known as total overlap conjecture for self-similar sets. A groundbreaking progress came through [**11**], where Hochman showed that $\dim_H(A) = \min\{s, 1\}$ in $\mathbb{R}$ holds with a weaker separation condition called exponential separation condition. As a corollary to [[**11**], Theorem 1.5], we may say that the conjecture is true for all IFS on $\mathbb{R}$ with similarity mappings defined by using algebraic parameters, that is, $f_i(x) = a_i x + b_i$, where $a_i, b_i$ are algebraic numbers. Recently, Rapaport [**21**] proved that the conjecture for IFS with algebraic contractions $a_i$'s and arbitrary translations $b_i$'s. We encourage the reader to see few recent advances done by Rapaport and Varjú [**22, 27**] in similar issue. For a self-similar measure $\mu_p$, we always have $\dim_H(\mu_p) \le \min\{\dim_S(\mu_p), m\}$, where $\dim_S(\mu_p)$ is the similarity dimension of $\mu_p$ given by

$$\dim_S(\mu_p) = \frac{\sum_{i=1}^{N} p_i \log(p_i)}{\sum_{i=1}^{N} p_i \log(c_i)}.$$

In 1991, Deliu et al. [**7**] proved that if an IFS consisting of similarity mappings satisfies the OSC, then $\dim_H(\mu_p) = \min\{\dim_S(\mu_p), m\}$. Due to equality of the support of $\mu_p$ and the attractor $A$, one may transfer results from measures to sets. Our current work is motivated by this dimension-based research. So, we mainly explore separation conditions, the Hausdorff measure, and the dimension of sets and Borel measures. We also construct some dense sets and show a weaker version of Hochman's main result for self-similar sets. We believe the remarks and examples included in the article will definitely make this article attractive and fragrant.

The paper is arranged as follows. In Section 2, we collect all required definitions and needed background. In Section 3, we prove our main results and provide some future aspects of our paper.

## 2. Preliminaries

In this section, we recall the basic definitions and refer the reader to [**2**, **4**, **8**, **14**] for a more detailed exposition.

DEFINITION 2.1. Let $A, B$ be nonempty compact sets in $\mathbb{R}^m$. The Hausdorff metric is defined as $h_d(A, B) = \inf\{\epsilon > 0 : A \subseteq B^\epsilon \text{ and } B \subseteq A^\epsilon\}$, where $A^\epsilon = \{x \in \mathbb{R}^m : d(x, y) < \epsilon \text{ for some } y \in A\}$. It is well-known [**2**, **8**] that the set of all nonempty compact subsets of $\mathbb{R}^m$, denoted by $\mathcal{K}(\mathbb{R}^m)$, is complete with $h_d$.

DEFINITION 2.2. The Hausdorff dimension of $A \subset \mathbb{R}^m$ is defined as follows:

$$\dim_H(A) = \inf\left\{\alpha > 0 : \text{for each } \epsilon > 0,\ \exists \text{ a cover } \{V_i\}_{i=1}^\infty \text{ of } A \text{ with } \sum_{i=1}^\infty |V_i|^\alpha < \epsilon\right\},$$

where $|V_i|$ denotes the diameter of $V_i$.

DEFINITION 2.3. Let $A$ be a non-empty bounded subset of $\mathbb{R}^m$. The box dimension of $A$ is defined as

$$\dim_B(A) = \lim_{\delta \to 0^+} \frac{\log N_\delta(A)}{-\log \delta},$$

where $N_\delta(A)$ denotes the smallest number of sets of diameter at most $\delta$ that can cover $A$, provided the limit exists. If this limit does not exist, then the upper and lower box dimensions, respectively, are defined by taking $\limsup_{\delta \to 0^+}$ and $\liminf_{\delta \to 0^+}$ in place of $\lim_{\delta \to 0^+}$.

DEFINITION 2.4. Let $\mu$ be a Borel finite measure on $\mathbb{R}^m$. The (lower) Hausdorff dimension of measure $\mu$ is defined by

$$\dim_H(\mu) = \inf\{\dim_H(A) : A \text{ is a Borel subset such that } \mu(A) > 0\}.$$

The Hausdorff dimension of a measure $\mu$ on $\mathbb{R}^m$ is defined as

$$\dim_H(\mu) = \inf\{\dim_H(A) : A \text{ is Borel, } \mu(\mathbb{R}^m \backslash A) = 0\}.$$

Since for self-similar measures, both the above values coincide (see [**11**, **12**] and the references cited therein) and both satisfy similar required properties in this paper, we will be proving the result for one, and the same statement will be true for the other without distinguishing them.

DEFINITION 2.5. Let $\mathcal{I} = \{\mathbb{R}^m; f_1, f_2, \ldots, f_N\}$ be an IFS and $A$ be the associated attractor.

(1) $\mathcal{I}$ satisfies the Strong Separation Condition (SSC) if $f_i(A) \cap f_j(A) = \emptyset$ for distinct $i, j \in \{1, 2, \ldots, N\}$.
(2) $\mathcal{I}$ is said to satisfy the Open Set Condition (OSC) if there exists a non-empty open set $U \subset X$ such that

$$\cup_{i=1}^N f_i(U) \subset U; \quad f_i(U) \cap f_j(U) = \emptyset, \text{ for } i \neq j.$$

(3) If the above open set $U$ satisfies $U \cap A \neq \emptyset$, then we say that $\mathcal{I}$ satisfies the Strong Open Set Condition (SOSC).
(4) Let $h_1, h_2$ be similarity maps on $\mathbb{R}$ such that $h_1(x) = c_1 x + a_1$ and $h_2(x) = c_2 x + a_2$. Then, we define a distance

$$d(h_1, h_2) = \begin{cases} |a_1 - a_2|, & \text{if } c_1 = c_2 \\ 1, & \text{otherwise.} \end{cases}$$

An IFS $\{\mathbb{R}; f_1, f_2, \dots, f_N\}$ is said to satisfy the exponential separation condition (ESC) [**11**] if there exists $\delta > 0$ such that $\triangle_n > \delta^n$ for infinitely many $n \in \mathbb{N}$, where $\triangle_n := \inf\{d(f_i, f_j) : i = (i_1, \dots, i_n), j = (j_1, j_2, \dots, j_n), i \neq j\}$ and $f_i = f_{i_1} \circ \cdots \circ f_{i_n}$.

Let $\mathcal{P}(\mathbb{R}^m)$ denote the set of all Borel probability measures on the Euclidean space $(\mathbb{R}^m, \|.\|_2)$. Then,

$$d_L(\mu, \nu) := \sup_{Lip(f) \leq 1} \Big\{ \Big| \int_{\mathbb{R}^m} f d\mu - \int_{\mathbb{R}^m} f d\nu \Big| \Big\}, \; (\mu, \nu) \in \mathcal{P}(\mathbb{R}^m) \times \mathcal{P}(\mathbb{R}^m),$$

defines a metric on $\mathcal{P}(\mathbb{R}^m)$, where $Lip(f)$ denotes the Lipschitz constant of $f$. Here $(\mathcal{P}(\mathbb{R}^m), d_L)$ is complete (see [**4**]).

Definition 2.6. Let $\mu, \lambda \in \mathcal{P}(\mathbb{R}^m)$. The convolution of $\mu$ and $\lambda$ is denoted by $\mu * \lambda$ and defined by the push-forward of the product measure $\mu \times \lambda$ under the mapping $(x, y) \to x + y$.

Definition 2.7. Let $\mu \in \mathcal{P}(\mathbb{R}^m)$ and $x \in \mathbb{R}^m$. The translation of $\mu$ by $x$ is defined as $(\mu + x)(A) = \mu(A + x)$, $A \subset \mathbb{R}^m$.

Let us define

$$\mathcal{L}(\mathbb{R}^m) = \Big\{ \mu \in \mathcal{P}(\mathbb{R}^m) : \mu = \sum_{i=1}^{k} a_i \delta_{x_i}, \, x_i \in \mathbb{R}^m, \, a_i > 0, \sum_{i=1}^{k} a_i = 1, \, k \in \mathbb{N} \Big\}.$$

Here $\delta_x$ denotes the Dirac measure at $x \in \mathbb{R}^m$. For computational purpose, let us note the following:

- For any $A \subset \mathbb{R}^m$, we have $\mu * \lambda(A) = (\mu \times \lambda)\big(\{(x, y) \in \mathbb{R}^m \times \mathbb{R}^m : x + y \in A\}\big)$.
- Let $f \in \mathcal{C}_0(\mathbb{R}^m)$, where $\mathcal{C}_0(\mathbb{R}^m)$ denotes the set of compactly supported and continuous functions on $\mathbb{R}^m$. Then

$$\int f d(\mu * \lambda) = \int \int f(x + y) d\mu(x) d\lambda(y).$$

- Let $\mu \in \mathcal{P}(\mathbb{R}^m)$ and $x \in \mathbb{R}^m$. Then $\mu * \delta_x = \mu - x$, and

$$\int f d(\mu * \delta_x) = \int f(y + x) d\mu(y) = \int f d(\mu - x), \; \forall \, f \in \mathcal{C}_0(\mathbb{R}^m).$$

Definition 2.8. The level-$n$ dyadic partitions $D_n$ of $\mathbb{R}$ is given by

$$D_n = \Big\{ \Big[ \frac{k}{2^n}, \frac{k+1}{2^n} \Big) : k \in \mathbb{Z} \Big\}.$$

The Shannon entropy of a probability measure $\mu$ with respect to a finite or countable partition $\mathcal{E}$ $(= D_n)$ is defined by $H(\mu, \mathcal{E}) = -\sum_{E \in \mathcal{E}} \mu(E) \log \mu(E)$. We define the entropy dimension [**11**] of $\mu \in \mathcal{P}(\mathbb{R})$ to be $\dim_e(\mu) = \lim_{n \to \infty} \frac{H(\mu, \mathcal{E})}{n}$, if the limit exists.

Before our next section, we would like to encourage the reader to see [**4, 6, 8, 12, 19, 23, 30**] for a better understanding of the concepts used in the paper.

## 3. Main Results

We are now ready for establishing our theorems. Note that most of the results except those involving the ESC will hold in $\mathbb{R}^m$ with obvious mild modifications in statements, but we will work in $\mathbb{R}$ for readability and for making this article understandable to new researchers and graduate students.

THEOREM 3.1. *For each $\beta \in [0,1]$, the set $\mathcal{D}_\beta := \{A \in \mathcal{K}(\mathbb{R}) : \dim_H(A) = \beta\}$ is dense in $\mathcal{K}(\mathbb{R})$.*

PROOF. Let $\beta \in [0,1]$, $C \in \mathcal{K}(\mathbb{R})$ and $\epsilon > 0$. Since $C$ is compact, there exist points $x_1, x_2, \dots, x_n$ in $C$ such that

$$C \subset \cup_{i=1}^n B(x_i, \epsilon).$$

Now, set $E = \{x_1, x_2, \dots, x_n\}$. If $\beta = 0$, then, in view of definition of the Hausdorff metric, it is easy to see that

$$h_d(C, E) \leq \epsilon \text{ and } \dim_H(E) = 0.$$

This shows the set $\mathcal{D}_\beta$ for $\beta = 0$ is dense in $\mathcal{K}(\mathbb{R})$. Now, suppose $\beta \in (0,1]$, we can construct a set $F \subset B(x_i, \epsilon)$ such that $\dim_H(F) = \beta$. Define $E' = E \cup F$. Then, by definition of the Hausdorff metric and finite stability of the Hausdorff dimension, we have

$$h_d(C, E') \leq \epsilon \text{ and } \dim_H(E') = \max\{\dim_H(E), \dim_H(F)\} = \beta,$$

completing the assertion. □

REMARK 3.2. Similarly, for each $\beta \in [0,1]$, the set $\mathcal{D}_\beta^B := \{A \in \mathcal{K}(\mathbb{R}) : \dim_B(A) = \beta\}$ is dense in $\mathcal{K}(\mathbb{R})$. However, we know that the lower box dimension does not hold the finite stability property for general sets, but the above result can also be proved for this dimension because $\underline{\dim}_B(E \cup F) = \underline{\dim}_B(F)$ holds for any finite set $E$.

It is worth noting the next result.

THEOREM 3.3. *The sets $\mathcal{D}_{LU}^B := \{A \in \mathcal{K}(\mathbb{R}) : \underline{\dim}_B(A) \neq \overline{\dim}_B(A)\}$, $\mathcal{D}^{BH} := \{A \in \mathcal{K}(\mathbb{R}) : \dim_H(A) \neq \dim_B(A)\}$ and $\mathcal{D}_{LU}^{BH} := \{A \in \mathcal{K}(\mathbb{R}) : \dim_H(A) < \underline{\dim}_B(A) < \overline{\dim}_B(A)\}$ are dense in $\mathcal{K}(\mathbb{R})$.*

PROOF. Falconer [8] hinted at a construction of a Cantor-type set $F$ such that $\underline{\dim}_B(F) < \overline{\dim}_B(F)$. So, $\mathcal{D}_{LU}^B$ is non-empty. Since for $F' = \{0, 1, \frac{1}{2}, \frac{1}{3}, \dots\}$, $0 = \dim_H(F') \neq \dim_B(F') = \frac{1}{2}$, thus the set $\mathcal{D}^{BH}$ is nonempty. See [8] for more details and non-emptyness of $\mathcal{D}_{LU}^{BH}$. Now, it is easy to obtain by definitions of the lower and upper box dimensions that $\underline{\dim}_B(E \cup F) = \underline{\dim}_B(F)$ and $\overline{\dim}_B(E \cup F) = \overline{\dim}_B(F)$ for any finite set $E$. Using the above observations, the proof follows on similar lines as the proof of Theorem 3.1. □

THEOREM 3.4. *For each $\beta \in (0,1]$, the set $\mathcal{D}_{\beta,0} := \{A \in \mathcal{K}(\mathbb{R}) : \dim_H(A) = \beta$ and $\mathcal{H}^\beta(A) = 0\}$ is dense in $\mathcal{K}(\mathbb{R})$.*

PROOF. Let $\beta \in [0,1]$, $C \in \mathcal{K}(\mathbb{R})$ and $\epsilon > 0$. Since $C$ is compact, there exist points $x_1, x_2, \dots, x_n$ in $C$ such that

$$C \subset \cup_{i=1}^n B(x_i, \epsilon).$$

Now, set $E = \{x_1, x_2, \ldots, x_n\}$. Now, since $\beta \in (0,1]$ then we can construct sets $F_m \subset B(x_i, \epsilon)$ such that $\dim_H(F_m) = \beta - \frac{1}{m}$ for all sufficiently large $m \in \mathbb{N}$. Define $F = \cup_m F_m$, then $\dim_H(F) = \dim_H(\cup_m F_m) = \sup_m \dim_H(F_m) = \beta$ and

$$\mathcal{H}^\beta(F) = \mathcal{H}^\beta(\cup_m F_m) \leq \sum_m \mathcal{H}^\beta(F_m) = 0$$

Set $E' = E \cup F$. Then, by definition of the Hausdorff metric and finite stability of the Hausdorff dimension, we have

$$h_d(C, E') \leq \epsilon, \ \dim_H(E') = \max\{\dim_H(E), \dim_H(F)\} = \beta, \ \text{and} \ \mathcal{H}^\beta(E') = 0$$

completing the assertion. □

REMARK 3.5. Note that $\mathcal{H}^0(A) \geq 1$, for all nonempty set $A$. This implies that $\mathcal{D}_{0,0} := \{A \in \mathcal{K}(\mathbb{R}) : \dim_H(A) = 0 \text{ and } \mathcal{H}^0(A) = 0\}$ is an empty set.

THEOREM 3.6. *For each $\beta \in [0,1)$, the set $\mathcal{D}_{\beta,\infty} := \{A \in \mathcal{K}(\mathbb{R}) : \dim_H(A) = \beta$ and $\mathcal{H}^\beta(A) = \infty\}$ is dense in $\mathcal{K}(\mathbb{R})$.*

PROOF. Let $\beta \in [0,1]$, $C \in \mathcal{K}(\mathbb{R})$ and $\epsilon > 0$. Since $C$ is compact, there exist points $x_1, x_2, \ldots, x_n$ in $C$ such that

$$C \subset \cup_{i=1}^n B(x_i, \epsilon).$$

Now, set $E = \{x_1, x_2, \ldots, x_n\}$. Now, since $\beta \in [0,1)$ then we can construct a set $F \subset B(x_i, \epsilon)$ such that $\dim_H(F) = \beta$ and $\mathcal{H}^\beta(F) = \infty$. Set $E' = E \cup F$. Then, by definition of the Hausdorff metric and finite stability of the Hausdorff dimension, we have

$$h_d(C, E') \leq \epsilon, \ \dim_H(E') = \max\{\dim_H(E), \dim_H(F)\} = \beta, \ \text{and} \ \mathcal{H}^\beta(E') = 0$$

completing the assertion. □

REMARK 3.7. Since all compact sets are bounded, we have

$$\mathcal{H}^1(A) < \infty, \ \forall A \in \mathcal{K}(\mathbb{R}).$$

In view of the above, the set $\mathcal{D}_{1,\infty} := \{A \in \mathcal{K}(\mathbb{R}) : \dim_H(A) = 1 \text{ and } \mathcal{H}^1(A) = \infty\}$ is an empty set.

THEOREM 3.8. *The set $\mathcal{D}_{\beta,*} := \{A \in \mathcal{K}(\mathbb{R}) : \dim_H(A) = \beta$ and $0 < \mathcal{H}^\beta(A) < \infty\}$ is dense in $\mathcal{K}(\mathbb{R})$.*

PROOF. Let $\beta \in [0,1]$, $C \in \mathcal{K}(\mathbb{R})$ and $\epsilon > 0$. Since $C$ is compact, there exist points $x_1, x_2, \ldots, x_n$ in $C$ such that

$$C \subset \cup_{i=1}^n B(x_i, \epsilon).$$

Now, set $E = \{x_1, x_2, \ldots, x_n\}$. If $\beta = 0$ then in view of the definition of the Hausdorff metric, it is easy to see that

$$h_d(C, E) \leq \epsilon \text{ and } \dim_H(E) = 0.$$

This shows the set $\mathcal{D}_{\beta,*}$ for $\beta = 0$ is dense in $\mathcal{K}(\mathbb{R})$. Now, since $\beta \in (0,1]$ then we can construct a set $F \subset B(x_i, \epsilon)$ such that $\dim_H(F) = \beta$ and $0 < \mathcal{H}^\beta(F) < \infty$. Set $E' = E \cup F$. Then, by definition of the Hausdorff metric and finite stability of the Hausdorff dimension, we have

$$h_d(C, E') \leq \epsilon, \ \dim_H(E') = \max\{\dim_H(E), \dim_H(F)\} = \beta, \ \text{and} \ \mathcal{H}^\beta(E') = 0$$

completing the assertion.. □

THEOREM 3.9. *The set* $V = \{f \in \mathcal{C}([0,1]) : \dim_H(Gr(f+g)) = \dim_H(Gr(g)), \forall g \in \mathcal{C}([0,1])\}$ *is a dense vector-subspace of* $\mathcal{C}([0,1])$*, where* $Gr(g)$ *denotes the graph of* $g$.

PROOF. Using the fact that the set of Lipschitz functions is dense in $\mathcal{C}([0,1])$ and by Lemma 3.1 of [**28**], it follows that $V$ is dense in $\mathcal{C}([0,1])$. Now, let $f, h \in V$ and $\alpha \in \mathbb{R}$. Then we have

$$\dim_H(Gr((f+h)+g)) = \dim_H(Gr(f+(h+g))) = \dim_H(Gr(h+g)) = \dim_H(Gr(g)),$$

and $\dim_H(Gr((\alpha f)+g)) = \dim_H(Gr(f+g)) = \dim_H(Gr(g)), \forall g \in \mathcal{C}([0,1])$, which implies $f+h, \alpha f \in V$. This completes our proof. □

THEOREM 3.10. *For a self-similar set* $A$ *in* $\mathbb{R}$ *with* $s_* = \dim_H(A)$*, we have all the following cases:*

- $0 < \mathcal{H}^{s_*}(A) < +\infty$.
- $\mathcal{H}^{s_*}(A) = 0$.
- $\mathcal{H}^{s_*}(A) = +\infty$.

PROOF. We know [**1, 13, 24**] that every self-similar set with the OSC satisfies $0 < \mathcal{H}^{s_*}(A) < +\infty$. Also, in view of [**24**, Theorem 2.2], one can conclude that if the generating IFS of a self-similar set satisfies the ESC but not the OSC then $\mathcal{H}^{s_*}(A) = 0$. Choose an IFS $\mathcal{I} = \left\{\mathbb{R}; f_1(x) = \frac{x}{2}, f_2(x) = 1\right\}$ then $A = \{0, 1, \frac{1}{2}, \frac{1}{4}, \frac{1}{8}, \dots\}$ and $\dim_H(A) = 0$ with $\mathcal{H}^0(A) = +\infty$, where $\mathcal{H}^0(.)$ denotes the counting measure. This proves the assertion. □

REMARK 3.11. Let us understand via the following examples:

- Let $\mathcal{I} = \left\{\mathbb{R}; f_1(x) = \frac{x}{2}, f_2(x) = \frac{x}{3}\right\}$ be an IFS consisting of similarity maps. Then we have $A = \{0\}$ and $0 < \mathcal{H}^0(A) = 1 < \infty$.
- Let $\mathcal{I} = \left\{\mathbb{R}; f_1(x) = \frac{x}{2}, f_2(x) = \frac{x}{3}, f_3(x) = \frac{x}{2} + \frac{1}{2}\right\}$ be an IFS consisting of similarity maps. Then we have $A = [0,1]$ and $0 < \mathcal{H}^1(A) = 1 < \infty$.

The following theorem is a weaker version of Hochman's groundbreaking result on self-similar sets. We may introduce notation $c_{\max} = \max\{c_i : 1 \le i \le N\}$, $I = \{1, 2, \dots, N\}$, and the collection of all finite alphabets of $I$ as $I^* = \cup_{n \in \mathbb{N}} I^n$.

THEOREM 3.12. *Let* $\mathcal{I} = \{\mathbb{R} : f_1, f_2, \dots, f_N\}$ *be an IFS consisting of similarity mappings. If there exists an alphabet* $i \in I^*$ *such that* $\{f_{ji} : j \in I^n\}$ *satisfies the ESC for each* $n \in \mathbb{N}$*, then* $\dim_H(A) = \dim_B(A) = \min\{s, 1\}$*, where* $s$ *is the similarity dimension of the IFS* $\mathcal{I}$.

PROOF. Using the assumption that $\mathcal{I}_n = \{f_{ji} : j \in I^n\}$ satisfies the ESC for any $n \in \mathbb{N}$, Hochman's theorem [**11**] yields the following:

$$\dim_H(A_n) = s_n, \text{ where } A_n \text{ is the attractor of } \mathcal{I}_n,$$

where $\sum_{j \in I^n} c_{ji}^{s_n} = 1$. Note that $A_n \subset A$, hence, $s_n = \dim_H(A_n) \le \dim_H(A)$. Suppose $\dim_H(A) < s$. Then

$$\begin{aligned} c_i^{-s_n} &= \sum_{j \in I^n} c_j^{s_n} \ge \sum_{j \in I^n} c_j^{\dim_H(A)} = \sum_{j \in I^n} c_j^{\dim_H(A)-s} c_j^s \\ &\ge \sum_{j \in I^n} c_{\max}^{n(\dim_H(A)-s)} c_j^s = c_{\max}^{n(\dim_H(A)-s)}. \end{aligned}$$

From the above, we arrive at a contradiction, as $n \to \infty$. Therefore, $\dim_H(A) = s$, completing the assertion. $\square$

We pose an open question to construct an IFS that satisfies our given assumption of the above theorem but not the ESC. The following corollary is simple. Hence, it is mentioned without proof.

COROLLARY 3.13. *Let* $\mathcal{I} = \{\mathbb{R} : f_1, f_2, \dots, f_N\}$ *be an IFS consisting of similarity mappings. If there exists an alphabet* $i \in I^*$ *such that* $\{f_{ji} : j \in I\}$ *satisfies the SSC, then* $\dim_H(A) = \dim_B(A) = \min\{s, 1\}$, *where* $s$ *is the similarity dimension of the IFS* $\mathcal{I}$.

THEOREM 3.14. *Let* $\mathcal{I} = \{\mathbb{R} : f_1, f_2, \dots, f_N\}$ *be an IFS consisting of similarity mappings, and*

(1) *The IFS satisfies OSC.*
(2) *The IFS satisfies SOSC.*
(3) *There exists* $i \in I^*$ *such that* $\{f_{ji} : j \in I\}$ *satisfies the SSC.*
(4) $\dim_H(A) = \min\{s, 1\}$.

*Then, we have*

$$(1) \iff (2) \implies (3) \implies (4).$$

PROOF. Bandt and Graf in [**1**] proved the equivalence $(1) \iff (2)$ in a more general setting; see also a similar result by Schief in [**24**, Theorem 2.2]. By Corollary 3.13, we have $(3) \implies (4)$. So, we only need to prove $(2) \implies (3)$. Since the IFS $\mathcal{I}$ satisfies SOSC, there exists an open set $V$ of $\mathbb{R}$ such that

$$\cup_{k=1}^{N} f_k(U) \subset U,\ U \cap A \neq \emptyset,\ f_k(U) \cap f_j(U) = \emptyset\ \forall\, k \neq j,\ 1 \leq k, j \leq N.$$

Since $U \cap A \neq \emptyset$, there exists an index $i \in I^* = \{1, 2, \dots, N\}^*$ such that $f_i(A) \subset U$. We denote $f_k(A)$ by $A_k$ for any $k \in \{1, 2, \dots, N\}^*$. Now, by using the condition $f_k(U) \cap f_j(U) = \emptyset\ \forall\, k \neq j,\ 1 \leq k, j \leq N$ and $f_i(A) \subset U$, it is clear that, the sets $\{A_{ji} : j \in I\}$ are pairwise disjoint. The IFS defined by $\mathcal{L} = \{f_{ji} : j \in I\}$ satisfies the SSC. Thus, the proof is complete. $\square$

Now, let us note that $(3) \implies (2)$ and $(4) \implies (3)$ may not hold in general through the next remark.

REMARK 3.15. Let $\mathcal{I} = \{\mathbb{R} : f_1, f_2, f_3\}$ be an IFS, where

$$f_1(x) = \frac{x}{4},\ f_2(x) = \frac{x}{4} + t,\ f_3(x) = \frac{x}{4} + \frac{3}{4},\ \text{for a fixed constant } \frac{1}{16} < t < \frac{11}{16}.$$

Then $\{f_{11}, f_{21}, f_{31}\}$ will satisfy the SSC:

$$f_{11}([0,1]) = \left[0, \frac{1}{16}\right];\ f_{21}([0,1]) = \left[t, t + \frac{1}{16}\right];\ f_{31}([0,1]) = \left[\frac{3}{4}, \frac{13}{16}\right].$$

But the IFS does not satisfy the OSC (equivalently, the SOSC) (see Hochman [**11**]). Let $\mathcal{I} = \{\mathbb{R} : f_1, f_2, f_3\}$ be an IFS, where

$$f_1(x) = \frac{x}{2},\ f_2(x) = \frac{x}{4},\ f_3(x) = \frac{x}{2} + \frac{1}{2}.$$

Then $\dim_H(A) = 1 = \min\{s, 1\}$ but for each $i \in I^*$ the IFS $\{f_{ji} : j \in I\}$ does not satisfy the SSC.

REMARK 3.16. Note that OSC and SOSC are not equivalent for bi-Lipschitz IFS. For example: let $\mathcal{I} = \left\{\mathbb{R}^2; f_1(x,y) = (\frac{x}{2}, \frac{x}{2} + \frac{y}{2}), f_2(x,y) = (\frac{x}{2} + \frac{y}{2}, \frac{y}{2})\right\}$ be an IFS consisting of bi-Lipschitz mappings. Then $A = \{(0,0)\}$ is the attractor of this IFS, and if $U = (0,1) \times (0,1)$ then $f_1(U) \cup f_2(U) \subset U$ and $f_1(U) \cap f_2(U) = \emptyset$, that is, the IFS satisfies the OSC. However, it does not satisfy the SOSC.

REMARK 3.17. We should remark that we get the equality of box and Hausdorff dimensions for self-similar sets without any separation conditions [**9**], but for bi-Lipschitz IFS, we may not get the equality of box and Hausdorff dimensions, that is, $\dim_H(A) \neq \dim_B(A)$ where $A$ is the attractor of the IFS. For example: see the famous Bedford-McMullen carpets [**3**, **16**], let $\mathcal{I} = \left\{[0,1]^2; f_i(x,y) = \left(\frac{x}{n} + \frac{k_i}{n}, \frac{y}{m} + \frac{l_i}{n}\right), \text{ for some fixed } 0 < m < n, \text{ for some fixed } k_i, l_i\right\}$ be an IFS consisting of finite number of bi-Lipschitz mappings (also affine transformations). Note that

$$\frac{1}{n}\|(x,y) - (z,w)\|_2 \leq \|f_i(x,y) - f_i(z,w)\|_2 \leq \frac{1}{m}\|(x,y) - (z,w)\|_2,$$

where $\|.\|_2$ is the Euclidean norm. Bedford [**3**] and McMullen [**16**] showed that $\dim_H(A) < \dim_B(A)$.

We note the coming remark to highlight the issue of not considering the case when all $p_i = 0$ and $c_i = 0$ for studying invariant measures in fractal geometry literature.

REMARK 3.18. In the first case, consider $p_1 = 1$ and remaining $p_i = 0$, for all $i = 2, 3, \ldots, N$. Clearly, $\mu_p = \delta_{x_0}$, a Dirac measure supported on $x_0$. Here, $x_0$ is the fixed point of $f_1$. Secondly, if some $p_i$ are zero, for example, $p_1 = 0$ and others are positive then $\mu_p$ will be supported on $A_*$, where $A_* = \cup_{i=2}^N f_i(A_*)$. Third, consider for each $i$, $f_i : \mathbb{R} \to \mathbb{R}$ the constant map given as $f_i(x) = x_i, \forall x \in \mathbb{R}$, and $p_i > 0$. Then, it is straightforward to verify that $\mu_p$ is the convex combination of Dirac measures supported on $\{x_1, x_2, x_3, \ldots, x_N\}$, that is, $\mu_p = \sum_{i=1}^N p_i \delta_{x_i}$. These previous cases convince us to consider $p_i > 0$ and $c_i > 0$ for all $i = 1, 2, \ldots, N$, to avoid the above trivial cases. Further, for $p_i = c_i^s$, then in case of similarity IFS under the OSC, [**17**] confirms that $\mu_p$ is equal to normalized $s$-dimensional Hausdorff measure supported on the attractor of the associated IFS.

The next example shows the dimension of measures (hence the dimension of fractal sets) cannot be continuous in general.

EXAMPLE 3.19. In view of [**17**], we may choose a sequence of IFSs and a sequence of probability vectors such that the associated sequence of invariant measures $\mu_p^n$ is the normalized Lebesgue measure supported on $[0, \frac{1}{n}]$, which converges to the Dirac measure $\delta_0$ supported on $\{0\}$. So, we have $\dim_H(\mu_p^n) = 1$ for all $n$ and $\dim_H(\delta_0) = 0$. Now, choose a sequence of IFSs $\mathcal{I}_n = \{\mathbb{R}; f_{n,1}, f_{n,2}\}$, where

$$f_{n,1}(x) = \frac{x}{7n}, \quad f_{n,2}(x) = \frac{x}{7n} + \frac{1}{7n} \text{ for all } x \in \mathbb{R}.$$

One may compute that the similarity ratio of $f_{n,i}$ is $c_{n,i} = \frac{1}{7n}$, $i = 1, 2$. Next, let us choose an IFS $\mathcal{I} = \{\mathbb{R}; f_1, f_2\}$ with $f_1(x) = f_2(x) = 0$, $\forall x \in \mathbb{R}$ and corresponding similarity ratios $c_1 = c_2 = 0$. With a sequence of probability vectors $(p_{n,1}, p_{n,2}) = (\frac{1}{2}, \frac{1}{2}) = p$, and a probability vector $(p_1, p_2) = (\frac{1}{2}, \frac{1}{2}) = p$, we have $f_{n,i} \to f_i$, $p_{n,i} \to p_i$, and $c_{n,i} \to c_i$ as $n \to \infty$ for $i = 1, 2$. Note that the sequence of invariant measures $(\mu_p^n)$ associated with $\mathcal{I}_n$ converges to the invariant measure $\mu_p$ which is associated with $\mathcal{I}$. Since $\mathcal{I}_n$ holds the SOSC, we deduce [**7**] that $\dim_H(\mu_p^n) = \frac{\ln(2)}{\ln(7)+\ln(n)} \to \dim_H(\mu_p) = 0$. In this example, the limit IFS $\mathcal{I}$ does not satisfy the SOSC.

LEMMA 3.20. *Let $\mu$ be a finite Borel measure. For a fixed $\nu \in \mathcal{L}(\mathbb{R})$, we have that*

$$\dim_H(\mu * \nu) = \dim_H(\mu) \text{ and } \dim_e(\mu * \nu) = \dim_e(\mu),$$

*provided* $\dim_e(\mu)$ *exists.*

PROOF. Let $\nu \in \mathcal{L}(\mathbb{R})$. By definition of $\mathcal{L}(\mathbb{R})$, we write $\nu = \sum_{i=1}^k a_i \delta_{x_i}$ for some $x_i \in \mathbb{R}$ and $a_i > 0$ such that $\sum_{i=1}^k a_i = 1$. Now, note that

$$\mu * \nu = \mu * \Big( \sum_{i=1}^k a_i \delta_{x_i} \Big) = \sum_{i=1}^k a_i \mu * \delta_{x_i}.$$

Using finite-stability of the Hausdorff dimension of measures [**14**], we have

$$\dim_H(\mu * \nu) = \dim_H \Big( \sum_{i=1}^k a_i \mu * \delta_{x_i} \Big) = \min_{1 \le i \le k} \{\dim_H(a_i \mu * \delta_{x_i})\} = \min_{1 \le i \le k} \{\dim_H(\mu * \delta_{x_i})\}.$$

From the above, it is enough to prove that $\dim_H(\mu * \delta_{x_i}) = \dim_H(\mu)$. We proceed as follows:

$$\begin{aligned} \dim_H(\mu * \delta_{x_i}) &= \inf \Big\{ \dim_H(A) : A \text{ Borel subset of } \mathbb{R}, \mu * \delta_{x_i}(A) > 0 \Big\} \\ &= \inf \Big\{ \dim_H(A) : A \text{ Borel subset of } \mathbb{R}, (\mu - x_i)(A) > 0 \Big\} \\ &= \inf \Big\{ \dim_H(A) : A \text{ Borel subset of } \mathbb{R}, \mu(A - x_i) > 0 \Big\} \\ &= \inf \Big\{ \dim_H(A - x_i) : A \text{ Borel subset of } \mathbb{R}, \mu(A - x_i) > 0 \Big\} \\ &= \dim_H(\mu). \end{aligned}$$

For the other part, applying a result of translation effect on entropy from [**12**], we get $\dim_e(\mu * \delta_{x_i}) = \dim_e(\mu)$. This with the fact [**12**] that the entropy is concave and almost convex yields

$$\dim_e(\mu * \nu) = \dim_e \Big( \sum_{i=1}^k a_i \mu * \delta_{x_i} \Big) = \sum_{i=1}^k a_i \dim_e(\mu * \delta_{x_i}) = \sum_{i=1}^k a_i \dim_e(\mu) = \dim_e(\mu).$$

This completes the proof of the lemma. □

It is well-known [**4**] that $\mathcal{L}(\mathbb{R})$ is a dense subset of $\mathcal{P}(\mathbb{R})$ where the latter is endowed with the metric $d_L$. We use this fact to prove the following theorem.

THEOREM 3.21. *Let $\alpha \in [0, 1]$. Then, the sets $\mathcal{S}_\alpha^H := \{\mu \in \mathcal{P}(\mathbb{R}) : \dim_H(\mu) = \alpha\}$ and $\mathcal{S}_\alpha^e := \{\mu \in \mathcal{P}(\mathbb{R}) : \dim_e(\mu) = \alpha\}$ are dense subsets of $\mathcal{P}(\mathbb{R})$.*

PROOF. Let $\mu \in \mathcal{P}(\mathbb{R})$ and $\epsilon > 0$. Using the density of $\mathcal{L}(\mathbb{R})$ in $\mathcal{P}(\mathbb{R})$, there exists $\nu$ in $\mathcal{L}(\mathbb{R})$ such that

$$d_L(\mu, \nu) < \frac{\epsilon}{2}.$$

Further, we consider a non-zero compactly supported measure $\lambda \in \mathcal{S}_\alpha$ with its support $K \subset \mathbb{R}$. Without loss of generality, we assume $K \subseteq [0, c]$ for some $c > 0$.

Let $\lambda' = \nu * \lambda_1$, where $\lambda_1(A) := \lambda(\frac{2|K|A}{\epsilon})$ and $|K| := \sup\{|x-y| : x, y \in K\}$. Note that Lemma 3.20 implies that, $\dim_H(\lambda') = \dim_H(\lambda) = \alpha$, that is, $\lambda' \in \mathcal{S}_\alpha$. By well-known properties of convolution and the Dirac measure as mentioned earlier, we have

$$\begin{aligned} d_L(\nu, \lambda') &= \sup_{Lip(f)\leq 1} \Big\{ \Big| \int_{\mathbb{R}} f d\nu - \int_{\mathbb{R}} f d\lambda' \Big| \Big\} \\ &= \sup_{Lip(f)\leq 1} \Big\{ \Big| \sum_{i=1}^{k} a_i f(x_i) - \sum_{i=1}^{k} a_i \int_{\mathbb{R}} f(y+x_i) d\lambda_1 \Big| \Big\} \\ &= \sup_{Lip(f)\leq 1} \Big\{ \Big| \sum_{i=1}^{k} a_i \int_{\mathbb{R}} f(x_i) d\lambda_1 - \sum_{i=1}^{k} a_i \int_{\mathbb{R}} f(y+x_i) d\lambda_1 \Big| \Big\} \\ &\leq \sup_{Lip(f)\leq 1} \Big\{ \sum_{i=1}^{k} a_i \int_{\mathbb{R}} |f(x_i) - f(y+x_i)| d\lambda_1 \Big\} \\ &\leq \sup_{Lip(f)\leq 1} \Big\{ \sum_{i=1}^{k} a_i \int_{\mathbb{R}} Lip(f)|y| d\lambda_1 \Big\} \\ &\leq \sup_{Lip(f)\leq 1} \Big\{ \sum_{i=1}^{k} a_i \int_{\mathbb{R}} \frac{Lip(f)\epsilon}{2} d\lambda_1 \Big\} \leq \frac{\epsilon}{2} \end{aligned}$$

Using the above estimate, we immediately get

$$d_L(\mu, \lambda') \leq d_L(\mu, \nu) + d_L(\nu, \lambda') < \frac{\epsilon}{2} + \frac{\epsilon}{2} = \epsilon.$$

Thus, $\mathcal{S}_\alpha^H$ is dense. On similar lines as above, using Lemma 3.20, we may show that $\mathcal{S}_\alpha^e$ is also dense. Therefore, the proof of the theorem is complete. □

THEOREM 3.22. *Let $\alpha \in [0,1]$. Then, the sets $\mathcal{S}_{\alpha,I}^H := \{\mu \in \mathcal{P}(\mathbb{R}) : \dim_H(\mu) = \alpha = \dim_H(A)$ for some set $A$ satisfying $\mu(A) > 0\}$ and $\mathcal{S}_{\alpha,II}^H := \{\mu \in \mathcal{P}(\mathbb{R}) : \dim_H(\mu) = \alpha \neq \dim_H(A)$ for all $A$ satisfying $\mu(A) > 0\}$ are dense subsets of $\mathcal{P}(\mathbb{R})$.*

PROOF. The proof may follow similar lines of the above theorem in view of the below points:

- Sets $\mathcal{S}_{\alpha,I}^H$ and $\mathcal{S}_{\alpha,II}^H$ are nonempty.
- If $\lambda' = \nu * \lambda_1$, where $\lambda_1(A) := \lambda(\frac{2|K|A}{\epsilon})$ satisfying $\dim_H(\lambda) = \alpha = \dim_H(C)$ and $\lambda(C) > 0$, then $\dim_H(\lambda') = \dim_H(\lambda) = \alpha = \dim_H(\frac{\epsilon}{2|K|}C)$ and $\lambda'(\frac{\epsilon}{2|K|}C) > 0$,

therefore, we omit the details. □

Now, in our last two theorems, we will try to study the separation conditions and invariant measures for the product of two IFSs. It will play an important role because several new fractal sets and measures can be constructed by taking product of simple and well-known sets and measures, respectively. To start with we may consider more general metric spaces. Let $(X, d_1)$ and $(Y, d_2)$ be two complete and separable metric spaces; then it is well known that the product space $(X \times Y, d)$ is a complete metric space with respect to the metric $d$ given by

$$d((x,y),(z,w)) = \max\{d_1(x,z), d_2(y,w)\}.$$

Now, for two IFSs $\mathcal{I}_1 = \{(X, d_1); f_1, f_2, \ldots, f_N\}$ and $\mathcal{I}_2 = \{(Y, d_2); g_1, g_2, \ldots, g_M\}$ we define product IFS $\mathcal{P} = \{X \times Y; \Phi_{ij}, 1 \le i \le N, 1 \le j \le M\}$, where $\Phi_{ij}(x, y) = (f_i(x), g_j(y))$. It is easy to observe that $\Phi_{ij}$ is contraction for all $(i, j) \in \{1, 2, \ldots, N\} \times \{1, 2, \ldots, M\}$. For any $(x, y), (z, w) \in X \times Y$, we have

$$\begin{aligned} d(\Phi_{ij}(x, y), \Phi_{ij}(z, w)) &= d((f_i(x), g_j(y)), (f_i(z), g_j(w))) \\ &= \max\{d_1(f_i(x), f_i(z)), d_2(g_j(y), g_j(w))\} \\ &\le \max\{c_i d_1(x, z), t_j d_2(y, w)\} \\ &\le \max\{c_i, t_j\} d((x, y), (z, w)). \end{aligned}$$

In the next theorem, we see the separation conditions on $\mathcal{I}_1$ and $\mathcal{I}_2$ can be transferred to the IFS $\mathcal{P}$. Similar results for fractal transformation and particular cases can be seen in [**20**].

THEOREM 3.23. *Let $\mathcal{I}_1$, $\mathcal{I}_2$ and $\mathcal{P}$ be the IFSs as defined above, then we have the following.*

1. *If IFSs $I_1$ and $I_2$ satisfy the SSC, then the IFS $\mathcal{P}$ satisfies the SSC.*
2. *If IFSs $I_1$ and $I_2$ satisfy the OSC, then the IFS $\mathcal{P}$ satisfies the OSC.*
3. *If IFSs $I_1$ and $I_2$ satisfy the SOSC, then the IFS $\mathcal{P}$ satisfies the SOSC*

PROOF. (1) Since $\mathcal{I}_1$ and $\mathcal{I}_2$ satisfy SSC, we have $f_i(A) \cap f_j(A) = \emptyset$ for all $i \neq j$ and $g_k(B) \cap g_l(B) = \emptyset$ for all $k \neq l$. This leads to

$$\Phi_{ij}(A \times B) \cap \Phi_{kl}(A \times B) = (f_i(A) \times g_j(B)) \cap (f_k(A) \times g_l(B)) = \emptyset,$$

for all $(i, j) \neq (k, l)$. Since $C = A \times B$, the above implies

$$\Phi_{ij}(C) \cap \Phi_{kl}(C) = \emptyset \text{ for all } (i, j) \neq (k, l).$$

Thus, the IFS $\mathcal{P}$ satisfies the SSC.

(2) Since $\mathcal{I}_1$ and $\mathcal{I}_2$ satisfy the OSC, we have open sets $U \subset X$ and $V \subset Y$ such that $\bigcup_{i=1}^{N} f_i(U) \subset U$, $f_i(U) \cap f_j(U) = \emptyset$ for all $i \neq j$, and $\bigcup_{k=1}^{M} g_k(V) \subset V$, $g_k(V) \cap g_l(V) = \emptyset$ for all $k \neq l$. This leads to

$$\Phi_{ij}(U \times V) \cap \Phi_{kl}(U \times V) = (f_i(U) \times g_j(V)) \cap (f_k(U) \times g_l(V)) = \emptyset$$

for all $(i, j) \neq (k, l)$. Now, define an open set $O = U \times V \subset X \times Y$, This yields the required result.

(3) Since $\mathcal{I}_1$ and $\mathcal{I}_2$ satisfy the SOSC, there exist open sets $U \subset X$ and $V \subset Y$ such that

$$\bigcup_{i=1}^{N} f_i(U) \subset U,\ U \cap A \neq \emptyset,\ f_i(U) \cap f_j(U) = \emptyset \text{ for all } i \neq j,$$

and

$$\bigcup_{k=1}^{M} g_k(V) \subset V,\ V \cap B \neq \emptyset,\ g_k(V) \cap g_l(V) = \emptyset \text{ for all } k \neq l.$$

Using $C = A \times B$ and $(A \times B) \cap (U \times V) \neq \emptyset$, one can prove the result on similar lines of item (2). □

It remains open to prove a similar result to the above theorem for the ESC. Let us now see evidence of a connection between invariant (fractal) measures of IFSs $\mathcal{I}_1$, $\mathcal{I}_2$, and $\mathcal{P}$.

THEOREM 3.24. *Let $\mathcal{I}_1$, $\mathcal{I}_2$ and $\mathcal{P}$ be the IFSs as defined above with probability vectors $p = (p_1, p_2, \ldots, p_N)$, $q = (q_1, q_2, \ldots, q_M)$ and $r = (r_{ij} = p_i q_j)_{1 \le i \le N, 1 \le j \le M}$, respectively. Let $\mu_p^{\mathcal{I}_1}, \mu_q^{\mathcal{I}_2}$ and $\mu_r^{\mathcal{P}}$ be the invariant measures associated with IFSs $\mathcal{I}_1, \mathcal{I}_2$ and $\mathcal{P}$, and their corresponding probability vector, respectively. Then*

$$\mu_r^{\mathcal{P}} = \mu_p^{\mathcal{I}_1} \times \mu_q^{\mathcal{I}_2},$$

*where $\mu_p^{\mathcal{I}_1} \times \mu_q^{\mathcal{I}_2}$ denotes the product measure on $X \times Y$.*

PROOF. With $r_{ij} = p_i q_j$, it is given that

$$\mu_p^{\mathcal{I}_1} = \sum_{i=1}^{N} p_i \mu_p^{\mathcal{I}_1} \circ f_i^{-1}, \; mu_q^{\mathcal{I}_2} = \sum_{j=1}^{M} q_j \mu_q^{\mathcal{I}_2} \circ g_j^{-1} \text{ and } \mu_r^{\mathcal{P}} = \sum_{i=1,j=1}^{N,M} r_{ij} \mu_r^{\mathcal{P}} \circ \Phi_{ij}^{-1}.$$

We know that the above measures $\mu_p^{\mathcal{I}_1}$ and $\mu_q^{\mathcal{I}_2}$ are Borel probability measures, hence they are $\sigma$-finite. By the well-known product measure theorem, the existence and uniqueness of the product measure $\mu_p^{\mathcal{I}_1} \times \mu_q^{\mathcal{I}_2}$ follow. Now, let $E$ and $F$ be measurable subsets of $X$ and $Y$, respectively. Then we have

$$\begin{aligned}
(\mu_p^{\mathcal{I}_1} \times \mu_q^{\mathcal{I}_2})(E \times F) &= \Big(\sum_{i=1}^{N} p_i \mu_p^{\mathcal{I}_1} \circ f_i^{-1}(E)\Big)\Big(\sum_{j=1}^{M} q_j \mu_q^{\mathcal{I}_2} \circ g_j^{-1}(F)\Big) \\
&= \sum_{i=1,j=1}^{N,M} r_{ij} \mu_p^{\mathcal{I}_1} \circ f_i^{-1}(E) \mu_q^{\mathcal{I}_2} \circ g_j^{-1}(F) \\
&= \sum_{i=1,j=1}^{N,M} r_{ij} (\mu_p^{\mathcal{I}_1} \times \mu_q^{\mathcal{I}_2})(f_i^{-1}(E), g_j^{-1}(F)) \\
&= \sum_{i=1,j=1}^{N,M} r_{ij} (\mu_p^{\mathcal{I}_1} \times \mu_q^{\mathcal{I}_2}) \circ \Phi_{ij}^{-1}(E \times F).
\end{aligned}$$

Since $\mu_r^{\mathcal{P}}$ is the unique measure satisfying the equation, $\mu_r^{\mathcal{P}} = \sum_{i=1,j=1}^{N,M} r_{ij} \mu_r^{\mathcal{P}} \circ \Phi_{ij}^{-1}$, the above equation yields that $\mu_{\mathcal{P}} = \mu_p^{\mathcal{I}_1} \times \mu_q^{\mathcal{I}_2}$, establishing the assertion. □

## Statements and Declarations

The authors have no competing interests to declare that are relevant to the content of this article. There is no data associated with this paper.

## Acknowledgements

Some parts of the paper have been presented at AMS Spring Eastern Virtual Sectional Meeting, Special Session on "Fractal Geometry and Dynamical Systems", USA, April 01-02, 2023.

## References

[1] C. Bandt and S. Graf, *Self-similar sets. VII. A characterization of self-similar fractals with positive Hausdorff measure*, Proc. Amer. Math. Soc. **114** (1992), no. 4, 995–1001, DOI 10.2307/2159618. MR1100644

[2] M. Barnsley, *Fractals everywhere*, Academic Press, Inc., Boston, MA, 1988. MR977274

[3] T. Bedford, *Crinkly curves, Markov partitions and box dimensions in self-similar sets*, PhD thesis, University of Warwick, 1984.

[4] P. Billingsley, *Convergence of probability measures*, John Wiley & Sons, Inc., New York-London-Sydney, 1968. MR233396
[5] S. Chandra and S. Abbas, *On fractal dimensions of fractal functions using function spaces*, Bull. Aust. Math. Soc. **106** (2022), no. 3, 470–480, DOI 10.1017/S0004972722000685. MR4510139
[6] J. B. Conway, *A course in functional analysis*, 96 Springer, 2019.
[7] A. Deliu, J. S. Geronimo, R. Shonkwiler, and D. Hardin, *Dimensions associated with recurrent self-similar sets*, Math. Proc. Cambridge Philos. Soc. **110** (1991), no. 2, 327–336, DOI 10.1017/S0305004100070407. MR1113431
[8] K. J. Falconer, *Fractal Geometry: Mathematical Foundations and Applications*, John Wiley Sons Inc., New York, 1999.
[9] K. Falconer, *Techniques in fractal geometry*, John Wiley & Sons, Ltd., Chichester, 1997. MR1449135
[10] Gurubachan, V. V. M. S. Chandramouli, and S. Verma, *Fractal dimension of $\alpha$-fractal functions without endpoint conditions*, Mediterr. J. Math. **21** (2024), no. 3, Paper No. 71, 23 pp., DOI 10.1007/s00009-024-02610-7. MR4719420
[11] M. Hochman, *On self-similar sets with overlaps and inverse theorems for entropy*, Ann. of Math. (2) **180** (2014), no. 2, 773–822, DOI 10.4007/annals.2014.180.2.7. MR3224722
[12] M. Hochman, *Some problems on the boundary of fractal geometry and additive combinatorics*, Recent developments in fractals and related fields, Trends Math., Birkhäuser/Springer, Cham, 2017, pp. 129–174, DOI 10.1007/978-3-319-57805-7_7. MR3775463
[13] J. E. Hutchinson, *Fractals and self-similarity*, Indiana Univ. Math. J. **30** (1981), no. 5, 713–747, DOI 10.1512/iumj.1981.30.30055. MR625600
[14] P. Mattila, *Fourier analysis and Hausdorff dimension*, Cambridge Studies in Advanced Mathematics, vol. 150, Cambridge University Press, Cambridge, 2015, DOI 10.1017/CBO9781316227619. MR3617376
[15] R. D. Mauldin and S. C. Williams, *On the Hausdorff dimension of some graphs*, Trans. Amer. Math. Soc. **298** (1986), no. 2, 793–803, DOI 10.2307/2000650. MR860394
[16] C. McMullen, *The Hausdorff dimension of general Sierpiński carpets*, Nagoya Math. J. **96** (1984), 1–9, DOI 10.1017/S0027763000021085. MR771063
[17] M. Morán and J.-M. Rey, *Singularity of self-similar measures with respect to Hausdorff measures*, Trans. Amer. Math. Soc. **350** (1998), no. 6, 2297–2310, DOI 10.1090/S0002-9947-98-02218-1. MR1475691
[18] P. A. P. Moran, *Additive functions of intervals and Hausdorff measure*, Proc. Cambridge Philos. Soc. **42** (1946), 15–23, DOI 10.1017/s0305004100022684. MR14397
[19] K. R. Parthasarathy, *Probability measures on metric spaces*, Probability and Mathematical Statistics, No. 3, Academic Press, Inc., New York-London, 1967. MR226684
[20] A. Priyadarshi, M. Verma, and S. Verma, *Fractal dimensions of fractal transformations and quantization dimensions for bi-Lipschitz mappings*, J. Fractal Geom. **12** (2025), no. 1-2, 1–33, DOI 10.4171/jfg/157. MR4875617
[21] A. Rapaport, *Proof of the exact overlaps conjecture for systems with algebraic contractions* (English, with English and French summaries), Ann. Sci. Éc. Norm. Supér. (4) **55** (2022), no. 5, 1357–1377. MR4517689
[22] A. Rapaport and P. P. Varjú, *Self-similar measures associated to a homogeneous system of three maps*, Duke Math. J. **173** (2024), no. 3, 513–602, DOI 10.1215/00127094-2023-0019. MR4729827
[23] W. Rudin, *Principles of mathematical analysis*, 3rd ed., International Series in Pure and Applied Mathematics, McGraw-Hill Book Co., New York-Auckland-Düsseldorf, 1976. MR385023
[24] A. Schief, *Separation properties for self-similar sets*, Proc. Amer. Math. Soc. **122** (1994), no. 1, 111–115, DOI 10.2307/2160849. MR1191872
[25] W. Shen, *Hausdorff dimension of the graphs of the classical Weierstrass functions*, Math. Z. **289** (2018), no. 1-2, 223–266, DOI 10.1007/s00209-017-1949-1. MR3803788
[26] K. Simon, *Overlapping cylinders: the size of a dynamically defined Cantor-set*, Ergodic theory of $\mathbf{Z}^d$ actions (Warwick, 1993), London Math. Soc. Lecture Note Ser., vol. 228, Cambridge Univ. Press, Cambridge, 1996, pp. 259–272, DOI 10.1017/CBO9780511662812.009. MR1411222

[27] P. P. Varjú, *On the dimension of Bernoulli convolutions for all transcendental parameters*, Ann. of Math. (2) **189** (2019), no. 3, 1001–1011, DOI 10.4007/annals.2019.189.3.9. MR3961088

[28] S. Verma and P. R. Massopust, *Dimension preserving approximation*, Aequationes Math. **96** (2022), no. 6, 1233–1247, DOI 10.1007/s00010-022-00893-3. MR4514041

[29] M. Verma, A. Priyadarshi, and S. Verma, *Vector-valued fractal functions: fractal dimension and fractional calculus*, Indag. Math. (N.S.) **34** (2023), no. 4, 830–853, DOI 10.1016/j.indag.2023.03.005. MR4593109

[30] M. Verma, A. Priyadarshi, and S. Verma, *Analytical and dimensional properties of fractal interpolation functions on the Sierpiński gasket*, Fract. Calc. Appl. Anal. **26** (2023), no. 3, 1294–1325, DOI 10.1007/s13540-023-00148-1. MR4599235

Department of Applied Sciences, IIIT Allahabad, Prayagraj, India 211015
*Email address*: saurabhverma@iiita.ac.in

Department of Mathematics, Indian Institute of Technology Delhi, New Delhi, India 110016
*Email address*: priyadarshi@maths.iitd.ac.in

SELECTED PUBLISHED TITLES IN THIS SERIES

825 **Saurabh Verma, María A. Navascués, and Amit Priyadarshi, Editors,** Advances in Dimension Theory, Fractal Functions and Measures, 2025

822 **Ahmad Barhoumi, Roozbeh Gharakhloo, and Andrei Martinez-Finkelshtein, Editors,** Recent Developments in Orthogonal Polynomials, 2025

821 **Bogdan D. Suceavă, Adara M. Blaga, Marian Ioan Munteanu, Cezar Oniciuc, Shoo Seto, Gabriel-Eduard Vîlcu, and Lihan Wang, Editors,** Recent Advances in Differential Geometry and Related Areas, 2025

820 **Sudeb Mitra, Rakesh Kumar Parmar, Saminathan Ponnusamy, and Toshiyuki Sugawa, Editors,** Geometric Function Theory and Related Topics, 2025

819 **Howard S. Cohl, Roberto S. Costas-Santos, and Robert S. Maier, Editors,** Applications and $q$-Extensions of Hypergeometric Functions, 2025

818 **Howard S. Cohl, Roberto S. Costas-Santos, and Robert S. Maier, Editors,** Classical Hypergeometric Functions and Generalizations, 2025

817 **Mikhail Khovanov, Joshua Sussan, and Anton Zeitlin, Editors,** From Representation Theory to Mathematical Physics and Back, 2025

816 **Luca F. Di Cerbo and Laurenţiu G. Maxim, Editors,** Geometry and Topology of Aspherical Manifolds, 2025

815 **Daniel Orr and Joshua Jeishing Wen, Editors,** Macdonald Theory and Beyond, 2025

814 **Susan Montgomery and Siu-Hung Ng, Editors,** Quantum Groups, Hopf Algebras, and Applications, 2025

813 **Thomas Creutzig, Julia Plavnik, and David Ridout, Editors,** Quantum Symmetries, 2025

812 **Ross Geoghegan, Craig R. Guilbault, and Kim Ruane, Editors,** Topology at Infinity of Discrete Groups, 2025

811 **Krishnendu Gongopadhyay and Shashikant A. Katre, Editors,** Geometry, Groups and Mathematical Philosophy, 2025

810 **Robert J. Berman and Yanir A. Rubinstein, Editors,** Convex and Complex: Perspectives on Positivity in Geometry, 2025

809 **Fabrice Baudoin and Luca Rizzi, Editors,** New Trends in Sub-Riemannian Geometry, 2025

808 **Loring W. Tu, Editor,** Group Actions and Equivariant Cohomology, 2024

807 **Galina Filipuk, Editor,** Recent Progress in Special Functions, 2024

806 **Maximilian Engel, Hildeberto Jardón-Kojakhmetov, and Cinzia Soresina, Editors,** Topics in Multiple Time Scale Dynamics, 2024

805 **Anthony Iarrobino, Pedro Macias Marques, Maria Evelina Rossi, and Jean Vallès, Editors,** Deformation of Artinian Algebras and Jordan Type, 2024

804 **Mahir Bilen Can and Jörg Feldvoss, Editors,** A Glimpse into Geometric Representation Theory, 2024

803 **Peter Gothen, Margarida Melo, and Montserrat Teixidor i Bigas, Editors,** Moduli Spaces and Vector Bundles—New Trends, 2024

802 **Ralph M. Kaufmann, Martin Markl, and Alexander A. Voronov, Editors,** Higher Structures in Topology, Geometry, and Physics, 2024

801 **K. A. Brown, T. J. Hodges, M. Vancliff, and J. J. Zhang, Editors,** Recent Advances in Noncommutative Algebra and Geometry, 2024

800 **Avinoam Mann, Louis H. Rowen, David J. Saltman, Aner Shalev, Lance W. Small, and Uzi Vishne, Editors,** Amitsur Centennial Symposium, 2024

799 **Alberto A. Condori, Elodie Pozzi, William T. Ross, and Alan A. Sola, Editors,** Recent Progress in Function Theory and Operator Theory, 2024

For a complete list of titles in this series, visit the
AMS Bookstore at **www.ams.org/bookstore/conmseries/**.